AF831179

Phytohormones in Abiotic Stress

Plants are continuously exposed to different environmental stresses that negatively impact their physiology and morphology, resulting in production reduction. As a result of constant pressure, plants evolve different mechanisms for sustenance and survival. Hormones play a major role in defences against the stresses and stimulate regulatory mechanisms. One of the ways through which they mitigate stress is via the production of hormones like auxins, ethylene, jasmonic acid, etc. The phytohormones help in signaling and enhance the chances of their survival. Plant hormones play many vital roles from integrating developmental events, physiological and biochemical processes to mediating both abiotic and biotic stresses. This book aims to highlight these issues and provide scope for the development of tolerance in crops against abiotic stresses to maximize yield for the growing population. There is an urgent need for the development of strategies, methods and tools for the broad-spectrum tolerance in plants supporting sustainable crop production under hostile environmental conditions.

The salient features are as follows:

- It includes both traditional and non-traditional phytohormones and focuses on the latest progress emphasizing the roles of different hormones under abiotic stresses.
- It provides a scope of the best plausible and suitable options for overcoming these stresses and puts forward the methods for crop improvement.
- It is an amalgamation of the biosynthesis of phytohormones and also provides molecular intricacies and signalling mechanisms in different abiotic stresses.
- This book serves as a reference book for scientific investigators from recent graduates, academicians and researchers working on phytohormones and abiotic stresses.

Phytohormones in Abiotic Stress

Edited by
Dhandapani Raju
R Ambika Rajendran
Ayyagari Ramlal
Virendra Pal Singh

CRC Press
Taylor & Francis Group
Boca Raton London New York

CRC Press is an imprint of the
Taylor & Francis Group, an **informa** business

First edition published 2024
by CRC Press
2385 NW Executive Center Drive, Suite 320, Boca Raton, FL 33431

and by CRC Press
4 Park Square, Milton Park, Abingdon, Oxon, OX14 4RN

CRC Press is an imprint of Taylor & Francis Group, LLC

© 2024 selection and editorial matter, Dhandapani Raju, R Ambika Rajendran, Ayyagari Ramlal and Virendra Pal Singh; individual chapters, the contributors.

Reasonable efforts have been made to publish reliable data and information, but the author and publisher cannot assume responsibility for the validity of all materials or the consequences of their use. The authors and publishers have attempted to trace the copyright holders of all material reproduced in this publication and apologize to copyright holders if permission to publish in this form has not been obtained. If any copyright material has not been acknowledged please write and let us know so we may rectify in any future reprint.

Except as permitted under U.S. Copyright Law, no part of this book may be reprinted, reproduced, transmitted, or utilized in any form by any electronic, mechanical, or other means, now known or hereafter invented, including photocopying, microfilming, and recording, or in any information storage or retrieval system, without written permission from the publishers.

For permission to photocopy or use material electronically from this work, access www.copyright.com or contact the Copyright Clearance Center, Inc. (CCC), 222 Rosewood Drive, Danvers, MA 01923, 978-750-8400. For works that are not available on CCC please contact mpkbookspermissions@tandf.co.uk

Trademark notice: Product or corporate names may be trademarks or registered trademarks and are used only for identification and explanation without intent to infringe.

ISBN: 978-1-032-37193-1 (hbk)
ISBN: 978-1-032-37194-8 (pbk)
ISBN: 978-1-003-33578-8 (ebk)

DOI: 10.1201/9781003335788

Typeset in Times
by MPS Limited, Dehradun

Contents

SECTION II Non-Traditional Phytohormones

Foreword

In an ever-changing world, where environmental uncertainties are bound, the remarkable tenacity of plant life takes the central stage. Plants, as the silent sentinels of nature, continually confront various challenges that test their growth and productivity. They stand as nature, unwavering soldiers, facing adversities head-on. Unlike mobile creatures, plants are sessile. They cannot escape harsh climatic conditions or environmental stresses. Instead, they have evolved ingenious strategies to withstand these challenges. Yet, these stresses, which encompass waterlogging, searing heatwaves, relentless droughts, cold and salt-riddled soils, continue to thwart their full potential. Abiotic stresses remain among the leading causes of global crop losses. Amid these trials, plants deploy a remarkable defence mechanism – phytohormones. These natural chemical messengers, including auxins, ethylene, gibberellins, cytokinins and jasmonic acid, emerge as the plants' allies. Phytohormones serve as signals of distress and hope, enhancing the plant's chances of survival.

Phytohormones play pivotal roles in the intricate tapestry of plant life. They integrate developmental events, orchestrate physiological processes and steer biochemical reactions. Moreover, they serve as mediators in the face of both abiotic and biotic stresses, offering a multi-faceted defence mechanism. As our world grapples with the urgent need for sustainable agriculture, the quest for strategies, methods and tools to bolster broad-spectrum resistance in plants becomes paramount. Crop failures loom large and the demand for increased food production to meet the needs of a growing global population continues to surge. It is a race against time, exacerbated by the diverse abiotic stresses triggered by a fluctuating climate, often driven by human activities. Studies have unveiled the pivotal roles of various plant hormones, such as brassinosteroids, strigolactones and jasmonates, in orchestrating responses to environmental stimuli. These hormones act as the plant's vigilant guardians, ready to trigger adaptive measures when adversity strikes.

In this context, *Phytohormones in Abiotic Stress* takes root. This book seeks to provide an extensive exploration into the world of phytohormones and their utility in bolstering plant tolerance to abiotic stresses. Tailored for agriculturalists and plant physiologists, it navigates the latest discoveries and insights into the roles of hormones in mediating responses to environmental challenges. In addition to unravelling the discoveries, this book offers a spotlight on the most promising approaches to combat the challenges that lie ahead. It serves as a beacon of knowledge in an era where resilience in agriculture is non-negotiable. The book will provide a wide coverage of phytohormones and their utility in the development of abiotic stress tolerance in plants in light of the current knowledge on the subject for agriculturalists and plant physiologists. The book deals with recent and cutting-edge updates about the discoveries and information about the role of hormones in mediating environmental stresses. It will also highlight and provide updated information on the most plausible approaches to combat the challenges.

Dr Viswanathan Chinnusamy
Joint Director (Research)
Indian Agricultural Research Institute (IARI), New Delhi, India

Preface

Plants are continuously facing and exposed to different environmental stresses. Plants, being sessile, need to develop strategies to overcome these harsh climatic conditions and stresses such as waterlogging, heat, drought, cold and salinity are some of the primary causes of crop losses globally. One of the ways through which they mitigate stress is via the production of hormones like auxins, ethylene, gibberellins, cytokinins, jasmonic acid and others. The phytohormones help in signalling and their production enhances their chances of survival. Apart from that, these hormones play many vital roles from integrating developmental events and physiological and biochemical processes to mediating both abiotic and biotic stresses. There is an urgent need for the development of strategies, methods and tools for the broad-spectrum resistance in plants supporting sustainable crop production under hostile environmental conditions.

Keeping the above objective, this book has been designed to provide wide coverage on phytohormones and their utility in the development of abiotic stress tolerance in plants. The efforts are made to include the sheer breadth of plant physiology with special reference to hormones and abiotic stress and the rapidly increasing knowledge in the field makes it impossible to include all of the relevant material in an entry-level text. Consequently, in light of the current knowledge on the subject, the broad scope of the field, ranging from molecular, environmental physiology, agronomy and genetics, makes it useful for agriculturalists and plant physiologists.

This edition is comprised of three units: *Introduction*, *traditional phytohormones* and *non-traditional phytohormones*. All efforts are made to retain the readability and overall approach.

Dhandapani Raju
R Ambika Rajendran
Ayyagari Ramlal
Virendra Pal Singh
Editors

Editor Biography

Dr Dhandapani Raju is a senior scientist of plant physiology in the Division of Physiology, Indian Agricultural Research Institute (ICAR-IARI), Delhi, India. He is a post-doctoral fellow at the agricultural laboratory, College of Agriculture, Food and Natural Resources, University of Missouri. He completed his undergraduate in agriculture from Tamil Nadu Agricultural University (TNAU), Tamil Nadu, India. He pursued his post-graduate in agriculture with a specialization in plant physiology from C. S. Azad University of Agriculture and Technology, Uttar Pradesh, India. He completed his doctoral degree in plant physiology from the Indian Agricultural Research Institute. He has been awarded "young scientist" by the Indian Society of Plant Physiology, New Delhi, India, for his outstanding work on phenotyping banana fruits and identifying vitamin A enrichment gene (phytoene synthase gene) for the first time in bananas. His academic achievements are evident from possessing various awards, scholarships and merit medals (book prizes, ICAR-JRF, IARI-SRF and ICAR-ARS) at the national level during post-graduate education. He is the recipient of several best poster and oral presentation awards at the national level. Through agricultural research and development, he has invented five technologies, registered three international patents and published more than 15 peer-reviewed articles in both international and national journals of repute. He has currently been working on high throughput phenotyping of rice plants for drought tolerance for the last four years. He has hands-on experience in the phenotyping of whole-plant transpiration and developed the software for automatic and near real-time estimation of WUE in rice. He has hands-on experience in both structural and functional genomics techniques for nucleic acid isolation, quality testing and gene expression analysis. He is also involved in teaching and guiding students at the post-graduation level. He has guided nine MSc students to date. He has conducted three national-level trainings for professionals at state and central levels. Currently, he is working on a national-level multi-institutional project on plant phenomics to develop rice and wheat varieties with improved water use efficiency and nitrogen use efficiency.

Dr R Ambika Rajendran is a senior scientist of plant breeding in the Division of Genetics, Indian Agricultural Research Institute (ICAR-IARI), Delhi, India. She earned her bachelor's from Tamil Nadu Agricultural University (TNAU), Tamil Nadu, India in 2003, her master's from Kerala Agricultural University (KAU), Kerala, India in 2005 and obtained her doctorate in 2010 from Tamil Nadu Agricultural University (TNAU). She joined the Directorate of Maize Research (DMR), New Delhi (currently referred to as Indian Institute of Maize Research IIMR, Punjab, India) in 2009 till 2016; later, she joined the Division of Genetics, IARI as a scientist in a senior scale in 2016. She has over 15 years of research experience. She currently works in abiotic stress tolerance and doubled haploidy in soybeans. She worked in maize breeding for high oil along with baby corn. Her significant research includes heterotic grouping and patterning of quality protein maize using molecular markers, in vitro mutagenesis in rice and identification of sources of high oil post-*Fusarium* stalk rot resistance genetic stocks in maize and identification of tolerant sources for pre-germination anaerobic stress tolerance in soybean. She has edited two books, *Genetics and Breeding in Pulses* and *Handbook of Cereals*, and also served as an editor for a book on *Soybean*. She has many peer-reviewed research articles published in journals of international and national repute and book chapters to her credit. She is also serving as a reviewer in various national and international journals.

Mr Ayyagari Ramlal is a doctoral scholar at the Plant Biotechnology Laboratory, School of Biological Science, Universiti Sains Malaysia, Malaysia. He completed his bachelor's and master's in botany from the University of Delhi, Delhi, India in 2017 and 2020, respectively. He has worked as a junior research fellow at the Division of Genetics, Indian Agricultural Research Institute

(ICAR-IARI), Delhi, India and Banaras Hindu University (BHU), Uttar Pradesh, India. He also worked on an innovation research project initiated by the University of Delhi. He has received the Young and Best Author awards from different societies/organizations in India. He has published more than 20 peer-reviewed research and review articles in international journals of repute and many book chapters. He served as an editor for a book on *Soybeans* and is also serving as a reviewer for many international and national journals and as a Review Editor for the section Biodiversity of Frontiers in Young Minds.

Dr Virendra Pal Singh completed his PhD in plant physiology and has over 35 years of research and 27 years of teaching experience. His areas of specialization include post-harvest and seed physiology, abiotic stress and climate change. He is a fellow at the Indian Society for Plant Physiology and served as the editor of the *Indian Journal of Plant Physiology* for nine years and national consulting editor for the *Advances in Plant Physiology* series (ten volumes). He is a gold medallist in MSc in botany with a specialisation in plant physiology. He was awarded a Council of Scientific and Industrial Research (CSIR) fellowship for PhD and post-doctoral. He was awarded the Emeritus Professorship by the Indian Council of Agricultural Research (ICAR), Delhi, India. He worked on an Indo-U.S. project on tropical climate and he has supervised nine PhD and three MSc students and trained many agricultural research service (ARS) scientists. He served as an advisor in various committees of Agricultural Universities and Institutes and Staff Selection Commission, India. He has more than a hundred publications to his credit.

Contributors

Ananya Anurag Anand
Indian Institute of Information Technology Allahabad
Prayagraj, Uttar Pradesh, India

A Arunachalam
Tree Improvement Research Division
Central Agroforestry Research Institute (ICAR-CAFRI)
Jhansi, Uttar Pradesh, India

Rohit Babar
ICAR-National Institute of Abiotic Stress Management
Malegaon, Baramati, Pune, Maharashtra, India

Diksha Bagal
Department of Botany
School of Life Sciences, Central University of Jammu
Samba, Jammu and Kashmir (UT), India

Pooja Baweja
Department of Botany
Maitreyi College, University of Delhi
Delhi, India

Isha Bhat
Department of Biosciences
Jamia Millia Islamia
Jamia Nagar, New Delhi, India

Varsha C
Tree Improvement Research Division
Central Agroforestry Research Institute (ICAR-CAFRI)
Jhansi, Uttar Pradesh, India

Sangram B Chavan
ICAR-National Institute of Abiotic Stress Management
Malegaon, Baramati, Pune, Maharashtra, India

Vaishnavi Dahiya
Department of Botany
University of Delhi
Delhi, India

Chandrasekaran Deepika
Department of Genetics and Plant Breeding
Tamil Nadu Agricultural University
Coimbatore, Tamil Nadu, India

Sourav Singh Deo
Department of Botany
Deshbandhu College, University of Delhi
New Delhi, India

Rajkumari Sanayaima Devi
Department of Botany
Deen Dayal Upadhyaya College
New Delhi, India

Rajkumar Dhakar
Division of Agriculture Physics
ICAR-Indian Agricultural Research Institute (IARI)
Pusa Campus, New Delhi, India

Nguyen Trung Duc
Vietnam National University of Agriculture
Hanoi, Vietnam

Naveen Goel
Department of Botany
University of Delhi
New Delhi, India

A K Handa
Tree Improvement Research Division
Central Agroforestry Research Institute (ICAR-CAFRI)
Jhansi, Uttar Pradesh, India

Amooru Harika
Division of Plant Physiology
ICAR-Indian Agricultural Research Institute (IARI)
Pusa Campus, New Delhi, India
Department of Plant and Environmental Sciences
Clemson University, Clemson, SC, US

Farheen Islam
Department of Botany
University of Delhi
Delhi, India

Meenakshi Jadhav
ITM University
Gwalior, Jhansi Rd, Turari, Gwalior, Lakhnotikhurd
Madhya Pradesh, India

Richa Jain
Department of Botany
University of Delhi
New Delhi, Delhi, India

Sourabh Karwa
Division of Plant Physiology
ICAR - Indian Agricultural Research Institute (IARI)
Pusa Campus, New Delhi, India

Kamaraj Keerthana
Department of Genetics and Plant Breeding
Tamil Nadu Agricultural University
Coimbatore, Tamil Nadu, India

Deepesh Kumar
National Institute for Plant Biotechnology (ICAR-NIPB)
Pusa Campus, New Delhi, India

Rahul Kumar
Division of Genetics
ICAR-Indian Agricultural Research Institute (IARI)
Pusa Campus, New Delhi, India
ICAR-Research Complex for North Eastern Hill Region (RC-NEH)
Tripura Centre, Lembucherra, Agartala, India

Sudhir Kumar
Division of Plant Physiology
ICAR-Indian Agricultural Research Institute (IARI)
Pusa Campus, New Delhi, India

Kishori Lal
Department of Botany
Institute of Science, Banaras Hindu University (BHU)
Varanasi, Uttar Pradesh, India

S K Lal
Division of Genetics
ICAR-Indian Agricultural Research Institute (IARI)
Pusa Campus, New Delhi, India

Rohit Kumar Mahto
Division of Genetics
ICAR-Indian Agricultural Research Institute (IARI)
Pusa Campus, New Delhi, India
School of Biotechnology
Institute of Science, Banaras Hindu University (BHU)
Varanasi, Uttar Pradesh, India

Pravin Mane
ICAR-National Institute of Abiotic Stress Management
Malegaon, Baramati, Pune, Maharashtra, India

Ashlesha Manta
Department of Botany
University of Delhi
New Delhi, Delhi, India

Shashi Meena
Division of Plant Physiology
ICAR-Indian Agricultural Research Institute (IARI)
Pusa Campus, New Delhi, India

Naina Miglani
Department of Botany
University of Rajasthan
Jaipur, Rajasthan, India

Kirti Nain
Centre of Medical Biotechnology
Maharshi Dayanand University
Rohtak, Haryana, India

Shivani Nagar
Division of Plant Physiology
ICAR-Indian Agricultural Research Institute (IARI)
Pusa Campus, New Delhi, India

Aparna Nautiyal
Department of Botany
Deshbandhu College, University of Delhi
Delhi, India

Riya Pakhre
Department of Environmental Studies
University of Delhi
New Delhi, Delhi, India

Ankita Rajendra Parab
School of Biological Sciences
Universiti Sains Malaysia (USM)
Georgetown, Penang, Malaysia

Raman Pushpa
Tamil Nadu Rice Research Institute
Tamil Nadu Agricultural University (TNAU)
Aduthurai, Tamil Nadu, India

Adeeb Rahman
Department of Botany
University of Delhi Delhi, India
Plant RNAi Biology Group
International Centre for Genetic engineering and Biotechnology (ICGEB)
New Delhi, Delhi, India

Roshni Rajamohan
Department of Botany
Deshbandhu College, University of Delhi
Delhi, India

K Rajarajan
Tree Improvement Research Division
Central Agroforestry Research Institute (ICAR-CAFRI)
Jhansi, Uttar Pradesh, India

Ambika Rajendran
Division of Genetics
ICAR-Indian Agricultural Research Institute (IARI)
Pusa Campus, New Delhi, India

Anjali Rajhans
Department of Botany
University of Delhi
Delhi, India

Dhandapani Raju
Division of Plant Physiology
ICAR-Indian Agricultural Research Institute (IARI)
Pusa Campus, New Delhi, India
Agriculture Laboratory, Division of Plant Science & Technology
College of Agriculture, Food and Natural Resources
University of Missouri, Columbia, MO

Ayyagari Ramlal
Plant Biotechnology Laboratory, School of Biological Sciences
Universiti Sains Malaysia (USM)
Georgetown, Penang, Malaysia
Division of Genetics
ICAR-Indian Agricultural Research Institute (IARI)
Pusa Campus, New Delhi, India

Madhu Rani
University Institute of Engineering & Technology
Maharshi Dayanand University
Rohtak, Haryana, India

Sumit Sagar
Department of Botany
University of Delhi
Delhi, India

Sakshi Sahu
Tree Improvement Research Division
Central Agroforestry Research Institute (ICAR-CAFRI)
Jhansi, Uttar Pradesh, India

Manisha Saini
Division of Genetics
ICAR-Indian Agricultural Research Institute (IARI)
Pusa Campus, New Delhi, India

Savita
Department of Botany
Hansraj College, University of Delhi
Delhi, India

Vidisha Saxena
Department of Botany
University of Delhi
Delhi, India

Nallamuthu Ramya Selvi
Department of Genetics and Plant Breeding
Tamil Nadu Agricultural University
Coimbatore, Tamil Nadu, India

Daniya Shahid
Department of Botany
University of Delhi
Delhi, India

Aravindan Shanmugam
Department of Genetics and Plant Breeding
Tamil Nadu Agricultural University
Coimbatore, Tamil Nadu, India

Anshul Sharma
University Institute of Engineering & Technology
Maharshi Dayanand University
Rohtak, Haryana, India

Shiv Shankar Sharma
School of Biotechnology
Institute of Science, Banaras Hindu University (BHU)
Varanasi, Uttar Pradesh, India
National Institute for Plant Biotechnology (ICAR-NIPB)
Pusa Campus, New Delhi, Delhi, India

Vasundhara Sharma
Division of Plant Physiology
ICAR-National Research Centre on Seed Spices (NRCSS)
Ajmer, Rajasthan, India

Vrinda Sharma
Department of Botany
School of Life Sciences, Central University of Jammu
Samba, Jammu and Kashmir (UT), India

Prashant Shaw
Department of Botany
University of Delhi, New Delhi
Delhi, India

Devendra Pratap Singh
Division of Genetics
ICAR-Indian Agricultural Research Institute (IARI)
Pusa Campus, New Delhi, India
Department of Botany
Institute of Science, Banaras Hindu University (BHU)
Varanasi, Uttar Pradesh, India

Neha Singh
Department of Botany
Gargi College, Siri Fort Road, University of Delhi
Delhi, India

Shubham Kumar Singh
Division of Genetics
ICAR-Indian Agricultural Research Institute (IARI)
Pusa Campus, New Delhi, India
Department of Botany, Institute of Science
Banaras Hindu University (BHU)
Varanasi, Uttar Pradesh, India

Virendra Pal Singh
Division of Plant Physiology
ICAR-Indian Agricultural Research Institute (IARI)
Pusa Campus, New Delhi, India

Nisha Sogan
Department of Botany
Deshbandhu College, University of Delhi
New Delhi, India

Renu Soni
Department of Botany
Gargi College, Siri Fort Road, University of Delhi
Delhi, India

Sreeramanan Subramaniama
School of Biological Sciences
Universiti Sains Malaysia (USM)
Georgetown, Penang, Malaysia
Chemical Centre Biology (CCB)
Universiti Sains Malaysia (USM)
Bayan Lepas, Penang, Malaysia
School of Chemical Engineering Technology
Universiti Malaysia Perlis (UNIMAP)
Arau, Perlis, Malaysia
National Poison Centre
Universiti Sains Malaysia (USM)
Georgetown, Penang, Malaysia

Sakshi Suman
Department of Botany
University of Delhi
Delhi, India

Ramalingam Suresh
Department of Rice
Tamil Nadu Agricultural University (TNAU)
Coimbatore, Tamil Nadu, India

Sukumar Taria
Division of Plant Physiology
ICAR-Indian Agricultural Research Institute (IARI)
Pusa Campus, New Delhi, India

Sachchidanand Tripathi
Department of Botany
Deen Dayal Upadhyaya College
New Delhi, India

Shikha Tripathi
National Institute for Plant Biotechnology (ICAR-NIPB)
Pusa Campus, New Delhi, Delhi, India
Department of Botany
Institute of Science, Banaras Hindu University (BHU)
Varanasi, Uttar Pradesh, India

Deepika Kumar Umesh
Central Sericultural Research and Training Institute
Central Silk Board, Berhampore, West Bengal, India

Apoorva Verma
Department of Botany,
University of Delhi, New Delhi, Delhi, India

Ritika Vishnoi
Department of Biosciences and Bioengineering
Indian Institute of Technology Roorkee
Roorkee, India

Sakthivel Viswabharathy
Department of Genetics and Plant Breeding
Tamil Nadu Agricultural University
Coimbatore, Tamil Nadu, India

Sheel Yadav
National Bureau of Plant Genetic Resources (ICAR-NBPGR)
Pusa Campus, New Delhi, India

Unit I

Introduction

1 Introduction to Phytohormones

Ayyagari Ramlal, Ambika Rajendran, Dhandapani Raju, and Virendra Pal Singh

1.1 INTRODUCTION

A hormone is described as "a chemical which, being created in any one area of the organism, is transported to another portion and there regulates a specific physiological process" by Went and Thimann in 1937. These chemical messengers work in coordination through which the activity of specific organs is coordinated with that of other organs (Went & Thimann, 1937; Ghorbel & Brini, 2021). The term "hormone" and its concept came into existence based on the observation made by Sachs 1880–1893. He stated, "Morphological differences between plant organs are due to differences in their material composition" (Sachs, 1893; Went & Thimann, 1937; Davies, 2010). The majority of plant cells seem to be able to create most hormones with varied transport routes, and they can impact both the cell from where they originated and distant cells. The term "hormone" has become widely accepted to describe these molecules in plants because there are significant similarities between the roles of hormones in animals and plants, including that they are active in very small quantities and serve as chemical signals (as opposed to having nutritional or catalytic functions). According to Went and Thimann (1937), "To prevent any potential misunderstanding with animal systems, the word phytohormones has been created for similar compounds in plants" (Weyers & Paterson, 2001). The prefix, however, may be dropped as this book will exclusively cover the plant kingdom. The five traditional plant hormones that were discovered at the start of the mid-20th century are widely mentioned, namely ethylene (1901 by D. Neljubow (Abeles & Heggestad, 1973), auxins (discovered in 1926 by Fr. Went (Moore & Moore, 1979)), gibberellins (1926 by E. Kurosawa (Armstrong, 1958)), cytokinins (discovered in the 1950s by F. Skoog (Skoog et al., 1966)) and abscisic acid (ABA; 1950s, T. Bennett-Clark and N. Kefford (Sacher, 1980)). Other substances that fit the definition of hormones have been discovered over the past 50 years. Brassinosteroids (BRs) (Fujioka & Sakurai, 1997; Fujioka & Yokota, 2003), jasmonates (JAs) (Howe, 2004), salicylic acid (SA) (Raskin, 1992), oxylipins (Blee, 2002), strigolactones (SLs) (Xie et al., 2010), karrikins (Nelson et al., 2009), indolamines (IAs) (Dowling & Ehinger, 1978), nitric oxide (NO) (Herbette et al., 2003) and peptide hormone (PH) (Gancheva et al., 2019) are some of the most recently discovered non-traditional phytohormones. Phytohormones serve a variety of purposes, including cell division, elongation, differentiation, and overall growth; additionally, they impact seed germination, root and shoot development, flowering, fruiting and stress responses, and they all have a significant impact on growth and development. From seed to seed, hormones have an impact on the life cycle of plants and how they react and interplay with the biotic and abiotic factors. Unraveling the roles of phytohormones has been challenging due to their pleiotropic effects, and it is still one of the most active areas of plant biology study. The study of plant hormones and the genes that govern their synthesis, transport and after-effects has discovered several new tools for agricultural advancements because of their core functions as integrators and regulators. This chapter briefly discusses the biosynthesis, evolution and purposes of both conventional and non-traditional plant hormones. Their functions and roles will also be discussed. The molecular mechanisms and proteins that underlie the

DOI: 10.1201/9781003335788-2

production, activity and interactions between each of the key hormones will be covered in greater detail in the following chapters of the book, with special reference to abiotic stress factors.

1.2 TRADITIONAL PHYTOHORMONES

1.2.1 Auxins

Auxins were the first essential plant hormone identified. Auxins are a class of plant hormones (or plant growth regulators) and they possess distinct morphogenetic characteristics that contribute significantly to various aspects of plant development (George et al., 2008). Darwin (1880), initially made his observations about the presence of some chemical entity that led to the bending of grass coleoptile (Darwin, 1880). Dutch biologist Frits Warmolt Went initially described auxins and their role in plant growth in the 1920s and isolated auxin based on the decapitated coleoptile and named it *Wuchsstoff* (and later changed to auxin from auxien meaning to grow) (Davies, 2010; Jiang et al., 2017). Went isolated for the first time one of these phytohormones and determined their chemical structure as indole-3-acetic acid (IAA). Auxin occurs in all parts of a plant but at varying concentrations and is closely linked through metabolism and transport (Simon et al., 2011). Auxins play a crucial role in coordinating growth and behavioral processes in plant life cycles and are essential for the development of the plant body. The distribution pattern of auxin in the plant is a critical factor for the plant's growth, its reaction to the environment and, in particular, the development of the plant organs. All auxins are made of an aromatic ring and a carboxylic acid group at the molecular level (Taiz & Zeiger et al., 1998). The most prominent member of the auxin family, indole-3-acetic acid (IAA), causes most of the auxin effects in intact plants. It is the most potent native auxin (Benková et al., 2003) and synthesized from the precursor, tryptophan (Trp) via the Shikimate pathway.

1.2.2 Gibberellic Acid

Gibberellic acid (also gibberellins A3, GA and GA3) is a hormone found in plants and fungi (Silva et al., 2013). GA promotes elongation and germination of stems and, in some plants, flowering and development of fruits. Manipulating the GA levels through genetic variation or chemical application of synthetic GA or GA inhibitors are standard farming practices to optimize plant growth and yield. The development of semi-dwarf cereal varieties with an attenuated response to stem elongation was the most important achievement of agricultural practices of the 20th century. It significantly increased the yield of global crops (Claeys et al., 2014). The characteristic response to GA-induced stem lengthening was first investigated in rice plants infected by the fungus *Gibberella fujikuroi*, which produces GA. Subsequently, the compound responsible for stem elongation was purified and characterized, and shortly thereafter, GAs were found in non-infected plant extracts. The ability of purified GAs to restore wild-like growth in dwarf pea mutants and corn mutants was essential to their recognition as endogenously produced plant hormones. These mutations and radioactive tracers have been used to gradually identify the biosynthetic pathway for GA production (Hedden et al., 2012). This process converts trans-geranylgeranyl diphosphate into bioactive GA (GGDP). Terpene synthases (TPSs), cytochrome P450 monooxygenases (P450s) and 2-oxoglutarate-dependent dioxygenases are employed in the MEP route to produce GA from GGDP (2ODDs). Gibberellins play a role in germination, including the natural breaking of dormancy. The GA produced in the scutellum diffuses to the aleurone cells, where it stimulates the secretion of alpha-amylase, as shown by the model for the gibberellin-induced synthesis of alpha-amylase. When a plant is subjected to low temperatures, a larger quantity of gibberellins is produced. They encourage germination, fruit ripening without seeds, breaking and blossoming and cell elongation (Yamaguchi et al., 2008). It produces an enzyme to promote growth in the embryo when its hormone interacts with a receptor, calcium activates the protein calmodulin and the complex binds to DNA.

1.2.3 Cytokinins

A group of plant hormones known as cytokinins (CKs) encourages cytokinesis, or cell division, in the roots and shoots of plants (Aina et al., 2012). Expanding on Haberlandt's results, Jablonski and Skoog observed in 1954 that the chemical that triggered cell division in the pith cells was also present in the vascular tissue (Haberlandt, 1913; Jablonski & Skoog, 1954; Mok et al., 2000). Miller et al. (1955) extracted autoclaved herring fish sperm DNA in crystalline form and discovered and improved the cell division component. This active ingredient was named "Kinetin" and became the first cytokinin, due to its ability to promote cell division (Miller et al., 1955; Schaller et al., 2015). Eventually, 6-furfuryl-amino purine was found to be kinetin. Later, it was suggested to include kinetin and other substances with comparable properties under the umbrella word *kinin*. By simultaneously isolating and crystallizing zeatin from the milky endosperm of corn (*Zea mays* L.) in the 1960s, Miller and Lethum isolated the first naturally occurring cytokinin (Lethum, 1967; Miller, 1971; Großkinsky & Petrášek, 2019). In the production of isoprene cytokinins, the first reaction is catalyzed by the adenosine phosphate-isopentenyl transferase (IPT). As prenyl donors, it may utilize dimethylallyl pyrophosphate (DMAPP) or hydroxy methylbutenyl pyrophosphate (HMBPP) in addition to ATP, ADP, or AMP as substrates. The rate-limiting step in the production of cytokinin is this process. The methylerythritol phosphate route generates DMADP and HMBDP, which are necessary for cytokinin production (MEP). Recycled tRNAs can also be used by bacteria and plants to make cytokinins. tRNAs with anticodons that begin with uridine and carry prenylated adenosine close to the anticodon release when the adenosine is broken down as a cytokinin. These adenines are prenylated by tRNA-isopentenyl transferase (Koprna et al., 2016). CKs were later discovered to delay leaf senescence, which has been reported to improve plant drought tolerance.

1.2.4 Ethylene

Unique among plant hormones, ethylene (C_2H_4) is a simple two-carbon gaseous phytohormone. Being gaseous, ethylene can freely travel between cells and is often produced at the site of action. It can gather in specific places and spread out quickly (Wang et al., 2018). These gaseous qualities are believed to influence their roles in coordinating fruit ripening and reactions to injury, mechanical impedance and stress (Polko et al., 2019). Through complex positive and negative feedback networks, transcriptional and posttranscriptional mechanisms and response pathways, ethylene synthesis and levels are closely regulated. Working together, these regulatory systems allow for a rapid and precise ethylene response. These regulatory systems work together to make a very quick and accurate ethylene response (Pandey et al., 2021). By enhancing fruit ripening, postponing senescence and producing plants that are more tolerant to stress, understanding and controlling ethylene biosynthesis and actions can significantly affect agriculture and horticulture (Hartman et al., 2019).

The enzyme methionine adenosyl transferase catalyzes the first reaction in the ethylene biosynthesis from the methionine to S-adenosyl-L-methionine (SAM). The enzyme 1-aminocyclopropane-1-carboxylic acid (ACC) synthase (ACS) then converts SAM to ACC and 5′-methylthioadenosine (MTA). The rate at which ethylene is produced depends on ACS activity. Therefore, controlling this enzyme is essential for the production of ethylene (Hartman et al., 2021; Pattyn et al., 2021). The penultimate stage involves the ACC-oxidase (ACO), formerly known as the ethylene-forming enzyme, and necessitates oxygen. Either endogenous or exogenous ethylene can stimulate the production of ethylene (Cho et al., 2022). High amounts of auxins, particularly IAA and CKs, promote ACC production. The majority of plants fight salinity through the hormone ethylene. Through a complicated signaling mechanism, ethylene controls plant growth and development and helps plants adapt to stressful situations (Maric et al., 2022).

1.2.5 Abscisic Acid

The primary hormone that regulates a plant's capacity for survival in a harsh, fluctuating environment is the abscisic acid (ABA). Plants create more ABA when under water stress, which causes them to respond by closing their stomata to stop sweating and expressing genes that make osmoprotectants (Zhang et al., 2021). ABA triggers similar genes in mature seeds, and it is assumed that these genes endow the developing embryo with desiccation tolerance (Jones et al., 2014). In addition, ABA controls seed dormancy and germination, assists in controlling development and takes part in biotic stress reactions (Wang et al., 2016). Since the 1980s, molecular genetics techniques in *Arabidopsis thaliana* and other plants have added to the ongoing efforts to elucidate the pathway for ABA production. At the beginning of the 21st century, after more than 40 years since ABA (abscisic acid) was first studied, scientists had completely figured out how it's made and used in plants. It's hard to highlight Jan Zeevart's exact role in understanding ABA because many scientists from around the world, working in different places, have contributed a lot to this knowledge (Waadt et al., 2014). ABA is an isoprenoid plant hormone that is made in the plastidial 2-C-methyl-D-erythritol-4-phosphate (MEP) pathway. In contrast to sesquiterpenes, which are structurally related but are made from the mevalonic acid-derived precursor farnesyl diphosphate (FDP), the C15 backbone of ABA is developed after cleavage of C40 carotenoids in MEP. Zeaxanthin is the first committed ABA precursor; subsequent oxidation of xanthoxin to ABA occurs as a result of a sequence of enzyme-catalyzed epoxidations and isomerizations via violaxanthin and final cleavage of the C40 carotenoid by a deoxygenation process through ABA (Bassaganya et al., 2011).

1.3 NON-TRADITIONAL PHYTOHORMONES

1.3.1 Jasmonates

Jasmonates (or jasmonic acid; JA) are oxylipins that are produced when a lipoxygenase oxidizes a polyunsaturated fatty acid (Caarls et al., 2015). The biological activity of the jasmonate family of chemicals, which are produced from JA, varies. Since JA itself has a very modest level of activity, the term "prohormone" has been used occasionally to describe it. Methyl jasmonate (MeJA) is the volatile methyl ester of JA (Acosta & Farmer, 2010). In 1962, MeJA was purified to describe jasmonate hormones for the first time. MeJA has a pleasant aroma and is volatile, like many mild esters. The jasmine blossom, *Jasminum grandiflorum* L., was used to extract and purify it (Demole et al., 1962; Sharma & Laxmi, 2016). Later, the metabolic process for the synthesis of jasmonates as well as the biological functions of jasmonates was discovered. Its function in plant defensive responses was discovered in the 1980s, and subsequently, the receptor protein and signal transduction pathway were discovered (Shitan et al., 2013). Plastid-localized lipases are necessary for JA production and to be quickly triggered by injury. Jasmonates are derived from a-linolenic acid, an octadecanoid fatty acid with three unsaturated double bonds, by the majority of higher plants; OPDA is converted to jasmonate in the peroxisome (Hsieh et al., 2014). JA controls a wide range of physiological functions in plants, but its impact on wound healing is best understood. After being mechanically injured or being eaten by a herbivore, JA biosynthesis is quickly stimulated, which causes the production of the necessary response genes. After playing a defense-related role, JAs have also been linked to cell death and leaf senescence (Schaller et al., 2009). JA can interact with a wide variety of kinases and transcriptional regulators linked to senescence. By causing the buildup of reactive oxygen species (ROS), JA can potentially cause mitochondrial apoptosis. These substances cause apoptosis, or programmed cell death, which compromises cells by damaging mitochondrial membranes (Reinbothe et al., 2009). The functions of JAs in these processes point to strategies by which the plant protects itself against biotic threats and restricts the spread of diseases.

1.3.2 Brassinosteroids

The sixth class of plant hormones has been identified as the brassinosteroids (BRs, or less frequently, BS), a subclass of polyhydroxysteroids (Khripach et al., 2000). They might be useful as an anticancer medication for malignancies that respond to endocrine hormones to trigger apoptosis and stop growth. Mitchell et al. (1970) observed that treating organic extracts of rapeseed pollen resulted in stem lengthening and cell proliferation, and sparked the first investigation into brassinosteroids. The biologically active molecule, called brassinolide, was discovered in 1979 when stem elongation and cell divisions were observed in rapeseed pollens (Grove et al., 1979; Karssen et al., 2012). All plant species contain brassinosteroids involved in disease resistance, vascular differentiation, development, light responses and growth. Brassinosteroids were initially discovered in 1979. Campesterol is used in the biosynthesis of BR. The BR biosynthesis mutants in *Arabidopsis*, tomatoes, and peas were studied to see whether the biosynthetic route that Japanese scientists had described was accurate. No experimental proof has been provided to back up the sites of BR synthesis in plants. The expression of BR biosynthetic and signal transduction genes in various plant organs lends credence to the idea that all tissues produce BRs. Additionally, this is supported by the hormones' short-range action (Lalarukh et al., 2022). The flow is from the base to the tips, and experiments have demonstrated that long-distance transfer is feasible (acropetal). If this movement is meaningful in terms of biology is yet unknown (Tossi et al., 2015). Under deficit irrigation system, BR effectively reduces drought stress and enhances wheat growth. Due to its crucial role in lowering oxidative stress indicators, it also exhibited additional favorable effects on raising plant development metrics. BRs may be found in plants like the bean using some bioassays.

1.3.3 Karrikins

A class of plant growth regulators known as karrikins is discovered in the smoke produced while burning plant matter (Flematti et al., 2015). Karrikins imitate strigolactone, a signaling hormone that aids in promoting seed germination and plant growth. Strigolactones are hormones that support the formation of symbiotic arbuscular mycorrhizal fungus in the soil, which in turn promotes plant growth and branching (Andreo-Jimenez et al., 2015). The heating or burning of carbohydrates, including sugars and polysaccharides, mostly cellulose, results in the formation of karrikins (Barickman et al., 2013). These sugars become karrikins when plant material burns. Karrikins' pyran moiety is most likely directly generated from pyranose sugar. Although karrikins are not known to naturally occur in plants, it has been hypothesized that karrikin-like compounds do exist in plants (Gutjahr et al., 2015). Karrikins are said to promote seedling vitality and stimulate seed germination. Karrikins affect the photomorphogenesis of seedlings in *Arabidopsis*, resulting in shorter hypocotyls and bigger cotyledons (Van Staden et al., 2006). Such reactions could be advantageous for seedlings when they emerge into the post-fire landscape. Since the KAI2 protein is also necessary for the formation of leaves, karrikins may impact several other aspects of plant development.

1.3.4 Oxylipins

A series of oxygenated natural products known as oxylipins are created from fatty acids by chemical reactions, including at least one stage of dioxygen-dependent oxidation (Wasternack et al., 2007). Plant growth and biotic/abiotic stress responses are known to be regulated by oxylipins (Wasternack & Feussner, 2018; Wasternack & Strnad, 2018). Polyunsaturated fatty acids (PUFAs) are converted into oxylipins by cyclooxygenases (COX), lipoxygenases (LOX) or cytochrome P450 epoxygenase (Barquissau et al., 2017). Numerous aerobic creatures, such as fungi, mammals and plants, produce oxylipins. Numerous oxylipins are important physiologically (Bolwell et al., 2002). Dioxygenases or monooxygenases initiate the biosynthesis of oxylipins. The production of oxylipins is started by dioxygenases or monooxygenases. Dioxygenases include heme-dependent fatty acid oxygenases (plants, fungi), lipoxygenases (plants, animals, and fungi), and cyclooxygenases (animals). Members of the cytochrome P450 superfamily, involved in the oxylipin biosynthesis process, may oxidize double bonds to produce epoxides or saturated carbons to produce alcohols. The C18 and C20 polyenoic fatty acids are the primary oxylipin precursors in both the plant and animal worlds. Oxylipins may have a variety of biological functions due to their wide structural diversity, including those of second messengers, antibacterial, insecticidal and antifungal substances (Blée, 2002; Howe & Jander, 2008; Howe et al., 2018).

1.3.5 Polyamines

In plant tissue culture, organic polycations with low molecular weight known as polyamines (PAs) are known to exhibit substantial biological activity (Chen et al. 2019). By assisting in plant growth, development, molecular signaling, cell division, differentiation, totipotency, and other stress responses, these aliphatic amines which have two or more amino groups attached play a crucial role in a variety of physiological processes (Xu & Wu 2009; Vuosku et al., 2018). In plants, PAs have a variety of uses, from callus induction to the exclusion of the plant's secondary metabolites, firmly establishing itself as a powerful elicitor in plant tissue culture (Rakesh et al., 2021).

1.3.6 Strigolactones

A class of chemical substances known as strigolactones is created by roots. These substances have been categorized as phytohormones or plant hormones based on how they work (Umehara et al., 2015). Three distinct physiological mechanisms have been linked to strigolactones thus far: First, they promote the growth of parasitic organisms that reside in the roots of the host plant, including *Striga lutea* and other members of the genus *Striga*. Second, strigolactones are essential for symbiotic fungi to recognize the plant, particularly arbuscular mycorrhizal fungi since they form a mutualistic interaction with these plants and supply them with phosphate and other soil nutrients. Thirdly, strigolactones have been shown to act as branching inhibiting hormones in plants; when present, these substances block excessive bud growth in the terminal stalks, interrupting the ramification process in plants. Although the strigolactones' biosynthesis route (refer Chapter 18 for more details) is still not completely understood, many steps have been discovered, including the enzymes needed to carry out the necessary chemical transition (Alder et al., 2012).

1.3.7 Salicylic Acid

Salicylic acid is a phenolic phytohormone in plants and plays a part in ion absorption and transport, photosynthesis, transpiration, growth and development (Vlot et al., 2009). In endogenous signaling, salicylic acid plays a role in modulating plant defense against pathogens (Hayat et al., 2007). By stimulating the synthesis of defense-related metabolites and proteins associated with pathogenesis, it contributes to the resistance to infections (i.e., systemic acquired resistance) (Van Huijsduijnen et al., 1986). Delaney et al. (1994), Gaffney et al. (1993), Lawton et al. (1995) and Vernooij et al. (1994) have demonstrated the alteration of SA levels using *Nicotiana tabacum* or *Arabidopsis* as model plants expressing *nahG* which encodes for SA-metabolizing enzyme salicylate hydroxylase, upon infection with pathogen. Thus, failing to show an increase in SA levels, development of SAR and no PR genes expression in the leaves. It was found that the plants were susceptible to both avirulent and virulent strains of the pathogens (Delaney et al., 1994; Gaffney et al., 1993; Lawton et al., 1995; Vernooij et al., 1994; Vlot et al., 2009). Through improved seed germination, bud blooming and fruit ripening, salicylic acid can help plants develop exogenously. However, salicylic acid concentrations that are too high can inhibit these developmental processes (Koo et al., 2020). Additionally, methyl salicylate (MeSA), a volatile salicylic acid methyl ester, can permeate through the atmosphere, promoting plant-plant communication (Taiz & Zeiger, 2002). When MeSA is converted back to salicylic acid by the stomata of the next plant, it can trigger an immunological response (Taiz & Zeiger, 2002; Kumar, 2014). Salicylic acid binding proteins (SABPs) and the putative receptor NPR genes (nonexpressor of pathogenesis-related genes) are two of the proteins that have been shown to interact with SA in plants (Kumar, 2014).

1.3.8 Nitric Oxide

A gaseous signaling molecule is called nitric oxide (NO) (Liu et al., 2017). Stomatal closure, root growth, germination, nitrogen fixation, cell death and stress response are just a few examples of nitric oxide's hormonal and defensive responses (Shapiro et al., 2005). NO is crucial for many biological activities in plants, including growth, development and responses to biotic and abiotic stresses (Shang et al., 2022). NO regulates the activities of plant cell organelles and can be created by an as-yet-unidentified NO synthase, a specific nitrite reductase, a nitrate reductase, mitochondrial cytochrome c oxidase, or nonenzymatic mechanisms (e.g., ATP synthesis in chloroplasts and mitochondria). It is an essential vertebrate biological messenger that participates in some biological processes (Weller et al., 2012). Almost all organisms, including bacteria, plants, fungi and mammal cells, produce it as a bioproduct (Rőszer, 2012).

1.3.9 Indoleamines

Tryptophan, an important amino acid, is used in biological processes to create indoleamines. With the help of the enzymes 5-HTP decarboxylase and tryptophan hydroxylase, tryptophan is converted into serotonin by adding a hydroxyl group and then removing the carboxyl group (Carlson, 2012). It is a bio-stimulator that controls plant growth and productivity. Melatonin (N-acetyl-5-methoxytryptamine) is an indoleamine found in plants and mammals. It reduces oxidative stress brought on by reactive oxygen species, potentially boosting photosynthetic rate, chlorophyll content and it has a role in controlling auxin and SA (Zhang et al., 2015; Pardo-Hernández et al., 2020; Zhao et al., 2022).

1.3.10 Peptide Hormones

All small secreted peptides involved in cell communication fall into the category of plant peptidic hormones. These tiny peptide hormones govern cell division and expansion, defense systems, pollen self-incompatibility, and other critical aspects of plant growth and development (Lindsey et al., 2001). Water stress detected in the roots is sent to the stomata in the leaves via a tiny peptide CLE25 as a long-distance signal (Porrini et al., 2019). Plant development and growth are significantly influenced by peptide signaling. Membrane-localized receptor kinases, the most well-known class of receptor-like molecules in plants, have been identified as specific receptors for various peptides (Matsubayashi et al., 2006).

1.4 CONCLUSION

Various traditional phytohormones, including auxins, cytokinins, ethylene, gibberellins and abscisic acid, and non-traditional phytohormones, such as brassinosteroids, jasmonates, oxylipins, polyamines and others, play important roles during plant growth and development and involved in maintenance and sustenace of the plants. Biosynthesis and functions of both the phytohormones are essential and known to be involved in providing tolerance and resistance to various kinds of stresses. In the subsequent chapters, traditional and non-traditional phytohormones will be discussed in detail, with special reference to abiotic stresses.

1.5 FUTURE PROSPECTS

The phytohormones play important roles in controlling all aspects of plants' growth, development and also interplay with each other. Given the significance of phytohormones in the middle of the 20th century, it is crucial to comprehend how these molecules function and can boost agricultural production to meet the growing demands of the population. Genomic and other high throughput techniques have improved the understanding acquired through biochemical, physiological and genetic investigations by identifying the key genes driving these signaling pathways as well as their broad interaction; thereby they can be manipulated for improving resistance and yield. Therefore, more emphasis should be given to the phytohormones, such that their full potential can be harnessed and utilized for crop improvement programs.

REFERENCES

Abeles, F. B., & Heggestad, H. E. (1973). Ethylene: an urban air pollutant. Journal of the Air Pollution Control Association, 23(6), 517–521.

Acosta, I. F., & Farmer, E. E. (2010). Jasmonates. The Arabidopsis Book, American Society of Plant Biologists, 8, e0129.

Aina, O., Quesenberry, K., & Gallo, M. (2012). Thidiazuron-induced tissue culture regeneration from quartered-seed explants of *Arachis paraguariensis*. Crop Science, 52(3), 1076–1083.

Alder, A., Jamil, M., Marzorati, M., Bruno, M., Vermathen, M., Bigler, P., Ghisla, S., Bouwmeester, H., Beyer, P., & Al-Babili, S. (2012). The path from b-carotene to carlactone, astrigolactone-like plant hormone. Science, 335, 1348–1351. 10.1126/science.1218094.

Andreo-Jimenez, B., Ruyter-Spira, C., Bouwmeester, H. J., & Lopez-Raez, J. A. (2015). Ecological relevance of strigolactones in nutrient uptake and other abiotic stresses, and in plant-microbe interactions below-ground. Plant and Soil, 394(1), 1–19.

Armstrong, J. L. (1958). Research on Gibberellin: I. Gibberellin Studies on Radishes. Bios, 29(3), 139–141.

Barickman, T. C., Kopsell, D. A., & Sams, C. E. (2013). Selenium influences glucosinolate and isothiocyanates and increases sulfur uptake in Arabidopsis thaliana and rapid-cycling *Brassica oleracea*. Journal of Agricultural and Food Chemistry, 61(1), 202–209.

Barquissau, V., Ghandour, R. A., Ailhaud, G., Klingenspor, M., Langin, D., Amri, E. Z., & Pisani, D. F. (2017). Control of adipogenesis by oxylipins, GPCRs, and PPARs. Biochimie, 136, 3–11.

Bassaganya-Riera, J., Guri, A. J., Lu, P., Climent, M., Carbo, A., Sobral, B. W., et al. (2011). Abscisic acid regulates inflammation via ligand-binding domain-independent activation of peroxisome proliferator-activated receptor γ. Journal of Biological Chemistry, 286(4), 2504–2516.

Benková, E., Michniewicz, M., Sauer, M., Teichmann, T., Seifertová, D., Jürgens, G., & Friml, J. (2003). Local, efflux-dependent auxin gradients as a common module for plant organ formation. Cell, 115(5), 591–602.

Blée, E. (2002). Impact of phyto-oxylipins in plant defense. Trends in Plant Science, 7(7), 315–322.

Bolwell, G. P., Bindschedler, L. V., Blee, K. A., Butt, V. S., Davies, D. R., Gardner, S. L., ... & Minibayeva, F. (2002). The apoplastic oxidative burst in response to biotic stress in plants: a three-component system. Journal of Experimental Botany, 53(372), 1367–1376.

Caarls, L., Pieterse, C. M., & Van Wees, S. C. (2015). How salicylic acid takes transcriptional control over jasmonic acid signaling. Frontiers in Plant Science, 6, 170.

Carlson, N. R. (2012). Physiology of behavior, 11th edition. Pearson, Boston, MA.

Chen, D., Shao, Q., Yin, L., Younis, A., & Zheng, B. (2019). Polyamine function in plants: metabolism, regulation on development, and roles in abiotic stress responses. Frontiers in Plant Science, 9, 1945.

Cho, H. Y., Chou, M. Y., Ho, H. Y., Chen, W. C., & Shih, M. C. (2022). Ethylene modulates translation dynamics in *Arabidopsis* under submergence via GCN2 and EIN2. Science Advances, 8(22), eabm7863.

Claeys, H., De Bodt, S., & Inzé, D. (2014). Gibberellins and DELLAs: central nodes in growth regulatory networks. Trends in Plant Science, 19(4), 231–239.

Darwin, C. (1880). The power of movement in plants. John Murray, London.

Davies, P. J. (2010). The plant hormones: their nature, occurrence, and functions. In: P. J. Davies (ed.), Plant hormones (pp. 1–15). Springer, Dordrecht. 10.1007/978-1-4020-2686-7_1

Delaney, T. P., et al. (1994). A central role of salicylic acid in plant disease resistance. Science, 266, 1247–1250. 10.1126/science.266.5188.1247.
Demole, E., Lederer, E., & Mercier, D. (1962). Isolement et détermination de la structure du jasmonate de méthyle, constituant odorant caractéristique de l'essence de jasmin. Helvetica Chimica Acta, 45, 675–685. 10.1002/hlca.19620450233.
Dowling, J. E., & Ehinger, B. (1978). Synaptic organization of the dopaminergic neurons in the rabbit retina. Journal of Comparative Neurology, 180(2), 203–220.
Flematti, G. R., Dixon, K. W., & Smith, S. M. (2015). What are karrikins and how were they 'discovered by plants? BMC Biology, 13(1), 1–7.
Fujioka, S., & Sakurai, A. (1997). Biosynthesis and metabolism of brassinosteroids. Physiologia Plantarum, 100(3), 710–715.
Fujioka, S., & Yokota, T. (2003). Biosynthesis and metabolism of brassinosteroids. Annual Review of Plant Biology, 54(1), 137–164.
Gaffney, T., et al (1993). Requirement of salicylic acid for the induction of systemic acquired resistance. Science, 261, 754–756. 10.1126/science.261.5122.754.
Gancheva, M. S., Malovichko, Y. V., Poliushkevich, L. O., Dodueva, I. E., & Lutova, L. A. (2019). Plant peptide hormones. Russian Journal of Plant Physiology, 66, 171–189.
George, E. F., Hall, M. A., & Klerk, G. J. D. (2008). Plant Growth Regulators I: Introduction; Auxins, their Analogues and Inhibitors. In: E. F. George et al. (eds), Plant Propagation by Tissue Culture. Dordrecht: Springer. https://doi.org/10.1007/978-1-4020-5005-3_5
Ghorbel, M., & Brini, F. (2021). Hormone mediated cell signaling in plants under changing environmental stress. Plant Gene, *28*, 100335.
Großkinsky, D. K., & Petrášek, J. (2019). Auxins and cytokinins–the dynamic duo of growth-regulating phytohormones heading for new shores. The New Phytologist, 221(3), 1187–1190.
Grove, M. D., Spencer, G. F., Rohwedder, W. K., Mandava, N., Worley, J. F., Warthen, J. D., et al. (1979). Brassinolide, a plant growth-promoting steroid isolated from *Brassica napus* pollen. Nature, 281(5728), 216–217.
Gutjahr, C., Gobbato, E., Choi, J., Riemann, M., Johnston, M. G., Summers, W., et al. (2015). Rice perception of symbiotic arbuscular mycorrhizal fungi requires the karrikin receptor complex. Science, 350(6267), 1521–1524.
Haberlandt, G. (1913). Zur physiologie der zellteilung. Sitzungsberichte der Preussischen Akademie der Wissenschaften, Physikalisch-Mathematische Klasse I (pp. 318–345).
Hartman, S., Liu, Z., Van Veen, H., Vicente, J., Reinen, E., Martopawiro, S., et al. (2019). Ethylene-mediated nitric oxide depletion pre-adapts plants to hypoxia stress. Nature Communications, 10(1), 1–9.
Hartman, S., Sasidharan, R., & Voesenek, L. A. C. J. (2021). The role of ethylene in metabolic acclimations to low oxygen. New Phytologist, 229, 64–70. 10.1111/nph.16378.
Hayat, S., Ali, B., & Ahmad, A. (2007). Salicylic acid: biosynthesis, metabolism and physiological role in plants. In: S. Hayat, & A. Ahmad (eds.), Salicylic acid: A plant hormone (pp. 1–14). Springer, Dordrecht. 10.1007/1-4020-5184-0_1
Hedden, P., & Thomas, S. G. (2012). Gibberellin biosynthesis and its regulation. Biochemical Journal, 444(1), 11–25.
Herbette, S., Lenne, C., De Labrouhe, D. T., Drevet, J. R., & Roeckel-Drevet, P. (2003). Transcripts of sunflower antioxidant scavengers of the SOD and GPX families accumulate differentially in response to downy mildew infection, phytohormones, reactive oxygen species, nitric oxide, protein kinase and phosphatase inhibitors. Physiologia Plantarum, 119(3), 418–428.
Howe, G. A. (2004). Jasmonates. In: P. J. Davies (ed.), Plant hormones: biosynthesis, signal transduction, action! (pp. 646–680). Springer Netherlands, Dordrecht. 10.1007/978-1-4020-2686-7_28.
Howe, G. A., & Jander, G. (2008). Plant immunity to insect herbivores. Annual Review of Plant Biology, 59, 41–66. 10.1146/annurev.arplant.59.032607.092825.
Howe, G. A., Major, I. T., & Koo, A. J. (2018). Modularity in jasmonate signaling for multi-stress resilience. Annual Review in Plant Biology, 69(1), 387–415.
Hsieh, H. L., & Okamoto, H. (2014). Molecular interaction of jasmonate and phytochrome A signaling. Journal of Experimental Botany, 65(11), 2847–2857.
Jablonski, J. R., & Skoog, F. (1954). Cell enlargement and cell division in excised tobacco pith tissue 1. Physiologia Plantarum, 7, 16–24. 10.1111/j.1399-3054.1954.tb07552.x.
Jiang, Z., Li, J., & Qu, L. J. (2017). Auxins. In: J. Li, C. Li & S. M. Smith (eds.), Hormone metabolism and signaling in plants (pp. 39–76). Cambridge, MA: Academic Press.
Jones, A. M., Danielson, J. Å., ManojKumar, S. N., Lanquar, V., Grossmann, G., & Frommer, W. B. (2014). Abscisic acid dynamics in roots detected with genetically encoded FRET sensors. Life, 3, e01741.

Karssen, C. M., van Loon, L. C., & Vreugdenhil, D. (Eds.). (2012). Progress in plant growth regulation (Vol. 13). Dordrecht: Springer Science & Business Media.
Khripach, V., Zhabinskii, V., & de Groot, A. (2000). Twenty years of brassinosteroids: steroidal plant hormones warrant better crops for the XXI century. Annals of Botany, 86(3), 441–447.
Koo, Y. M., Heo, A. Y., & Choi, H. W. (2020). Salicylic acid is a safe plant protector and growth regulator. The Plant Pathology Journal, 36(1), 1.
Koprna, R., De Diego, N., Dundálková, L., & Spíchal, L. (2016). Use of cytokinins as agrochemicals. Bioorganic & Medicinal Chemistry, 24(3), 484–492.
Kumar, D. (2014). Salicylic acid signaling in disease resistance. Plant Science, 228, 127–134.
Lalarukh, I., Amjad, S. F., Mansoora, N., Al-Dhumri, S. A., Alshahri, A. H., Almutari, M. M., et al. (2022). The integral effects of brassinosteroids and timber waste biochar enhance the drought tolerance capacity of a wheat plant. Scientific Reports, 12(1), 1–10.
Lawton, K. (1995). Systemic acquired resistance in Arabidopsis requires salicylic acid but not ethylene. Molecular Plant-Microbe Interactions, 8, 863. 10.1094/mpmi-8-0863.
Letham, D. S. (1967). Chemistry and physiology of kinetin-like compounds. Annual Review of Plant Physiology, 18(1), 349–364.
Lindsey, K. (2001). Plant peptide hormones: The long and the short of it. Current Biology, 11(18), R741–R743.
Liu, H., Weng, L., & Yang, C. (2017). A review on nanomaterial-based electrochemical sensors for H2O2, H2S, and NO inside cells or released by cells. Microchimica Acta, 184(5), 1267–1283.
Maric, A., & Hartman, S. (2022). Ethylene controls translational gatekeeping to overcome flooding stress in plants. The EMBO Journal, 41(19), e112282.
Matsubayashi, Y., & Sakagami, Y. (2006). Peptide hormones in plants. Annual Review of Plant Biology, 57, 649–674.
Miller, C. O. (1971). Cytokinin production by mycorrhizal fungi. Mycorrhizae, 3(2), 168–174.
Miller, C. O., Skoog, F., Okumura, F. S., Von Saltza, M. H., & Strong, F. M. (1955). Isolation, structure and synthesis of kinetin, a substance promoting cell division. Journal of the American Chemical Society, 78, 1375–1380. 10.1021/ja01588a032.
Mitchell, J. W., Mandava, N., Worley, J. F., Plimmer, J. R., & Smith, M. V. (1970). Brassins—a new family of plant hormones from rape pollen. Nature, 225(5237), 1065–1066.
Mok, M. C., Martin, R. C., & Mok, D. W. (2000). Cytokinins: Biosynthesis metabolism and perception. In Vitro Cellular & Developmental Biology - Plant, 36, 102–107. 10.1007/s11627-000-0021-7.
Moore, T. C., & Moore, T. C. (1979). Auxins. Biochemistry and physiology of plant hormones (pp. 32–89). Springer, Verlag.
Nelson, D. C., Riseborough, J. A., Flematti, G. R., Stevens, J., Ghisalberti, E. L., Dixon, K. W., & Smith, S. M. (2009). Karrikins discovered in smoke trigger Arabidopsis seed germination by a mechanism requiring gibberellic acid synthesis and light. Plant Physiology, 149(2), 863–873.
Pandey, B. K., Huang, G., Bhosale, R., Hartman, S., Sturrock, C. J., Jose, L., et al. (2021). Plant roots sense soil compaction through restricted ethylene diffusion. Science, 371(6526), 276–280.
Pardo-Hernández, M., López-Delacalle, M., & Rivero, R. M. (2020). ROS and NO regulation by melatonin under abiotic stress in plants. Antioxidants, 9(11), 1078.
Pattyn, J., Vaughan-Hirsch, J., & Van de Poel, B. (2020). The regulation of ethylene biosynthesis: a complex multilevel control circuitry. New Phytologist, 229, 770–782. 10.1111/nph.16873.
Polko, J. K., & Kieber, J. J. (2019). The regulation of cellulose biosynthesis in plants. The Plant Cell, 31(2), 282–296.
Porrini, E., Ruggenenti, P., Luis-Lima, S., Carrara, F., Jiménez, A., de Vries, A. P., et al. (2019). Estimated GFR: time for a critical appraisal. Nature Reviews Nephrology, 15(3), 177–190.
Rakesh, B., Sudheer, W. N., & Nagella, P. (2021). Role of polyamines in plant tissue culture: An overview. Plant Cell, Tissue and Organ Culture, 145(3), 487–506.
Raskin, I. (1992). Role of salicylic acid in plants. Annual Review of Plant Biology, 43(1), 439–463.
Reinbothe, C., Springer, A., Samol, I., & Reinbothe, S. (2009). Plant oxylipins: role of jasmonic acid during programmed cell death, defense and leaf senescence. The FEBS Journal, 276(17), 4666–4681.
Rőszer, T. (2012). Nitric oxide synthesis in the mitochondria of animal cells. In: T. Rőszer (ed.), The biology of subcellular nitric oxide (pp. 169–178). Springer, Dordrecht.
Sacher, R. F. (1980). Physiological age of potato seed tubers. University Microfilms International, MI, USA: Washington State University.
Sachs, J. (1893). Ueber Wachsthurnsperioden und Bildungsreize. Marburg: Physiologische Notizen VI.
Schaller, A., & Stintzi, A. (2009). Enzymes in jasmonate biosynthesis–structure, function, regulation. Phytochemistry, 70(13-14), 1532–1538.

Schaller, G. E., Bishopp, A., & Kieber, J. J. (2015). The yin-yang of hormones: cytokinin and auxin interactions in plant development. The Plant Cell, 27(1), 44–63.
Shang, J. X., Li, X., Li, C., & Zhao, L. (2022). The role of nitric oxide in plant responses to salt stress. International Journal of Molecular Sciences, 23(11), 6167.
Shapiro, A. D. (2005). Nitric oxide signaling in plants. Vitamins & Hormones, 72, 339–398.
Sharma, M., & Laxmi, A. (2016). Jasmonates: emerging players in controlling temperature stress tolerance. Frontiers in Plant Science, 6. 10.3389/fpls.2015.01129.
Shitan, N., Sugiyama, A., & Yazaki, K. (2013). Functional analysis of jasmonic acid-responsive secondary metabolite transporters. In: A. Goossens, & L. Pauwels (eds.), Jasmonate Signaling (pp. 241–250). Humana Press, Totowa. 10.1007/978-1-62703-414-2_19.
Silva, A. L. L., Rodrigues, C., Costa, J. D. L., Machado, M. P., Penha, R. D. O., Biasi, L. A., et al. (2013). The gibberellic acid fermented extract obtained by solid-state fermentation using citric pulp by *Fusarium moniliforme*: Influence on Lavandula angustifolia Mill. cultivated in vitro. Pakistan Journal of Botany, 45, 2057–2064.
Simon, S., & Petrášek, J. (2011). Why plants need more than one type of auxin. Plant Science, 180(3), 454–460.
Skoog, F., Armstrong, D. J., Cherayil, J. D., Hampel, A. E., & Bock, R. M. (1966). Cytokinin activity: localization in transfer RNA preparations. Science, 154(3754), 1354–1356.
Taiz, L., & Zeiger, E. (2002). Plant physiology. Sinauer Associates, Sunderland. p. 306. ISBN 0-87893-823-0. OCLC 50002466.
Taiz, L., & Zeiger, E. (1998). Plant physiology and development, 2nd edition. Sinauer Associates, Inc., Publ. Sunderland. 792p.
Tossi, V. E., Acebedo, S. L., Cassia, R. O., Lamattina, L., Galagovsky, L. R., & Ramírez, J. A. (2015). A bioassay for brassinosteroid activity based on the in vitro fluorimetric detection of nitric oxide production. Steroids, 102, 46–52.
Umehara, M., Cao, M., Akiyama, K., Akatsu, T., Seto, Y., Hanada, A., et al. (2015). Structural requirements of strigolactones for shoot branching inhibition in rice and *Arabidopsis*. Plant and Cell Physiology, 56(6), 1059–1072.
Van Huijsduijnen, R. H., Alblas, S. W., De Rijk, R. H., & Bol, J. F. (1986). Induction by salicylic acid of pathogenesis-related proteins and resistance to alfalfa mosaic virus infection in various plant species. Journal of General Virology, 67(10), 2135–2143.
Van Staden, J., Sparg, S. G., Kulkarni, M. G., & Light, M. E. (2006). Post-germination effects of the smoke-derived compound 3-methyl-2H-furo [2, 3-c] pyran-2-one, and its potential as a preconditioning agent. Field Crops Research, 98(2-3), 98–105.
Vernooij, B., et al (1994). Salicylic acid is not the translocated signal responsible for inducing systemic acquired resistance but is required in signal transduction. The Plant Cell, 6, 959. 10.2307/3870006.
Vlot, A. C., Dempsey, D. M. A., & Klessig, D. F. (2009). Salicylic acid, a multifaceted hormone to combat disease. Annual Review of Phytopathology, 47, 177–206.
Vuosku, J., Karppinen, K., Muilu-Mäkelä, R., Kusano, T., Sagor, G. H. M., Avia, K., et al. (2018). Scots pine aminopropyltransferases shed new light on the evolution of the polyamine biosynthesis pathway in seed plants. Annals of Botany, 121(6), 1243–1256.
Waadt, R., Hitomi, K., Nishimura, N., Hitomi, C., Adams, S. R., Getzoff, E. D., & Schroeder, J. I. (2014). FRET-based reporters for the direct visualization of abscisic acid concentration changes and distribution in *Arabidopsis*. Life, 3, e01739.
Wang, D., Gao, Z., Du, P., Xiao, W., Tan, Q., Chen, X., et al. (2016). Expression of ABA metabolism-related genes suggests similarities and differences between seed dormancy and bud dormancy of peach (*Prunus persica*). Frontiers in Plant Science, 6, 1248.
Wang, H., Chang, X., Lin, J., Chang, Y., Chen, J. C., Reid, M. S., & Jiang, C. Z. (2018). Transcriptome profiling reveals regulatory mechanisms underlying corolla senescence in petunia. Horticulture Research, 5, 1–13.
Wasternack, C. (2007). Jasmonates: an update on biosynthesis, signal transduction and action in plant stress response, growth and development. Annals of Botany, 100(4), 681–697.
Wasternack, C., & Feussner, I. (2018). The oxylipin pathways: biochemistry and function. Annual Review of Plant Biology, 69, 363–386.
Wasternack, C., & Strnad, M. (2018). Jasmonates: News on occurrence, biosynthesis, metabolism, and action of an ancient group of signaling compounds. International Journal of Molecular Sciences, 19(9), 2539.
Weller, R., & Richard, W. (2012). Could the sun be good for your heart? TedxGlasgow March, 6073–6078.
Went, F. W., & Thimann, K. V. (1937). Phytohormones. MacMillan Company, New York.

Weyers, J. D., & Paterson, N. W. (2001). Plant hormones and the control of physiological processes. New Phytologist, 152(3), 375–407.
Xie, X., Yoneyama, K., & Yoneyama, K. (2010). The strigolactone story. Annual Review of Phytopathology, 48, 93–117.
Xu, C., Wu, X., & Zhang, H. (2009). Impact of D-Arg on drought resistance and endogenous polyamines in mycorrhizal *Pinus massoniana*. Journal of Nanjing Forestry University (Natural Sciences Edition), 33(4), 19–23.
Yamaguchi, S. (2008). Gibberellin metabolism and its regulation. Annual Review of Plant Biology, 59, 225–251. 10.1146/annual.plant.59.032607.092804.
Zhang, Y., Kilambi, H. V., Liu, J., Bar, H., Lazary, S., Egbaria, A., et al. (2021). ABA homeostasis and long-distance translocation are redundantly regulated by ABCG ABA importers. Science Advances, 7(43), eabf6069.
Zhang, N., Sun, Q., Zhang, H., Cao, Y., Weeda, S., Ren, S., & Guo, Y. D. (2015). Roles of melatonin in abiotic stress resistance in plants. Journal of Experimental Botany, 66, 647–656. 10.1093/jxb/eru336.
Zhao, C., Nawaz, G., Cao, Q., & Xu, T. (2022). Melatonin is a potential target for improving horticultural crop resistance to abiotic stress. Scientia Horticulturae, 291, 110560.

2 Plant Hormones Mediated Alleviation of Abiotic Stress

Vasundhara Sharma and Manisha Saini

2.1 INTRODUCTION

Climate change due to various natural and anthropogenic activities causes biotic and abiotic stresses affecting agriculture worldwide. Abiotic stresses such as drought, flood, salinity, sodicity, nutrient imbalances and high or low temperature are detrimental to plant growth and development, which causes 51–82% annual yield reduction in the world (Oshunsanya et al., 2019). Plants face different stresses throughout the growing period, which may occur simultaneously, in sequence or as concurrent stresses, causing more severity compared to individual stress in plants (Lamaoui et al., 2018). For instance, combined heat and drought stress cause a reduction in photosynthesis, nutrient uptake and yield attributes due to the over-production of reactive oxygen species in maize (Hussain et al., 2019). Similarly, the combined salinity and drought stress have a greater adverse impact on plant photosynthesis, growth, ionic balance and oxidative balance than individual stress (Angon et al., 2022). The effects of abiotic stress on plants, such as qualitative and quantitative changes in protein synthesis, delay in germination, membrane denaturation, stomatal closure, disruption of flower bud formation, etc., can occur from the germination of seed till the harvest of the crop, resulting in poor crop yields (Oshunsanya et al., 2019).

Being sessile, plants have to adopt different strategies to cope with these stresses and maintain their survival under adverse conditions. Plant responses to abiotic stress depend on the extent and duration of particular stress as well as also vary with age, developmental stage and genotype of the plant (Peleg et al., 2012). Plants regulate their growth cycle and stimulate cellular, biochemical and physiological adaptive responses such as efficient signaling mechanisms, synthesis of plant growth hormones, compatible solutes (sugars, proline), activation of antioxidative defense system (superoxide dismutase, catalase, peroxidase, ascorbate peroxidase and reduced glutathione) and other stress-responsive genes and proteins to optimize their yield under stress conditions (Anjum et al., 2016). Among these responses, plant hormones are the key regulators of plants' responses to abiotic stresses. The major hormones in regulating abiotic stress responses in plants are abscisic acid (ABA), ethylene (ET), salicylic acid (SA), jasmonic acid (JA) and brassinosteroids (BRs) (Figure 2.1). Other than these, hormones such as auxin, gibberellins (GAs) and cytokinins (CKs) are central regulators of plant growth and development, however, their role in abiotic stress tolerance has begun to be elucidated and there are growing pieces of evidence that these are an integral part in plant adaptation to abiotic stress. Some other compounds like karrikins, oxylipins, nitric oxides, polyamines, indoleamines and peptides are also recently known to play a role in abiotic stress responses in plants (Savchenko et al., 2014, Shah et al., 2020, Kim et al., 2021, Antala, 2022). The roles and mechanisms of abiotic stress-regulating hormones (ABA, ET, SA, JA, BR, auxin, GAs and CKs) are discussed in this chapter for the identification of different mechanisms used by plants to maintain their growth and survival under abiotic stress conditions.

DOI: 10.1201/9781003335788-3

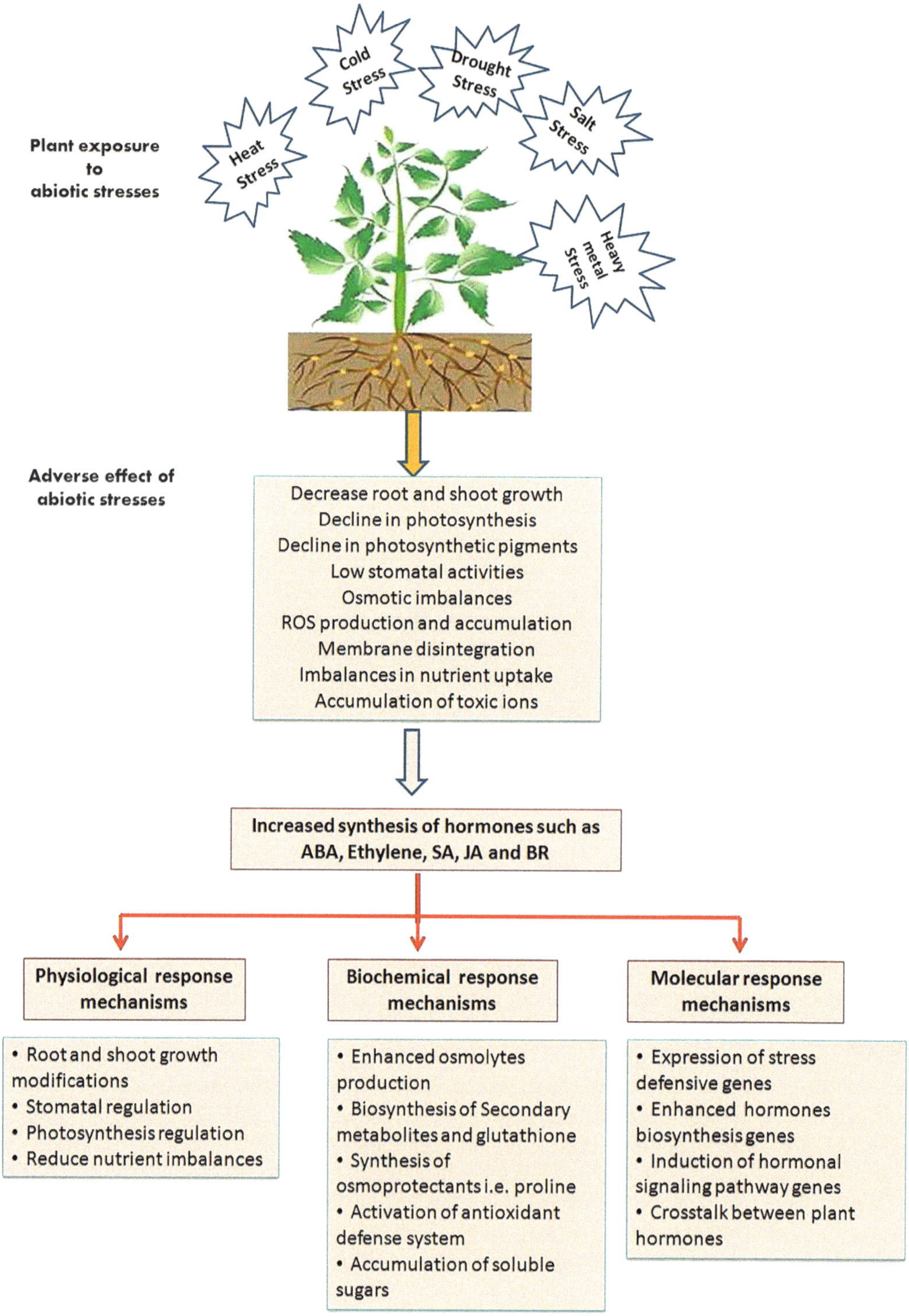

FIGURE 2.1 Role of hormones in plant defense responses against abiotic stresses.

2.2 ABSCISIC ACID

ABA is the central regulator of various abiotic stresses like drought, heavy metal stress, salinity stress, osmotic stress, low temperature, heat stress and radiation stress and also known as "stress hormone" (Wani & Kumar, 2015). ABA regulates various physiological processes such as seed germination, seed dormancy, stomatal closure, embryo morphogenesis, protein, lipid storage and leaf senescence under non-stress conditions. Increased endogenous ABA levels mainly control three important processes under abiotic stress conditions i.e., inhibition of seed germination to surpass stress, restriction of shoot and root growth and stomatal closure (Daszkowska-Golec, 2016). ABA regulates root system architecture under drought stress by promoting root growth, soil microbial communities and root exudation under limited water conditions (Aslam et al., 2022). ABA accumulation under drought stress conditions has been reported in many plants including wheat, maize, sorghum, barley, soybean and rice (Sah et al., 2016). ABA also helps in scavenging the reactive oxygen species (ROS) by initiating the ROS scavenging system, including various antioxidative enzymes such as superoxide dismutase (SOD), catalase (CAT), ascorbate peroxidase (APX), peroxidase (PDX) and glutathione reductase (GR); thus, reducing the oxidative damage in wheat (Zhou et al., 2019). ABA-induced and suppressed proteins are involved in the metabolic pathways of glycolysis, photosynthesis, protein synthesis and transport processes through membrane and stress responsiveness (Zhu et al., 2010). Osmotic stresses, including drought, cold and high salinity, can cause water loss from a cell in the vegetative stage (Fujita et al., 2011), and hence result in increased ABA biosynthesis (Nambara & Marion-Poll, 2005) and transport (Kuromori & Shinozaki, 2010). During heat stress, ABA accumulation upregulates the gene expression of heat shock proteins (HSPs), heat shock transcriptional factors (HSFs) and sucrose synthesis and transport genes to strengthen plant heat tolerance (Wang et al., 2017, Santiago & Sharkey, 2019). The exogenous ABA application was beneficial to protect membrane lipid peroxidation, maintaining a low Na^+/K^+ ratio, decreasing osmotic stress, enhancing the antioxidative defense system and hormonal balance under salt stress conditions; thus, reducing salt-induced damage (Chen et al., 2022). ABA mediates protection against metal stress such as Cd, Ni, Zn and Al by increasing growth and photosynthesis in plants (Hsu & Kao, 2003, Fediuc et al., 2005, Rizwan et al., 2017). Further, Xiong and Zhu (2003) also reported increased endogenous ABA under mineral and nutrient deficiency. ABA participates in the formation of the adventitious root, secondary aerenchyma and hyponastic growth through interaction with ethylene and GA under hypoxia due to floods in rice, soybean, tomato and other plants (Wang & Komatsu, 2022).

2.3 ETHYLENE

Ethylene is an endogenous gaseous hormone that controls various physiological and developmental processes such as ripening, senescence, cell elongation, adventitious rooting and multiple stress responses. Ethylene biosynthesis can be rapidly induced under abiotic stress conditions. Ethylene signaling due to abiotic stress results in physiological responses, which inhibit plant growth and confer stress tolerance that maximizes plant survival. Ethylene promotes adventitious root formation, regulates stem and petiole growth and closes stomata under salt stress conditions (Druege, 2006). Ethylene also mediates salinity stress tolerance through increased photosynthesis and growth (Nazar et al., 2014). In *Arabidopsis,* under salt stress, the ethylene biosynthesis gene (*ETO1*) increases ROS formation, with Na^+/K^+ homeostasis. During various stresses, ROS production and ethylene signaling are linked with each other (Yang et al., 2017). Flooding and heavy metal stress increase ROS production, which can be regulated by ethylene and the antioxidant defense system in plants (Steffens, 2014). The application of ethylene could increase the salinity tolerance in plants through the expression of ROS scavengers (Cao et al., 2007). Ethylene also regulates plant responses to nutrient deficiencies (K, N, P, Ca and Fe) through different metabolic pathways (García et al., 2015). Ethylene is involved in sulfur, nitrogen and proline metabolism; glycine betaine; secondary metabolites and glutathione synthesis to protect plants under adverse situations (Yoshida et al., 2009, Ma et al., 2021, Jahan et al., 2021, Gautam et al., 2022). Ethylene signaling and biosynthesis are also required for conferring thermotolerance and ameliorating heat-induced adverse effects on peas (Savada et al., 2017), tomatoes (Pan et al., 2019) and rice (Wu & Yang, 2019). Ethylene production has been linked to cold stress tolerance in grapes (Sun et al., 2016) and soybeans (Robison et al., 2019). Ethylene regulates the formation of aerenchyma, facilitates gas diffusion by adventitious roots, shoots elongation and elevates leaves above the water (Sasidharan & Voesenek, 2015). Increased levels of ethylene synthesis were observed in *Arabidopsis*, rice and soybeans under flooding stress (Wang & Komatsu, 2022).

2.4 SALICYLIC ACID

Salicylic acid (SA) is a phenolic compound that plays a crucial role in regulating plant defense responses against various pathogens, pests and abiotic stresses such as wounding, heat stress and exposure to ozone (Loake & Grant, 2007). SA regulates physiological processes in plants, such as photosynthesis, water relations, nitrogen, proline metabolism and antioxidant defense system under abiotic stress, thus enabling the plant to cope with abiotic stress (Khan et al., 2015, Wassie et al., 2020). SA application induced expression of stress-defensive genes such as HSPs, chaperones, cytochrome P450 and PR genes in *Arabidopsis* (Larkindale & Knight, 2002) and *Mitragyna speciosa* Korth. (Jumali et al., 2011). SA reduces the heat stress-induced plant membrane damage and improves plant growth, photosynthetic activities, proline production and antioxidant enzymes expression in soybeans (Wassie et al., 2020), tomatoes (Jahan et. al., 2019) and wheat (Khan et al., 2013). SA and melatonin application alleviates heat stress-induced damage in mint plants by restoring relative water content and increasing antioxidant enzymes (Haydari et al., 2019). Exogenous SA application improved the synthesis of total phenolics and the activity of PAL in cold-stored lemon to improve chilling tolerance (Siboza et al., 2014). Cold stress disturbs the ultrastructures in bananas which can be protected by SA application (Kang et al., 2007). Exogenous SA application to drought-stressed barley resulted in increased biomass and net CO_2 assimilation rate due to increased stomatal conductivity (Habibi, 2012). It also modulates the antioxidant defense system and decreases oxidative stress in drought-stressed plants (Alam et al., 2013). SA-induced stomatal closure and expression of PR genes were reported in SA-accumulating mutants in *Arabidopsis* to improve drought tolerance (Liu et al., 2022). Besides this, SA also governs plant tolerance to ozone and UV-B radiations, which significantly damage the plant defense system and metabolic activities. SA promotes molecular and physiological changes such as activating antioxidant response, maintaining cellular redox state and initiating processes against hypersensitive cell death in ozone-exposed plants (Pasqualini et al., 2002). Immersion in SA solutions improved post-harvest quality like the sensory quality, fresh weight and greenness, whereas ozone treatment suppressed the respiration rate and production of ethylene in parsley (Üner Öztürk & Koyuncu, 2021). SA treatment counteracts the effect of UV-A, UV-B and UV-C-induced damage on peppers by enhancing the photosynthetic pigment content and activities of antioxidant enzymes (Mahdavian et al., 2008). Application of SA was reported to increase growth and photosynthetic activities in different crops such as lead-exposed rice (Jing et al., 2007), Cd-exposed maize (Krantev et al., 2008) and Cu-exposed common bean (Zengin, 2014).

2.5 JASMONATES

Jasmonates are fatty acid derivates from plant hormones, including jasmonic acid (JAs), methyl jasmonates (MeJA) and jasmonate isoleucine conjugate (JA-Ile) (Ruan et al., 2019). JA signaling plays an important role in plant defense against abiotic stresses such as cold stress, salt stress, drought stress, heavy metal stress, ozone stress, light and circadian stress (Wang et al., 2020). JA mediates physiological responses including regulation of stomatal movement, accumulation of amino acids, soluble sugars and activation of the antioxidant system. It also induces the expression of JA-associated genes and transcription factors (Hu et al., 2017). JA interacts with other hormones such as ABA, SA, ET, GA, IAA and BR to induce physiological and molecular responses to mitigate the adverse effects of abiotic stress (Wang et al., 2020). In response to cold stress, JA enhances leaf senescence and positively regulates CBF (C-repeating binding factor) transcriptional pathways and, in turn, downstream cold-responsive gene expression, which improves cold tolerance in *Arabidopsis* (Hu et al., 2017). JA regulates stomatal movement to minimize water loss under drought stress conditions in *Arabidopsis* (Savchenko et al., 2014). Endogenous and exogenous JA participated in drought stress tolerance by expressing OsbHLH148 transcript expression in rice (Seo et al., 2011). JA improved plant growth, photosynthetic pigments, proline content and maintained ROS homeostasis in salt-stressed tomato plants (Abouelsaad & Renault, 2018). In salt-stressed wheat seedlings, exogenous application of JA decreased malondialdehyde (MDA) and hydrogen peroxide (H_2O_2) concentrations, whereas improved transcript expression of antioxidant enzymes improved the tolerance to salt stress (Qiu et al., 2014). Zhao et al. (2016) reported that JA deficiency mutant in tomatoes showed enhanced sensitivity to cadmium (Cd) stress by accumulation of more Cd in leaves and roots, increasing antioxidant enzymes activity and increasing accumulation of MDA and hydrogen peroxide in mutant plants compared to the control. Other than this, exogenous JA treatment also increases plant tolerance to high nickel (Ni) in *Zea mays*, *Glycine max* and other plants (Sirhindi et al., 2015, Azeem, 2018).

2.6 BRASSINOSTEROIDS

Brassinosteroids (BRs) are a group of steroid hormones with versatile roles in plant growth and development (Hafeez et al., 2021). BRs regulate processes such as cell division, cell elongation, xylem differentiation, photomorphogenesis, photosynthesis, nutrient metabolism, water status, plant reproduction and responses to biotic and abiotic stresses (Ahammed et al., 2020). BRs play a key role in initiating stress responses and promoting normal growth processes either independently or by crosstalk with other hormones (Planas-Riverola et al., 2019). BRs play a crucial role in environmental adaptation to abiotic stresses such as drought, heat, salinity, chilling, heavy metals, organic pollutants and pesticides (Kagale et al., 2007, Xia et al., 2018). Stress responses related to BR signaling pathways under abiotic stress conditions at the physiological level include an increase in photosynthesis, net CO_2 assimilation rate, stomatal conductance, photochemical efficiency of PSI, water use efficiency and increasing antioxidant enzymes that will minimize accumulation of ROS in the plant cell. It also increases the transcript-level expression of stress response and defense-related genes against adverse situations (Ahammed et al., 2020). BR treatment under salinity stress triggers ethylene biosynthesis and promotes BR-induced activities of antioxidant enzymes to enhance salt tolerance (Zhu et al., 2016). BR can induce nitric oxide production in a ROS-dependent manner, which is involved in BR-induced stress tolerance by mediating the induction of antioxidant genes and in turn induces antioxidant enzymes for coping with oxidative stress arising from abiotic stresses (Cui et al., 2011). Exogenous BR application counteracts growth inhibition and increases growth and plant survival exposed to heat and low-temperature stress (Sadura & Janeczko, 2018). BR promotes heat shock protein synthesis and protects the translation apparatus during prolonged heat stress in *Brassica napus* (Dhaubhadel et al., 2002). It also reduces the high temperature-induced photosynthesis inhibition by increasing antioxidant enzymes that minimize lipid peroxidation in brinjal (Wu et al., 2014), wheat and rice (Sonjaroon et al., 2018; Hussain et al., 2019). BR application also provides tolerance to drought stress conditions by improving photosynthesis, leaf water status, antioxidant defense and enhancing the ABA levels in grapevines (Wang et al., 2019) and tomatoes (Yuan et al., 2010). It has also been reported that BR mediates BR-associated signaling pathways to mitigate high zinc (Zn), mercury (Hg), Pb, arsenic (As) and Cd-induced metal stress in plants (Bücker-Neto et al., 2017). BR also improves pesticide and insecticide tolerance in plants (Hou et al., 2018; Sharma et al., 2019).

2.7 AUXIN

Auxin is regarded as an influential constituent of defense responses via the regulation of numerous genes and mediation of crosstalk between abiotic and biotic stress responses (Fahad et al., 2015). Auxin increases root and shoot growth of plants growing under salinity and heavy metal stress (Sheng & Xia, 2006; Egamberdieva, 2009). Elevated auxin levels by overexpressing auxin biosynthetic genes, such as *Arabidopsis YUC6* in potatoes and *iaaM* in *Arabidopsis*, modulate the expression of different abiotic stress-related genes (*DREB2A, DREB 2B, RAB 18, RD 22*), which regulates root architecture, ROS metabolism and antioxidant enzyme activities (Bielach et al., 2017). Overexpression of maize *ARGOS 1* and in *Arabidopsis*, *ARF5* homolog from sweet potato, whereas downregulation of the *ARF4* gene in tomato enhanced tolerance to drought and salinity (Nowicka, 2022). Auxin may contribute to crosstalk with other phytohormones, such as ABA and JA, to regulate the gene expression of TFs, such as WRKYs, improving cold stress tolerance in rice seedlings (Yang et al., 2015).

2.8 CYTOKININS (CKs)

Alteration of endogenous levels of CKs in response to stress indicates their involvement in abiotic stress, including drought and salinity (O'Brien & Benková, 2013). Mutants and transgenic cells/tissues with altered activity of cytokinin metabolic enzymes or perception machinery point towards their crucial involvement in several crop traits, including productivity and increased stress tolerance (Zalabák et al., 2013). CKs are often considered ABA antagonists, for instance, in water-stressed plants, decreased CK content and accumulation of ABA lead to an increased ABA/CK ratio. The reduced CK levels enhance apical dominance, which, together with the ABA regulation of the stomatal aperture, aids in adaptation to drought stress (O'Brien & Benková, 2013). CK acts antagonistically to ABA in regulating various processes in plants and helps in increasing salt tolerance. Recently, Wu et al. (2014) suggested that exogenous CK improved the salt resistance of brinjal, which was related to the increase in proline content. Furthermore, with increasing endogenous cytokinin in *Arabidopsis*, plants displayed an enhanced ability to survive freezing or dehydration (Kang et al., 2012). The role of CK in heat stress, nutrient stress and light stress has also been reported (Pavlů et al., 2018).

2.9 GIBBERELLINS (GAs)

There is increasing evidence for their vital roles in abiotic stress response and adaptation. GA reduction showed decreased growth under cold, salt and osmotic stress, whereas increased GA biosynthesis and signaling promote growth and plant escape responses to shading and submergence. The mutation of genes of GA metabolism and signaling (*GAI, RGA, RGL1* and *RGL2*) leads to DELLA-dependent reduced accumulation of bioactive GAs and regulates salinity tolerance in *Arabidopsis* (Achard et al., 2006). Overexpression of GA-related TFs such as DDF1, ERF6, CBF1, SNORKEL1, SNORKEL2 and SUB1A improved salinity tolerance, osmotic stress tolerance, cold tolerance and submergence tolerance, respectively, in *Arabidopsis* and rice (Vishal & Kumar, 2018). GA helps in improving plant detoxification mechanisms against heavy metal stress such as Pb and Cd with the activation of different genes, synthesis of catalase and accumulation of phytochelatins (Falkowska et al., 2011). GAs are known to interact with all other phytohormones, especially ABA and SA, in numerous developmental and stimulus-response processes (Vishal & Kumar, 2018, Emamverdian et al., 2020).

Other than these few studies, there are studies that support the role of nitric oxide, polyphenols and karrikins in the tolerance of plants to abiotic stress such as drought, salinity, low and high temperature, heavy metals, or nutrition deficiency (Siddiqui et al., 2011, Šamec et al., 2021, Antala, 2022).

2.10 CONCLUSION AND FUTURE PROSPECTS

Plant hormones play an important role in abiotic stress tolerance and maintaining the growth and survival of plants. In this chapter, we provided the functional attributes and mechanisms of plant hormones in the alleviation of major abiotic stress. This information may be useful for the application of hormones under stress conditions to maintain the yield potential of the plants. Furthermore, a comprehensive approach should be used to uncover the nature of underlying mechanisms, abiotic stress-induced upregulation of plant hormone biosynthesis genes as well as the precise description of hormonal crosstalk in influencing abiotic stress. Further research into genome editing and the use of computational tools for the integration of genome-scale mathematical modeling in systems biology, field and laboratory growth and development experiments and large-scale mutational analyses are important to provide deeper insight into plant growth and development in response to abiotic stress.

REFERENCES

Abouelsaad, I., & Renault, S. (2018). Enhanced oxidative stress in the jasmonic acid-deficient tomato mutant def-1 exposed to NaCl stress. Journal of Plant Physiology, 226, 136–144.

Achard, P., Cheng, H., De Grauwe, L., Decat, J., Schoutteten, H., Moritz, T.,& Harberd, N. P. (2006). Integration of plant responses to environmentally activated phytohormonal signals. Science, 311(5757), 91–94.

Ahammed, G. J., Li, X., Liu, A., & Chen, S. (2020). Brassinosteroids in plant tolerance to abiotic stress. Journal of Plant Growth Regulation, 39(4), 1451–1464.

Alam, M. M., Hasanuzzaman, M., Nahar, K., & Fujita, M. (2013). Exogenous salicylic acid ameliorates short-term drought stress in mustard (Brassica juncea L.) seedlings by up-regulating the antioxidant defense and glyoxalase system. Australian Journal of Crop Science, 7(7), 1053.

Angon, P. B., Tahjib-Ul-Arif, M., Samin, S. I., Habiba, U., Hossain, M. A., & Brestic, M. (2022). How do plants respond to combined drought and salinity stress?—A systematic review. Plants, 11(21), 2884.

Anjum, S. A., Tanveer, M., Ashraf, U., Hussain, S., Shahzad, B., Khan, I., & Wang, L. (2016). Effect of progressive drought stress on growth, leaf gas exchange, and antioxidant production in two maize cultivars. Environmental Science and Pollution Research, 23(17), 17132–17141.

Antala, M. (2022). Physiological roles of karrikins in plants under abiotic stress conditions. In Emerging Plant Growth Regulators in Agriculture (pp. 193–204). Academic Press.

Aslam, M. M., Waseem, M., Jakada, B. H., Okal, E. J., Lei, Z., Saqib, H. S. A., Yuan, W., Xu, W. and Zhang, Q. (2022). Mechanisms of Abscisic Acid-Mediated Drought Stress Responses in Plants. International Journal of Molecular Sciences, 23(3), 1084.

Azeem, U. (2018). Ameliorating Nickel stress by Jasmonic acid treatment in Zea mays L. Russian Agricultural Sciences, 44(3), 209–215.

Bielach, A., Hrtyan, M., & Tognetti, V. B. (2017). Plants under stress: involvement of auxin and cytokinin. International Journal of Molecular Sciences, 18(7), 1427.

Bücker-Neto, L., Paiva, A. L. S., Machado, R. D., Arenhart, R. A., & Margis-Pinheiro, M. (2017). Interactions between plant hormones and heavy metals responses. Genetics and Molecular Biology, 40, 373–386.

Cao, W. H., Liu, J., He, X. J., Mu, R. L., Zhou, H. L., Chen, S. Y., & Zhang, J. S. (2007). Modulation of ethylene responses affects plant salt-stress responses. Plant Physiology, 143(2), 707–719.

Chen, G., Zheng, D., Feng, N., Zhou, H., Mu, D., Zhao, L., Shen, X., Rao, G., Meng, F. & Huang, A. (2022). Physiological mechanisms of ABA-induced salinity tolerance in leaves and roots of rice. Scientific Reports, 12(1), 1–26.

Cui, J. X., Zhou, Y. H., Ding, J. G., Xia, X. J., Shi, K. A. I., Chen, S. C., Asami, T., Chen, Z. & Yu, J. Q. (2011). Role of nitric oxide in hydrogen peroxide-dependent induction of abiotic stress tolerance by brassinosteroids in cucumber. Plant, Cell & Environment, 34(2), 347–358.

Daszkowska-Golec, A. (2016). The role of abscisic acid in drought stress: how ABA helps plants to cope with drought stress. In Drought Stress Tolerance in Plants, Vol 2 (pp. 123–151). Springer.

Dhaubhadel, S., Browning, K. S., Gallie, D. R., & Krishna, P. (2002). Brassinosteroid functions to protect the translational machinery and heat-shock protein synthesis following thermal stress. The Plant Journal, 29(6), 681–691.

Druege, U. (2006). Ethylene and plant responses to abiotic stress. In Ethylene Action in Plants (pp. 81–118). Springer.

Egamberdieva, D. (2009). Alleviation of salt stress by plant growth regulators and IAA producing bacteria in wheat. Acta Physiologiae Plantarum, 31(4), 861–864.

Emamverdian, A., Ding, Y., & Mokhberdoran, F. (2020). The role of salicylic acid and gibberellin signaling in plant responses to abiotic stress with an emphasis on heavy metals. Plant Signaling & Behavior, 15(7), 1777372.

Fahad, S., Hussain, S., Matloob, A., Khan, F. A., Khaliq, A., Saud, S., Hassan, S., Shan, D., Khan, F., Ullah, N. & Faiq, M. (2015). Phytohormones and plant responses to salinity stress: a review. Plant Growth Regulation, 75, 391–404.

Falkowska, M., Pietryczuk, A., Piotrowska, A., Bajguz, A., Grygoruk, A., & Czerpak, R. (2011). The effect of gibberellic acid (GA_3) on growth, metal biosorption and metabolism of the green algae Chlorella vulgaris (Chlorophyceae) Beijerinck exposed to cadmium and lead stress. Polish Journal of Environmental Studies, 20(1), 53–59.

Fediuc, E., Lips, S. H., & Erdei, L. (2005). O-acetylserine (thiol) lyase activity in Phragmites and Typha plants under cadmium and NaCl stress conditions and the involvement of ABA in the stress response. Journal of Plant Physiology, 162(8), 865–872.

Fujita, Y., Fujita, M., Shinozaki, K., & Yamaguchi-Shinozaki, K. (2011). ABA-mediated transcriptional regulation in response to osmotic stress in plants. Journal of Plant Research, 124(4), 509–525.

García, M. J., Romera, F. J., Lucena, C., Alcántara, E., & Pérez-Vicente, R. (2015). Ethylene and the regulation of physiological and morphological responses to nutrient deficiencies. Plant Physiology, 169(1), 51–60.

Gautam, H., Fatma, M., Sehar, Z., Mir, I. R., & Khan, N. A. (2022). Hydrogen sulfide, ethylene, and nitric oxide regulate redox homeostasis and protect photosynthetic metabolism under high temperature stress in rice plants. Antioxidants, 11(8), 1478.

Habibi, G. (2012). Exogenous salicylic acid alleviates oxidative damage of barley plants under drought stress. Acta Biologica Szegediensis, 56(1), 57–63.

Hafeez, M. B., Raza, A., Zahra, N., Shaukat, K., Akram, M. Z., Iqbal, S., & Basra, S. M. A. (2021). Gene regulation in halophytes in conferring salt tolerance. In Handbook of Bioremediation (pp. 341–370). Academic Press.

Haydari, M., Maresca, V., Rigano, D., Taleei, A., Shahnejat-Bushehri, A. A., Hadian, J. Sorbo, S., Guida, M., Manna, C., Piscopo, M. & Basile, A. (2019). Salicylic acid and melatonin alleviate the effects of heat stress on essential oil composition and antioxidant enzyme activity in Mentha× piperita and Mentha arvensis L. Antioxidants, 8(11), 547.

Hou, J., Zhang, Q., Zhou, Y., Ahammed, G. J., Zhou, Y., Yu, J., Fang, H. & Xia, X. (2018). Glutaredoxin GRXS16 mediates brassinosteroid-induced apoplastic H2O2 production to promote pesticide metabolism in tomato. Environmental Pollution, 240, 227–234.

Hsu, Y. T., & Kao, C. H. (2003). Role of abscisic acid in cadmium tolerance of rice (Oryza sativa L.) seedlings. Plant, Cell & Eenvironment, 26(6), 867–874.

Hu, Y., Jiang, Y., Han, X., Wang, H., Pan, J., & Yu, D. (2017). Jasmonate regulates leaf senescence and tolerance to cold stress: crosstalk with other phytohormones. Journal of Experimental Botany, 68(6), 1361–1369.

Hussain, H. A., Men, S., Hussain, S., Chen, Y., Ali, S., Zhang, S., Zhang, K., Li, Y., Xu, Q., Liao, C. & Wang, L. (2019). Interactive effects of drought and heat stresses on morpho-physiological attributes, yield, nutrient uptake and oxidative status in maize hybrids. Scientific Reports, 9(1), 1–12.

Jahan, B., Iqbal, N., Fatma, M., Sehar, Z., Masood, A., Sofo, A., D'Ippolito, I. & Khan, N. A. (2021). Ethylene supplementation combined with split application of nitrogen and sulfur protects salt-inhibited photosynthesis through optimization of proline metabolism and antioxidant system in mustard (Brassica juncea L.). Plants, 10(7), 1303.

Jahan, M. S., Wang, Y., Shu, S., Zhong, M., Chen, Z., Wu, J., Sun, J.& Guo, S. (2019). Exogenous salicylic acid increases the heat tolerance in Tomato (Solanum lycopersicum L) by enhancing photosynthesis efficiency and improving antioxidant defense system through scavenging of reactive oxygen species. Scientia Horticulturae, 247, 421–429.

Jing, C. H. E. N., Cheng, Z. H. U., Li, L. P., Sun, Z. Y., & Pan, X. B. (2007). Effects of exogenous salicylic acid on growth and H2O2-metabolizing enzymes in rice seedlings under lead stress. Journal of Environmental Sciences, 19(1), 44–49.

Jumali, S. S., Said, I. M., Ismail, I., & Zainal, Z. (2011). Genes induced by high concentration of salicylic acid in'Mitragyna speciosa'. Australian Journal of Crop Science, 5(3), 296–303.

Kagale, S., Divi, U. K., Krochko, J. E., Keller, W. A., & Krishna, P. (2007). Brassinosteroid confers tolerance in *Arabidopsis* thaliana and Brassica napus to a range of abiotic stresses. Planta, 225(2), 353–364.
Kang, G. Z., Wang, Z. X., Xia, K. F., & Sun, G. C. (2007). Protection of ultrastructure in chilling-stressed banana leaves by salicylic acid. Journal of Zhejiang University Science B, 8, 277–282.
Kang, N. Y., Cho, C., Kim, N. Y., & Kim, J. (2012). Cytokinin receptor-dependent and receptor-independent pathways in the dehydration response of *Arabidopsis* thaliana. Journal of Plant Physiology, 169(14), 1382–1391.
Khan, M. I. R., Fatma, M., Per, T. S., Anjum, N. A., & Khan, N. A. (2015). Salicylic acid-induced abiotic stress tolerance and underlying mechanisms in plants. Frontiers in Plant Science, 6, 462.
Khan, M. I. R., Iqbal, N., Masood, A., Per, T. S., & Khan, N. A. (2013). Salicylic acid alleviates adverse effects of heat stress on photosynthesis through changes in proline production and ethylene formation. Plant Signaling & Behavior, 8(11), e26374.
Kim, J. S., Jeon, B. W., & Kim, J. (2021). Signaling peptides regulating abiotic stress responses in plants. Frontiers in Plant Science, 12, 704490.
Krantev, A., Yordanova, R., Janda, T., Szalai, G., & Popova, L. (2008). Treatment with salicylic acid decreases the effect of cadmium on photosynthesis in maize plants. Journal of Plant Physiology, 165(9), 920–931.
Kuromori, T., & Shinozaki, K. (2010). ABA transport factors found in *Arabidopsis* ABC transporters. Plant Signaling & Behavior, 5(9), 1124–1126.
Lamaoui, M., Jemo, M., Datla, R., & Bekkaoui, F. (2018). Heat and drought stresses in crops and approaches for their mitigation. Frontiers in Chemistry, 6, 26.
Larkindale, J., & Knight, M. R. (2002). Protection against heat stress-induced oxidative damage in *Arabidopsis* involves calcium, abscisic acid, ethylene, and salicylic acid. Plant Physiology, 128(2), 682–695.
Liu, J., Qiu, G., Liu, C., Li, H., Chen, X., Fu, Q., Lin Y. & Guo, B. (2022). Salicylic acid, a multifaceted hormone, combats abiotic stresses in plants. Life, 12(6), 886.
Loake, G., & Grant, M. (2007). Salicylic acid in plant defence—the players and protagonists. Current Opinion in Plant Biology, 10(5), 466–472.
Ma, W., Xu, L., Gao, S., Lyu, X., Cao, X., & Yao, Y. (2021). Melatonin alters the secondary metabolite profile of grape berry skin by promoting VvMYB14-mediated ethylene biosynthesis. Horticulture Research, 8, 1–15.
Mahdavian, K., Ghorbanli, M., & Kalantari, K. M. (2008). Role of salicylic acid in regulating ultraviolet radiation-induced oxidative stress in pepper leaves. Russian Journal of Plant Physiology, 55(4), 560–563.
Nambara, E., & Marion-Poll, A. (2005). Abscisic acid biosynthesis and catabolism. Annual Review of Plant Biology, 56, 165.
Nazar, R., Khan, M. I. R., Iqbal, N., Masood, A., & Khan, N. A. (2014). Involvement of ethylene in reversal of salt-inhibited photosynthesis by sulfur in mustard. Physiologia Plantarum, 152(2), 331–344.
Nowicka, B. (2022). Modifications of phytohormone metabolism aimed at stimulation of plant growth, improving their productivity and tolerance to abiotic and biotic stress factors. Plants, 11(24), 3430.
O'Brien, J. A., & Benková, E. (2013). Cytokinin cross-talking during biotic and abiotic stress responses. Frontiers in Plant Science, 4, 451.
Oshunsanya, S. O., Nwosu, N. J., & Li, Y. (2019). Abiotic stress in agricultural crops under climatic conditions. In Sustainable Agriculture, Forest and Environmental Management (pp. 71–100). Springer.
Pan, C., Zhang, H., Ma, Q., Fan, F., Fu, R., Ahammed, G. J., Yu, J. and Shi, K. (2019). Role of ethylene biosynthesis and signaling in elevated CO2-induced heat stress response in tomato. Planta, 250(2), 563–572.
Pasqualini, S., Della Torre, G., Ferranti, F., Ederli, L., Piccioni, C., Reale, L., & Antonielli, M. (2002). Salicylic acid modulates ozone-induced hypersensitive cell death in tobacco plants. Physiologia Plantarum, 115(2), 204–212.
Pavlů, J., Novák, J., Koukalová, V., Luklová, M., Brzobohatý, B., & Černý, M. (2018). Cytokinin at the crossroads of abiotic stress signalling pathways. International Journal of Molecular Sciences, 19(8), 2450.
Peleg, Z. V. I., Walia, H., & Blumwald, E. (2012). Integrating genomics and genetics to accelerate development of drought and salinity tolerant crops. In Plant Biotechnology and Agriculture (pp. 271–286). Academic Press.
Planas-Riverola, A., Gupta, A., Betegón-Putze, I., Bosch, N., Ibañes, M., & Caño-Delgado, A. I. (2019). Brassinosteroid signaling in plant development and adaptation to stress. Development, 146(5), dev151894.

Qiu, Z., Guo, J., Zhu, A., Zhang, L., & Zhang, M. (2014). Exogenous jasmonic acid can enhance tolerance of wheat seedlings to salt stress. Ecotoxicology and Environmental Safety, 104, 202–208.
Rizwan, M., Ali, S., Abbas, F., Adrees, M., Zia-ur-Rehman, M., Farid, M., Gill, R. A. and Ali, B., (2017). Role of organic and inorganic amendments in alleviating heavy metal stress in oil seed crops. In P. Ahmad (ed.), Oil Seed Crops: Yield and Adaptations under Environmental Stress (pp. 224–235). John Wiley and Sons.
Robison, J. D., Yamasaki, Y., & Randall, S. K. (2019). The ethylene signaling pathway negatively impacts CBF/DREB-regulated cold response in soybean (*Glycine max*). Frontiers in Plant Science, 10, 121.
Ruan, J., Zhou, Y., Zhou, M., Yan, J., Khurshid, M., Weng, W. Cheng, J. & Zhang, K. (2019). Jasmonic acid signaling pathway in plants. International Journal of Molecular Sciences, 20(10), 2479.
Sadura, I., & Janeczko, A. (2018). Physiological and molecular mechanisms of brassinosteroid-induced tolerance to high and low temperature in plants. Biologia Plantarum, 62(4), 601–616.
Sah, S. K., Reddy, K. R., & Li, J. (2016). Abscisic acid and abiotic stress tolerance in crop plants. Frontiers in Plant Science, 7, 571.
Šamec, D., Karalija, E., Šola, I., Vujčić Bok, V., & Salopek-Sondi, B. (2021). The role of polyphenols in abiotic stress response: The influence of molecular structure. Plants (Basel), 10(1), 118. doi:10.3390/plants10010118.
Santiago, J. P., & Sharkey, T. D. (2019). Pollen development at high temperature and role of carbon and nitrogen metabolites. Plant, Cell & Environment, 42(10), 2759–2775.
Sasidharan, R., & Voesenek, L. A. (2015). Ethylene-mediated acclimations to flooding stress. Plant Physiology, 169(1), 3–12.
Savada, R. P., Ozga, J. A., Jayasinghege, C., Waduthanthri, K. D., & Reinecke, D. M. (2017). Heat stress differentially modifies ethylene biosynthesis and signaling in pea floral and fruit tissues. Plant Molecular Biology, 95(3), 313–331.
Savchenko, T. V., Zastrijnaja, O. M., & Klimov, V. V. (2014). Oxylipins and plant abiotic stress resistance. Biochemistry (Moscow), 79(4), 362–375.
Seo, J. S., Joo, J., Kim, M. J., Kim, Y. K., Nahm, B. H., Song, S. I., Cheong, J. J., Lee, J. S., Kim, J. K. & Choi, Y. D. (2011). OsbHLH148, a basic helix-loop-helix protein, interacts with OsJAZ proteins in a jasmonate signaling pathway leading to drought tolerance in rice. The Plant Journal, 65(6), 907–921.
Shah, F. A., Wei, X., Wang, Q., Liu, W., Wang, D., Yao, Y., Hu, H., Chen, X., Huang, S., Hou, J. & Lu, R. (2020). Karrikin improves osmotic and salt stress tolerance via the regulation of the redox homeostasis in the oil plant Sapium sebiferum. Frontiers in Plant Science, 11, 216.
Sharma, A., Yuan, H., Kumar, V., Ramakrishnan, M., Kohli, S. K., Kaur, R., Thukral, A. K., Bhardwaj, R. & Zheng, B. (2019). Castasterone attenuates insecticide induced phytotoxicity in mustard. Ecotoxicology and Environmental Safety, 179, 50–61.
Sheng, X. F., & Xia, J. J. (2006). Improvement of rape (Brassica napus) plant growth and cadmium uptake by cadmium-resistant bacteria. Chemosphere, 64(6), 1036–1042.
Siboza, X. I., Bertling, I., & Odindo, A. O. (2014). Salicylic acid and methyl jasmonate improve chilling tolerance in cold-stored lemon fruit (Citrus limon). Journal of Plant Physiology, 171(18), 1722–1731.
Siddiqui, M. H., Al-Whaibi, M. H., & Basalah, M. O. (2011). Role of nitric oxide in tolerance of plants to abiotic stress. Protoplasma, 248, 447–455.
Sirhindi, G., Mir, M. A., Sharma, P., Gill, S. S., Kaur, H., & Mushtaq, R. (2015). Modulatory role of jasmonic acid on photosynthetic pigments, antioxidants and stress markers of Glycine max L. under nickel stress. Physiology and Molecular Biology of Plants, 21(4), 559–565.
Sonjaroon, W., Jutamanee, K., Khamsuk, O., Thussagunpanit, J., Kaveeta, L., & Suksamrarn, A. (2018). Impact of brassinosteroid mimic on photosynthesis, carbohydrate content and rice seed set at reproductive stage under heat stress. Agriculture and Natural Resources, 52(3), 234–240.
Steffens, B. (2014). The role of ethylene and ROS in salinity, heavy metal, and flooding responses in rice. Frontiers in Plant Science, 5, 685.
Sun, X., Zhao, T., Gan, S., Ren, X., Fang, L., Karungo, S. K., Wang, Y., Chen, L., Li, S. & Xin, H. (2016). Ethylene positively regulates cold tolerance in grapevine by modulating the expression of ETHYLENE RESPONSE FACTOR 057. Scientific Reports, 6(1), 1–14.
Üner Öztürk, K., & Koyuncu, M. A. (2021). Effects of ozone and salicylic acid on post-harvest quality of parsley during storage. Biological Agriculture & Horticulture, 37(3), 183–196.
Vishal, B., & Kumar, P. P. (2018). Regulation of seed germination and abiotic stresses by gibberellins and abscisic acid. Frontiers in Plant Science, 9, 838.
Wang, J., Song, L., Gong, X., Xu, J., & Li, M. (2020). Functions of jasmonic acid in plant regulation and response to abiotic stress. International Journal of Molecular Sciences, 21(4), 1446.

Wang, X., & Komatsu, S. (2022). The role of phytohormones in plant response to flooding. International Journal of Molecular Sciences, 23(12), 6383.
Wang, X., Zhuang, L., Shi, Y., & Huang, B. (2017). Up-regulation of HSFA2c and HSPs by ABA contributing to improved heat tolerance in tall fescue and *Arabidopsis*. International Journal of Molecular Sciences, 18(9), 1981.
Wang, Y. T., Chen, Z. Y., Jiang, Y., Duan, B. B., & Xi, Z. M. (2019). Involvement of ABA and antioxidant system in brassinosteroid-induced water stress tolerance of grapevine (Vitis vinifera L.). Scientia Horticulturae, 256, 108596.
Wani, S. H., & Kumar, V. (2015). Plant stress tolerance: engineering ABA: a potent Phytohormone. Transcriptomics 3, 1000113.
Wassie, M., Zhang, W., Zhang, Q., Ji, K., Cao, L., & Chen, L. (2020). Exogenous salicylic acid ameliorates heat stress-induced damages and improves growth and photosynthetic efficiency in alfalfa (Medicago sativa L.). Ecotoxicology and Environmental Safety, 191, 110206.
Wu, X., Yao, X., Chen, J., Zhu, Z., Zhang, H., & Zha, D. (2014). Brassinosteroids protect photosynthesis and antioxidant system of eggplant seedlings from high-temperature stress. Acta Physiologiae Plantarum, 36(2), 251–261.
Wu, Y. S., & Yang, C. Y. (2019). Ethylene-mediated signaling confers thermotolerance and regulates transcript levels of heat shock factors in rice seedlings under heat stress. Botanical Studies, 60(1), 1–12.
Xia, X. J., Fang, P. P., Guo, X., Qian, X. J., Zhou, J., Shi, K., Zhou, Y. H., & Yu, J. Q. (2018). Brassinosteroid-mediated apoplastic H2O2-glutaredoxin 12/14 cascade regulates antioxidant capacity in response to chilling in tomato. Plant, Cell & Environment, 41(5), 1052–1064.
Xiong, L., & Zhu, J. K. (2003). Regulation of abscisic acid biosynthesis. Plant Physiology, 133(1), 29–36.
Yang, C., Li, W., Cao, J., Meng, F., Yu, Y., Huang, J., Jiang, L., Liu, M., Zhang, Z., Chen, X. and Miyamoto, K. (2017). Activation of ethylene signaling pathways enhances disease resistance by regulating ROS and phytoalexin production in rice. The Plant Journal, 89(2), 338–353.
Yang, Y. W., Chen, H. C., Jen, W. F., Liu, L. Y., & Chang, M. C. (2015). Comparative transcriptome analysis of shoots and roots of TNG67 and TCN1 rice seedlings under cold stress and following subsequent recovery: insights into metabolic pathways, phytohormones, and transcription factors. PLoS One, 10(7), e0131391.
Yoshida, S., Tamaoki, M., Ioki, M., Ogawa, D., Sato, Y., Aono, M., Kubo, A., Saji, S., Saji, H., Satoh, S. and Nakajima, N. (2009). Ethylene and salicylic acid control glutathione biosynthesis in ozone-exposed *Arabidopsis* thaliana. Physiologia Plantarum, 136(3), 284–298.
Yuan, G. F., Jia, C. G., Li, Z., Sun, B., Zhang, L. P., Liu, N., & Wang, Q. M. (2010). Effect of brassinosteroids on drought resistance and abscisic acid concentration in tomato under water stress. Scientia Horticulturae, 126(2), 103–108.
Zalabák, D., Pospíšilová, H., Šmehilová, M., Mrízová, K., Frébort, I., & Galuszka, P. (2013). Genetic engineering of cytokinin metabolism: prospective way to improve agricultural traits of crop plants. Biotechnology Advances, 31(1), 97–117.
Zengin, F. (2014). Exogenous treatment with salicylic acid alleviating copper toxicity in bean seedlings. Proceedings of the National Academy of Sciences, India Section B: Biological Sciences, 84(3), 749–755.
Zhao, S., Ma, Q., Xu, X., Li, G., & Hao, L. (2016). Tomato jasmonic acid-deficient mutant spr2 seedling response to cadmium stress. Journal of Plant Growth Regulation, 35(3), 603–610.
Zhou, Y., He, R., Guo, Y., Liu, K., Huang, G., Peng, C., Liu, Y., Zhang, M., Li, Z. & Duan, L. (2019). A novel ABA functional analogue B2 enhances drought tolerance in wheat. Scientific Reports, 9(1), 1–9.
Zhu, Q., Zhang, J., Gao, X., Tong, J., Xiao, L., Li, W., & Zhang, H. (2010). The *Arabidopsis* AP2/ERF transcription factor RAP2. 6 participates in ABA, salt and osmotic stress responses. Gene, 457(1-2), 1–12.
Zhu, T., Deng, X., Zhou, X., Zhu, L., Zou, L., Li, P., Zhang, D. & Lin, H. (2016). Ethylene and hydrogen peroxide are involved in brassinosteroid-induced salt tolerance in tomato. Scientific Reports, 6(1), 1–15.

3 Recent Discoveries and Prospects of Phytohormones

Vaishnavi Dahiya, Sakshi Suman, Aparna Nautiyal, and Pooja Baweja

3.1 INTRODUCTION

Stress can be defined as an instant shift from normal environmental conditions, which disturbs the initial homeostatic state. Most of the time, stress is measured in terms of survival or crop productivity, which in turn is correlated to overall plant growth (Kranner et al., 2010; Baweja & Kumar, 2020). The plant mainly requires carbon, energy, light, mineral nutrients and water for their optimum growth, whenever any of these requirements are either limited or in excessive amounts, there is development of stress (Baweja & Kumar, 2020). Plants that experience Eu-stress undergo positive and growth-promoting stress, whereas the stress that affects plants negatively is called De-stress (Lichtenthaler, 1998). Plant stress can be divided into two broad categories: abiotic and biotic (Hopkins & Huner, 2009). Light, temperature, wind, water, flooding, drought, salinity, relative humidity nutrition, etc., are responsible for abiotic stress (Figure 3.1).

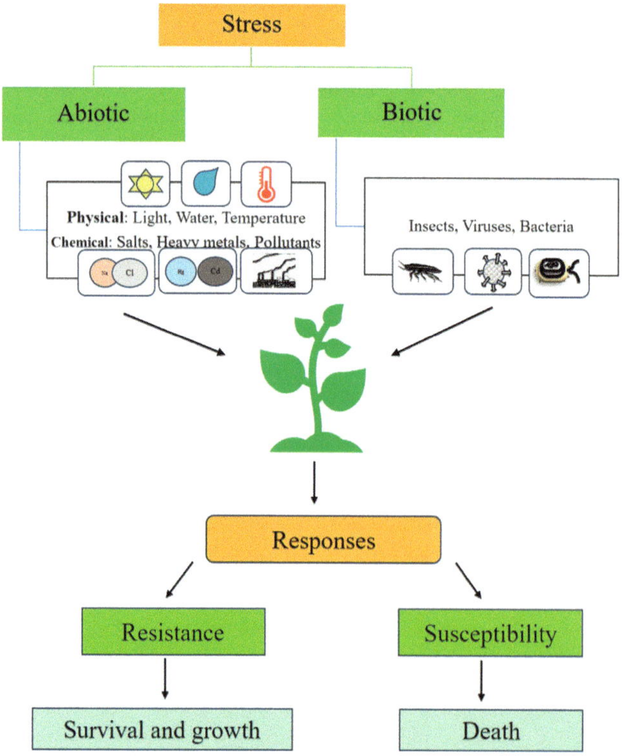

FIGURE 3.1 Biotic and abiotic stresses affecting plant growth and development (modified after Hopkins & Huner, 2009).

DOI: 10.1201/9781003335788-4

The stress condition directly results in a change in the structural properties, affects the signal transduction pathway and ultimately affects the physiological and morphological activity of plants and their adaptations to the response of stress (Vu et al., 2019). Plants that respond to various stresses are highly complex and involve changes at the transcriptome, cellular and physiological levels. Phytohormones play a crucial role in response to plants in abiotic and biotic stress, from which the plant may attempt to escape or survive under stressful conditions and focus on its resources to withstand the stress and develop specific protective mechanisms (Skirycz & Inze, 2010). Keeping this in view, some recent discoveries of phytohormones are discussed in this chapter.

3.2 SMALL RNAs (sRNA) AND PHYTOHORMONES

Small RNAs (sRNA) are approximately 21–24 nucleotides in size and mainly include small interfering RNAs (siRNA) and micro RNAs (miRNA) that are considered essential regulators of plant development under favorable as well as unfavorable conditions. Abiotic stress responses trigger the production of sRNAs, which sequentially regulate the expression of defense-related genes. Recent studies have identified emerging connections between sRNAs and plant hormones. sRNAs are involved in various plant hormonal crosstalk (Li et al., 2020).

ABA/Et are known to modify the levels of sRNAs. Overexpression of some ABA-responsive sRNAs e.g., ath-miR168/393/394, makes *Arabidopsis* more resistant to abiotic stress like drought and salinity, pointing towards a possible involvement of miRNAs in the abscisic acid–mediated stress responses (Song et al., 2013; Long et al., 2017; Li et al., 2020). The advancing network between sRNA and phytohormones implies that hormone biosynthesis steps are controlled by endogenous sRNAs, mainly for GA, auxin, CK and JA. At the signaling level, sRNA regulators either regulate hormone responses or target the genes involved in the signaling. Hormones shape plant phenotypic plasticity under optimal and stress conditions. sRNAs also regulate the plant's phenotypic plasticity under normal and stressful conditions along with hormones. High-throughput sRNA sequencing revealed that hormone treatment altered large sets of sRNAs in different species (Li et al., 2020).

sRNA biogenesis and miRNA precursors are regulated by HRFs (hormone-responsive factors). sRNAs are considered axes in hormonal networks. Given below are two sRNAs illustrating this:

1. miR159 levels in different species are controlled by hormones like BR, GA and ABA and miR159 action suggests that it is involved in many hormonal pathways. It elevates the biosynthesis of GA and BR and inhibits CK biosynthesis, while interfering with ABA in seed germination inhibition. miR159 targets a transcription factor of the MYB-like family (Alonso-Peral et al., 2010). To confirm its action, miR159 was suppressed in *Arabidopsis thaliana* L. and *Oryza sativa* L., which caused pleiotropic effects in plant growth suggesting that hormone levels are altered.
2. miR156 is a main player in age-to-phase transition. SPL expression is inhibited by miR156 as it suppresses GA or SL signaling, thus participating in flowering, branching and JA-dependent biotic defense.

These two miR159 and miR156 also crosstalk with each other because the deficiency of miR159 promotes levels of miR156 leading to delayed vegetative development.

3.3 PHYTOHORMONE PRIMING

Pre-treatment with exogenous application of phytohormones in prime plant cells is done towards developing defense strategies. The exogenous application activates crosstalk between phytohormones. Priming the seeds with hormonal solutions is known as hormonal priming and it has an important role

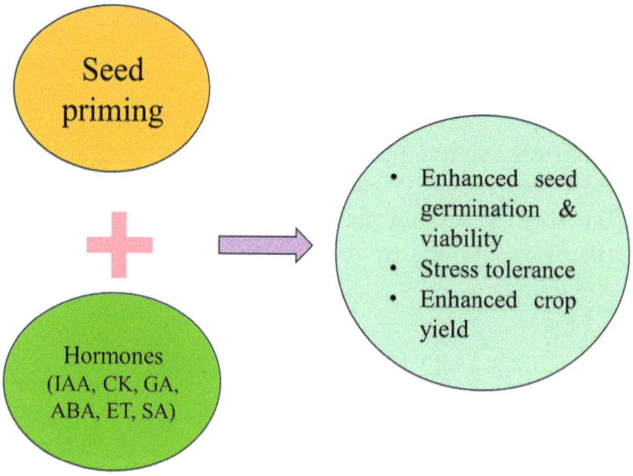

FIGURE 3.2 Seed priming with phytohormones and its effects (modified after Rhaman et al., 2020).

in maintaining healthy and productive plants and improving crop yield in adverse conditions (Hasanuzzaman & Fotopoulos, 2019). This results in enhanced seed germination, seedling growth, increased nutrient uptake and crop yield (Figure 3.2). Abiotic stress tolerance is conferred by seed priming by modulating the biochemical and molecular mechanisms involved in the abiotic defense of plants.

In *Agropyrone longatum* Host ex P.Beauv., seed priming with indole acetic acid (IAA) at 50 ppm improved tolerance to drought stress by amplifying the activity of antioxidant enzymes like catalase, superoxide dismutase and peroxidase (Eisvand et al., 2010). Seed priming with CK increased salt tolerance, germination, formation of chlorophyll and accumulation of biomass by decreasing ABA and increasing IAA concentration (Rhaman et al., 2020). The application of GA3 on the foliar region of *Solanum lycopersicum* L. increased relative water content in leaves, stomatal density and chlorophyll content by combating salinity stress (Jayasinghe et al., 2019).

Exogenous application of BR, Et and SA has shown to improve photosynthesis, level of antioxidants, decrease reactive oxygen species (ROS) and lipid peroxidation under heavy metal effect. It has been observed that exogenous application of BR on *Raphanus raphanistrum* subsp. *Sativus* (L.) Domin, decreased Cu toxicity, SA to *Arachis hypogaea* L. alleviated metal toxicity of cadmium (Cd), stimulated activities of antioxidant enzymes and increased content of non-enzymatic antioxidants (Xu et al., 2015). Jasmonic acid (JA) derivatives are also commonly used as priming agents to combat abiotic stress. Methyl jasmonate (MeJa) priming refined the growth of broccoli sprouts under salinity stress.

3.4 PHYTOHORMONE ENGINEERING

Genetic engineering is the upcoming most potent solution to improve the productivity of crops under challenging environmental conditions. This approach is greatly dependent on the transformation, integration and successful expression of genes in transgenic plants by vector-mediated or direct transfer. The hormonal metabolism and signaling pathways are potential targets for gene manipulation to mitigate biotic and abiotic stress (Kumar et al., 2016). Among all phytohormones, ABA might be the most fit for engineering due to its involvement in multiple functions under adverse conditions. Some of the vital enzymes involved in the ABA pathway are being studied for transgenics. Lu et al. (2013) recorded that overexpression of the MoCo sulfurase gene in soybeans leads to a yield with higher drought tolerance and higher production of biomass.

3.4.1 Cold Stress Tolerance and Phosphoprotein Cascade

Kolaksazov et al. (2013) suggested that stress phytohormones like ABA, SA and JA exert their action by triggering phosphoprotein cascade pathways, which leads to gene expression imparting tolerance to low temperatures. JA was reported in high concentrations in the three different populations of *Arabis alpina* Krock. ex Steud at a controlled temperature of 22°C, with a tenfold reduction in sensitive plants but no change in susceptible plants at 4°C.

3.4.2 Salicylic Acid and Chilling Stress

Gharib and Hegazi (2010) reported that SA lowered the negative impacts of low-temperature stress in common beans. SA is also known to stimulate the biosynthesis of growth-promoting and inhibiting substances. Experiments showed that cold stress slowed the process of germination and seedling growth of the six varieties of beans. Seed priming with SA improved the germination rate compared to seeds under control as well as cold stress.

3.4.3 Drought Tolerance and Putative Auxin Efflux Carrier

Auxin acts as an obligatory signal in abiotic stress response. Zhang et al. (2012) identified a putative carrier gene of auxin efflux *OsPIN3t* in *O. sativa,* which is associated with the polar transport of auxin as well as in the response to drought stress. Drought stress tolerance was enhanced when the *OsPIN3t* gene was overexpressed, whereas crown root abnormalities (seedling stage) were seen in the case of its knockdown.

3.5 MICROBIAL PHYTOHORMONES

Recent studies are focusing on the exogenous use of microbial phytohormones in combating heavy metal stress tolerance in plants (Ezeh et al., 2022). Microbial phytohormones take part in various salinity-induced root morphology changes, severe temperature, drought and heavy metal (HM) toxicity and also strike the metabolism of endogenously produced phytohormones (Sorty et al., 2016). Phytohormones produced by rhizobacteria are known to elevate stress tolerance and improve the development of plants in various stressful conditions (Sgroy et al., 2009; Liu et al., 2013). Auxins are reported to regulate plant growth and development by alternating plant gene expression patterns (Ljung, 2013). Although its production is reduced under stress, it still modulates the growth of plants by ABA induction (Kazan, 2013). Auxin application in sunflowers has proved to combat the harmful consequences of Pb. Outcomes observed were improved shoot biomass, indicating that auxin has a metal phytoextraction role (Fassler et al., 2010). Auxin produced by root-colonizing *B. licheniformis HSW-16, B. subtilis*, *Arthrobacter sp.* and *Enterobacter sp.* NIASMVII has the potential to enhance salt stress tolerance in wheat plants (Singh & Jha, 2016; Sorty et al., 2016). *Pseudomonas sp.* and *Serratia sp.*, which produce IAA, are known to induce drought tolerance and nutrient-limited tolerance in clover plants (Marulanda et al., 2009; Zaheer et al., 2016). SA, Et and BR lower ROS levels and lipid peroxidation by improving their antioxidant enzyme systems, thus enhancing photosynthesis in heavy metal-stressed plants (Bashar et al., 2019). In a study, plants under copper stress were treated with BR, which resulted in reduced Cu accumulation (Bashar et al., 2019). SA and JA multifunctional hormones generated in plants under abiotic stress are known to decrease HM toxicity (Dar et al., 2015).

3.6 PHYTOMELATONIN

Melatonin is especially known for its role as a neurohormone, with pleiotropic roles in activities mainly including the immune system, circadian rhythms and sleep and acting as an antioxidant. It is an endogenous synchronizer for endocrine as well as other rhythms. A recent study in *A. thaliana* indicated that melatonin could act as a phytohormone due to the identification of the first putative plant melatonin receptor (Sun et al., 2020; Arnao & Hernandez, 2020). Melatonin shows many actions that closely resemble hormonal actions. The most important action is its response in plants to biotic and abiotic stresses; others include regulation of growth and development in the root, seed germination, circadian rhythms, leaf senescence and postharvest fruit ripening. Phytomelatonin most likely functions in close crosstalk with other classical and modern phytohormones and its biosynthesis is tightly regulated by stress stimuli. The levels of phytomelatonin are controlled greatly by stressful conditions like drought, heavy metal stress, high light intensity and temperature variations. Phytomelatonin plays a major role in combating toxicity and promotes plant growth; therefore, its action might contribute to a potent phytoremediation method.

Phytomelatonin confers tolerance to abiotic stresses like salinity, drought, temperature and heavy metals via multiple mechanisms:

1. It is a scavenger of ROS as it stimulates antioxidant enzymes leading to a high number of antioxidants like AsA, GSH and phenolic compounds (Zhang et al., 2013; Li et al., 2015; Wei et al., 2015).
2. It interacts with ABA and mediates the movement of stomata. It also increases photosynthetic efficiency in stressed conditions by protecting chlorophyll from degradation (Qi et al., 2018; Chen et al., 2020).
3. It decreases the accumulation of toxic compounds like HMs. In *A. thaliana,* melatonin modulates the expression of many transporter genes of HM, e.g., ABC transporter genes and PCR2 and thereby decreases the accumulation of Cd (Gu et al., 2017; He et al., 2020).
4. It upregulates defense genes.
5. It improves the biosynthesis of secondary metabolites under stressful conditions.
6. It decreases cellular injury by modifying some components of the cell wall.
7. It served as a priming agent in *Medicago sativa* L., which provided drought stress tolerance by regulating osmoprotective and nitro-oxidative homeostasis (Antoniou et al., 2017).
8. In many plants like wheat, tomato, maize, cucumber and rice, phytomelatonin is known to enhance salinity stress tolerance (Liang et al., 2015; Zhou et al., 2016; Chen et al., 2018; Ke et al., 2018). It maintains homeostasis during salinity stress. Upon exposure to salinity stress, melatonin upregulated NHX1 and AKT1, encoding two ion channels that control the homeostasis of ions (Li et al., 2012).
9. Temperature stress inhibitory effects like germination and plant growth were reversed in plants like Bermudagrass, rice, *Lupinus albus* and tomato by exogenous treatment with melatonin (Wang et al., 2018).
10. Endogenous phytomelatonin levels are increased during high-temperature stress, which provides thermotolerance as melatonin is a potent antioxidant in plants (Liang et al., 2018; Ahammed et al., 2019).
11. High-temperature stress damage in tomato seedlings was alleviated by melatonin, which maintains redox homeostasis, polyamine and nitric oxide biosynthesis (Jahan et al., 2019).

3.7 NON-TRADITIONAL HORMONES

3.7.1 Brassinosteroids

BRs emerge as an essential phytohormone that maintains plant growth and development, even under adverse conditions along with normal conditions. BRs comprise a relatively new group of polyhydroxy steroidal plant hormones out of which brassinolide, 28-homobrassinolide and 24-epibrassinolide constitute the three most bioactive BRs. They play a critical role in developmental processes such as stem and root growth, floral initiation and development of flowers and fruits. BRs control several biological processes, such as growth, protein metabolism, cellular transport and signaling, biosynthesis of cell walls, formation of chromatin and cytoskeleton components, opening and closing of stomata and environmental responses (Peres et al., 2019). BRs show the stress impact mitigating roles in various abiotic stresses such as high temperature, chilling, soil salinity, light, drought, flooding, metals/metalloids and organic pollutants (Wani et al., 2016). In India, around 2% of the total geographical area of the country is salinity affected by enhancing the antioxidant enzymatic activity in plants. Ahanger et al. (2020) reported that treating the plants with BR alleviates the disastrous effects of salinity. It enhances the antioxidant enzyme activities and regulates the formation of ROS in the plant cell under salt stress conditions. It also acts as a signaling molecule during stress conditions and activates various enzymes involved in the defense mechanisms to face the stressed condition and regulates ion homeostasis. Recent research has shown the high efficiency of BRs in the modulation of antioxidant defense systems in response to and counteract abiotic stress (Vardhini et al., 2006). The BR application regulates the expression of different genes and transcriptional factors (Sharma & Sharma, 2022).

Effects of BRs under salinity stress:

1. Seed germination: Salinity negatively affects the germination process, as reported in *O. sativa* (Xu et al., 2011) and *Brassica juncea* (L.) Czern. (Ibrar et al., 2003). An experiment conducted on cucumbers reported that BRs reduce the harmful effects of salinity stress in the seed germination process by improving ethylene synthesis via enhancing the expression of *CsACO1* and *CsACO2* (Wang et al., 2011). However, high concentrations of BR can also be harmful (Liu et al., 2020).
2. Plant length and biomass: Due to the salinity stress, the growth of plants decreases because the energy is diverted to maintain homeostasis (Atkin & Macherel, 2009; Sarker & Oba, 2020). The epibrassinolide (EBL) application can neutralize the harmful impacts caused by salinity (Sharma et al., 2013). Improved growth by enhanced cell division and elongation in *Liriodendron tulipifera* L. increased shoot biomass in wheat and improved biomass accumulation in soybeans have been reported by EBL application.

3.7.2 Salicylic Acid

SA is one such plant hormone that helps in plant growth, development and responses to environmental stress. Recently, evidence has shown that SA plays a crucial role in mediating plant responses to various abiotic stresses such as chilling, drought, thermogenesis, osmotic stress and metal toxicity. SA acts as an important signaling molecule for the regulation of reactive oxygen species (ROS) production in plants. However, it depends on many factors, including plant species, application mode, the levels of SA (exogenous and endogenous) and level of stress faced by the plants. Research has shown that all abiotic stresses increase the level of endogenous SA, indicating that SA has a defensive role against abiotic stress by redox signaling (Liu et al., 2022) (Figure 3.3).

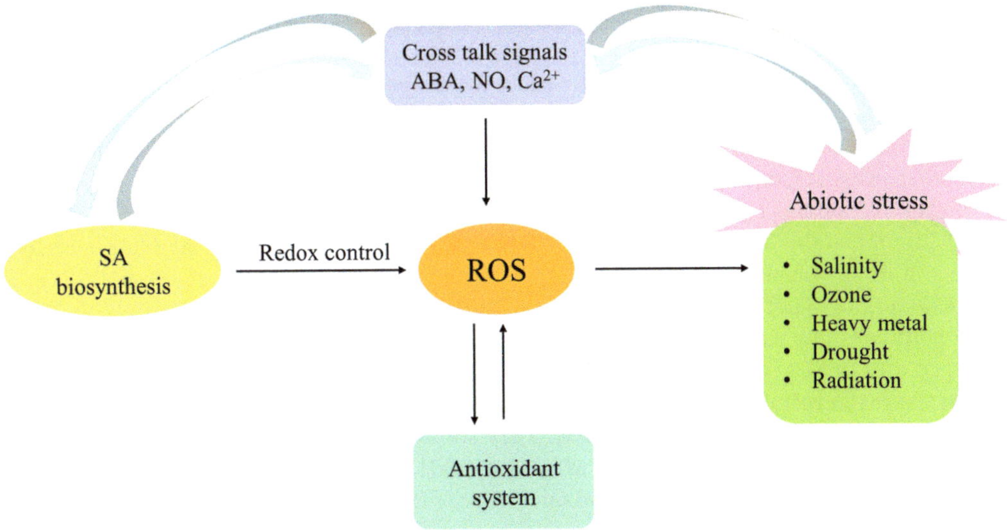

FIGURE 3.3 Modulation of SA signaling in plants under abiotic stress (Modified after Liu et al., 2022).

According to studies, pretreatment with SA induces H_2O_2 accumulation in rice roots is followed by an increase in levels of antioxidant molecules and enzymes. This leads to a decrease in oxidative injury by Cd stress (Guo et al., 2007). Exogenous application of SA induces the synthesis of HSPs, responsible for defense against heat stress in *Arabidopsis* (Clarke et al., 2004), tomato (Cronje et al., 1999; Snyman & Cronje, 2008) and *O. sativa* (Chang et al., 2007). Salicylic acid is involved in the regulation of pathogenesis-associated protein expression (Miura et al., 2014). Apart from a defense point of view, it plays a crucial role in the regulation of plant growth, ripening and development, as well as abiotic stresses. It consists of genes encoding chaperones, heat shock proteins, antioxidants and biosynthesis of secondary metabolites, such as alcohol dehydrogenase and cytochrome P450. It is also involved in the regulation of drought response along with ABA. Under drought stress, PR1 and PR2 genes (pathogenesis-related genes) are induced (Miura et al., 2013).

3.7.3 Strigolactones

Recent scientific research has identified a new class of phytohormones known as SLs that act as key regulators of plants. The first strigolactones, strigol, was isolated in 1966. All SLs have the basic structure that contains three annular rings, an ABC scaffold that is connected to a butenolide ring via an enol ether unit forming an ABC-D structure (Zwanenburg & Blanco, 2018). SLs are carotenoid-derived small signaling molecules that regulate multiple developmental processes and respond to various environmental signals. SLs also maintain the balance of resource distribution by strategic modification of the plant development, which allows plants to adapt to nutrient deficiency. The crosstalk between SL and other signaling pathways regulated by phytohormones such as auxin, CK, ET and ABA form the extensive signaling networks. The crosstalk of hormones in plant development and environmental response may occur in a fully concerted manner or as a cascade of sequential events (Wu et al., 2022). SLs regulate the coordinated development of roots and shoots, especially under N- and P-deficient conditions (Sun et al., 2014; Ito et al., 2015; Xi et al., 2015). Accordingly, SLs regulate above- and below-ground plant morphogenesis, shoot branching, leaf senescence, reproductive development, adventitious root formation and root hair density (Kretzschmar et al., 2012; Yamada et al., 2014; Sun et al., 2015; Tan et al., 2019; Mitra et al., 2021). SL biosynthesis-deficient mutants in *Arabidopsis* developed more lateral roots under optimal growth conditions, whereas an antagonist effect was observed under P-deficient conditions (Ruyter-Spira et al., 2011). However, continuous studies have suggested that SLs confer tolerance to various suboptimal growth conditions, especially drought and salinity (Saeed et al., 2017; Zhang et al., 2020). The significance of SLs is that they act as rhizosphere signaling molecules, play a vital role in the regulation of the architecture of the plant and enhance germination of root parasitic weeds that have fatal effects on plant growth. It also plays significant roles in plant biotic and abiotic stress responses. Recently, it emerged as an important biological target to study different signaling pathways, stress responses and various developmental stages of plants. Currently, two naturally occurring SL families have been reported; the stereochemical configuration of one of the forms is (+) strigol and the other has (−) orobanchol. The most prominent role of SLs has been found to help in the germination of the seeds in the *Orobanche* and *Striga*, parasitic weeds. The SLs are inhibitors of shoot branching and control architecture of the plants by crosstalk with other plant hormones in a cascade of events, although the interaction of details is still unknown. The practical uses of SLs in agriculture involve controlling parasitic weed seeds and amplification of arbuscular mycorrhizal fungi (Banerjee & Bhadra, 2020) (Table 3.1). It acts as a signaling agent for plant interactions with microorganisms. They stimulate the interaction of legume–rhizobium for nodulation (Soto et al., 2010; Foo & Davies, 2011). They can be used in agriculture for multipurpose work, such as inducers of suicidal seed germination of parasitic plants.

TABLE 3.1
SL Effects and Hormones Crosstalk on Various Plant Species

Plant Species	Plant Hormone	Type of Experiment	Effect of SL	Synergism or Antagonism	References
A. thaliana	ABA	SLs-response max2 mutant	Effect ABA import	Synergism	Ha et al., 2014; Ruiz-Lozano et al., 2016
	CK	GR24	The primary root gets elongated and inhibited	Synergism	Jiang et al., 2016
	ET	ET signaling deficient ein2 and etr1 mutants	Elimination of the SLs on the root morphogenesis	Synergism	Kapulnik et al., 2011
O. sativa	ABA	SL-deficient mutants D10 and D17 SL-perception mutant D13	Tolerance against drought	Synergism	Haider et al., 2018
	CK	SLs-insensitive tiller dwarfing mutants	The level of auxin increased	Synergism	Sun et al., 2019
	IAA	GR24	Distribution of IAA gets reduced and modulated AR formation	Antagonism	Sun et al., 2015

3.7.4 Jasmonic Acid

Plants themselves monitor and regulate their developmental status and external environment and adjust balance growth, differentiation and stress responses using a complex interconnected regulatory network composed of various signaling molecules and regulatory proteins. Jasmonic acid (JA) is a lipid-derived molecule that plays an important function in controlling various plant developmental and stress responses (Liu et al., 2021). JA and its metabolic derivatives such as JA isoleucine (JA-Ile) and methyl jasmonate (MeJA) are collectively known as jasmonates. JA signaling actively interacts with other hormone-signaling pathways, such as CK, BR and SA (Liu et al., 2021). JA has been involved in a wide range of abiotic stress responses like salt, drought, heavy metal, high and low temperature, light stress and ozone stress. Some evidence suggests that JA and its precursors, such as 12-oxo-phytodienoic acid (OPDA) participate in systemic signaling in plants, through which local stress can be perceived throughout the entire plant to induce a systemic defense response or systemic acquired acclimation (SAA) (Liu et al., 2022). It is involved in critical processes associated with plant survival and development including reproductive processes, flowering, fruiting, senescence, secondary metabolism and direct and indirect defense responses. MeJA activates plant defense machinery towards a pathogenic attack as well as environmental stresses like drought, salinity, low temperature and UV irradiation. It reduces heavy metal stress in plants by activating the antioxidant machinery (Yan et al., 2013). It confers tolerance in *A. thaliana* against Cu and Cd stress via the accumulation of phytochelatins (Maksymiec et al., 2007).

3.7.5 Other New Hormones

A. Karrikins (KARs) are discovered from burning plant material. It has the potential to break the dormancy of seeds of many species adapted to environments that regularly experience fire and smoke. The recent discovery shows its involvement in seed germination and growth in some taxa, indicating their role in fire ecology as well (Chiwocha et al., 2009). The plants defective in the KAR signaling pathway grow differently with several morphological changes (Antala et al., 2019).
B. Peptide hormones are involved in various aspects of growth and development. Short peptides act as ligands for receptor kinases that initiate a signaling cascade that regulates plant development in response to stimuli from an environment (Gancheva et al., 2019).
C. Indoleamines are ubiquitous tryptophan-derived indole alkaloids (Roshchina, 2010; Commisso et al., 2022). The ability of plants to produce indolamines, such as melatonin and serotonin, plays a vital role in alleviating abiotic stress (Ayyanath et al., 2023). Recent research shows evidence that plant indolamines are involved in various physiological processes, development and biotic/abiotic stress responses, leading to consideration as a new class of phytohormones (Sun et al., 2021).

3.8 CONCLUSION

Phytohormones are biochemical signaling molecules of key importance in regulating plant growth and development. Identifying the underlying principle by which plants respond to various stresses is a critical aspect. Although plant responses to abiotic stresses are dependent on various factors, phytohormones are still considered the most important endogenous substances for modulating these responses. Recently, extensive experiments have been conducted to establish clear connections between phytohormones and abiotic stresses. Small RNAs (sRNA), mainly siRNA and miRNAs, are essentially involved in phytohormonal crosstalk by regulating hormone responses or targeting the genes involved in signaling. Phytohormone priming or pretreatment with exogenous hormones like BR, Et JA and SA is being explored, known to develop defense responses and abiotic stress tolerance in plants. Phytohormone engineering or gene manipulation has the potential to combat biotic and abiotic stress. Recent studies show that phytomelatonin, which acts as an endogenous synchronizer, could act as a phytohormone. It is known to play a role in combating toxicity and conferring tolerance to abiotic stress. Brassinosteroids, salicylic acid, strigolactones, jasmonic acid and other newly discovered hormones like karrikins, peptide hormones and indoleamines are being regularly experimented with and studied to understand their roles in regulating plant growth, development and their crosstalk to combat stress.

REFERENCES

Ahammed, G. J., Xu, W., Liu, A., & Chen, S. (2019). Endogenous melatonin deficiency aggravates high temperature-induced oxidative stress in *Solanum lycopersicum* L. Environmental and Experimental Botany, 161, 303.

Ahanger, M. A., Mir, R. A., Alyemeni, M. N., & Ahmad, P. (2020). Combined effects of brassinosteroid and kinetin mitigates salinity stress in tomatoes through the modulation of antioxidant and osmolyte metabolism. Plant Physiology and Biochemistry, 147, 31–42. doi: 10.1016/j.plaphy.2019.12.007

Alonso-Peral, M. M., Li, J., Li, Y., Allen, R. S., Schnippenkoetter, W., Ohms, S., & Millar, A. A. (2010). The microRNA159-regulated GAMYB-like genes inhibit growth and promote programmed cell death in *Arabidopsis*. Plant Physiology, 154(2), 757–771.

Antala, M., Sytar, O., Rastogi, A., & Brestic, M. (2019). Potential of karrikins as novel plant growth regulators in agriculture. Plants, 9(1), 43.

Antoniou, C., Chatzimichail, G., Xenofontos, R., Pavlou, J. J., Panagiotou, E., Christou, A., & Fotopoulos, V. (2017). Melatonin systemically ameliorates drought stress-induced damage in *Medicago sativa* plants by modulating nitro-oxidative homeostasis and proline metabolism. Journal of Pineal Research, 62, e12401.

Arnao, M. B. & Hernández-Ruiz, J. (2020). Is phytomelatonin a new plant hormone?. Agronomy, 10(1), 95.

Atkin, O. K. & Macherel, D. (2009). The crucial role of plant mitochondria in orchestrating drought tolerance. Annals of Botany, 103(4), 581–597. doi: 10.1093/ aob/mcn094

Ayyanath, M. M., Shukla, M. R., & Saxena, P. K. (2023). Indoleamines impart abiotic stress tolerance and improve reproductive traits in Hazelnuts. Plants, 12(6), 1233.

Banerjee, P. & Bhadra, P. (2020). Mini-review on strigolactones: Newly discovered plant hormones. Bioresource Biotechnology Research Communications, 13(3), 1–7.

Bashar, K. K., Tareq, M. Z., Amin, M. R., Honi, U., Tahjib-Ul-Arif, M., Sadat, M. A., & Hossen, Q. M. M. (2019). Phytohormone-mediated stomatal response, escape and quiescence strategies in plants under flooding stress. Agronomy, 9(2), 43.

Baweja, P. & Kumar, G. (2020). Abiotic stress in plants: An overview, plant stress biology, 1–16. 10.1007/ 978-981-15-9380-2_1

Chang, P. F. L., Jinn, T. L., Huang, W. K., Chen, Y., Chang, H. M., & Wang, C. W. (2007). Induction of a cDNA clone from rice encoding a class II small heat shock protein by heat stress, mechanical injury, and salicylic acid. Plant Science, 172, 64–75.

Chen, K., Li, G. J., Bressan, R. A., Song, C. P., Zhu, J. K., & Zhao, Y. (2020). Abscisic acid dynamics, signaling, and functions in plants. Journal of Integrative Plant Biology, 62(1), 25–54.

Chen, Z., Gu, Q., Yu, X., Huang, L., Xu, S., Wang, R., Shen, W., & Shen, W. (2018). Hydrogen peroxide acts downstream of melatonin to induce lateral root formation. Annals of Botany 121, 1127–1136.

Chiwocha, S. D., Dixon, K. W., Flematti, G. R., Ghisalberti, E. L., Merritt, D. J., Nelson, D. C., & Stevens, J. C. (2009). Karrikins: a new family of plant growth regulators in smoke. Plant Science, 177(4), 252–256.

Clarke, S. M., Mur, L. A., Wood, J. E., & Scott, I. M. (2004). Salicylic acid dependent signaling promotes basal thermotolerance but is not essential for acquired thermotolerance in *Arabidopsis thaliana*. The Plant Journal, 38(3), 432–447.

Commisso, M., Negri, S., Gecchele, E., Fazion, E., Pontoriero, C., Avesani, L., & Guzzo, F. (2022). Indolamine accumulation and TDC/T5H expression profiles reveal the complex and dynamic regulation of serotonin biosynthesis in tomato (*Solanum lycopersicum* L.). Frontiers in Plant Science, 13, 975434.

Cronje, M. J. & Bornman, L. (1999). Salicylic acid influences Hsp70/Hsc70 expression in *Lycopersicon esculentum*: dose-and time-dependent induction or potentiation. Biochemical and Biophysical Research Communications, 265(2), 422–427.

Dar, T. A., Uddin, M., Khan, M. M. A., Hakeem, K. R., & Jaleel, H. (2015). Jasmonates counter plant stress: a review. Environmental and Experimental Botany, 115, 49–57.

Eisvand, H. R., Tavakkol-Afshari, R., Sharifzadeh, F., Maddah Arefi, H., & Hesamzadeh Hejazi, S. M. (2010). Effects of hormonal priming and drought stress on activity and isozyme profiles of antioxidant enzymes in deteriorated seed of tall wheatgrass (*Agropyron elongatum* Host). Seed Science and Technology, 38(2), 280–297.

Ezeh, C. C., Obi, C. J., & Moneke, A. N. (2022). Application of microbial synthesized phytohormones in the management of environmental impacts on soils. Bio-Research, 20(1), 1409–1425.

Fassler, E., Evangelou, M. W., Robinson, B. H., & Schulin, R. (2010). Effects of indole-3-acetic acid (IAA) on sunflower growth and heavy metal uptake in combination with ethylene diamine disuccinic acid (EDDS). Chemosphere, 80(8), 901–907.

Foo, E. & Davies, N. W. (2011). Strigolactones promote nodulation in pea. Planta, 234, 1073–1081.

Gancheva, M. S., Malovichko, Y. V., Poliushkevich, L. O., Dodueva, I. E., & Lutova, L. A. (2019). Plant peptide hormones. Russian Journal of Plant Physiology, 66, 171–189.

Gharib, F. A. & Hegazi, A. Z. (2010). Salicylic acid ameliorates germination, seedling growth, phytohormone and enzymes activity in bean (*Phaseolus vulgaris* L.) under cold stress. Journal of American Science, 6(10), 675–683.

Gu, Q., Chen, Z., Yu, X., Cui, W., Pan, J., Zhao, G., Xu, S., Wang, R., & Shen, W. (2017) Melatonin confers plant tolerance against cadmium stress via the decrease of cadmium accumulation and reestablishment of microRNA-mediated redox homeostasis. Plant Science, 261, 28–37.

Guo, B., Liang, Y. C., Zhu, Y. G., & Zhao, F. J. (2007). Role of salicylic acid in alleviating oxidative damage in rice roots (*Oryza sativa*) subjected to cadmium stress. Environmental Pollution, 147(3), 743–749.

Ha, C. V., Leyva-González, M. A., Osakabe, Y., Tran, U. T., Nishiyama, R., Watanabe, Y., & Tran, L. S. P. (2014). Positive regulatory role of strigolactone in plant responses to drought and salt stress. Proceedings of the National Academy of Sciences, 111(2), 851–856.

Haider, I., Andreo-Jimenez, B., Bruno, M., Bimbo, A., Floková, K., & Abuauf, H. (2018). The interaction of strigolactones with abscisic acid during the drought response in rice. Journal of Experimental Botany, 69, 2403–2414. doi: 10.1093/jxb/ery089

Hasanuzzaman, M. & Fotopoulos, V. (2019). Priming and pretreatment of seeds and seedlings: Implication in plant stress tolerance and enhancing productivity in crop plants. Springer, Singapore.

He, J., Zhuang, X., Zhou, J., Sun, L., Wan, H., Li, H., & Lyu, D. (2020). Exogenous melatonin alleviates cadmium uptake and toxicity in apple root-stocks. Tree Physiology, 40(6), 746–761. 10.1093/treephys/tpaa024

Hopkins, W. G. & Huner, N. P. (2009). Introduction to plant physiology, 4th edn. Wiley, Hoboken, NJ, p. 503.

Ibrar, M., Jabeen, M., Tabassum, J., Hussain, F., & Ilahi, I. (2003). Salt tolerance potential of *Brassica juncea* L. Journal of Science and Technology, 27(1-2), 79–84.

Ito, S., Ito, K., Abeta, N., Takahashi, R., Sasaki, Y., & Yajima, S. (2015). Effects of strigolactone signaling on *Arabidopsis* growth under nitrogen deficient stress conditions. Plant Signal Behaviour, 11, e1126031. doi: 10.1080/15592324.2015.1126031

Jahan, M. S., Shu, S., Wang, Y., Chen, Z., He, M., Tao, M., Sun, J., & Guo, S. (2019). Melatonin alleviates heat-induced damage of tomato seedlings by balancing redox homeostasis and modulating polyamine and nitric oxide biosynthesis. BMC Plant Biology, 19, 1–16.

Jayasinghe, T., Perera, P., & Wimalasekera, R. (2019). Effect of foliar application of gibberellin in mitigating salt stress in tomato (*Solanum Lycopersicum*), 'Thilina' variety. In Proceedings of the 6th International Conference on Multidisciplinary Approaches (iCMA), Faculty of Graduate Studies, University of Sri Jayewardenepura, Nugegoda, Sri Lanka, 26–27.

Jiang, L., Matthys, C., Marquez-Garcia, B., De Cuyper, C., Smet, L., & De Keyser, A. (2016). Strigolactones spatially influence lateral root development through the cytokinin signaling network. Journal of Experimental Botany, 67, 379–389. doi: 10.1093/jxb/erv478

Kapulnik, Y., Delaux, P. M., Resnick, N., Mayzlish-Gati, E., Wininger, S., & Bhattacharya, C. (2011). Strigolactones affect lateral root formation and root-hair elongation in *Arabidopsis*. Planta, 233, 209–216. doi: 10.1007/s00425-010-1310-y

Kazan, K. (2013). Auxin and the integration of environmental signals into plant root development. Annals of Botany, 112(9), 1655–1665.

Ke, Q., Ye, J., Wang, B., Ren, J., Yin, L., Deng, X., & Wang, S. (2018). Melatonin mitigates salt stress in wheat seedlings by modulating polyamine metabolism. Frontier of Plant Science, 9, 914.

Kranner, I., Minibayeva, F. V., Beckett, R. P., & Seal, C. E. (2010). What is stress? Concepts, definitions and applications in seed science. New Phytologist, 188(3), 655–673.

Kolaksazov, M., Laporte, F., Ananieva, K., Dobrev, P., Herzog, M., & Ananiev, E. D. (2013). Effect of chilling and freezing stresses on jasmonate content in *Arabis alpina*. Bulgarian Journal of Agricultural Science, 19, 15–17.

Kretzschmar, T., Kohlen, W., Sasse, J., Borghi, L., Schlegel, M., & Bachelier, J. B. (2012). A petunia ABC protein controls strigolactone-dependent symbiotic signalling and branching. Nature, 483, 341–344. doi: 10.1038/nature10873

Kumar, V., Sah, S. K., Khare, T., Shriram, V., & Wani, S. H. (2016). Engineering phytohormones for abiotic stress tolerance in crop plants. In: G. Ahammed, J. Q. Yu (eds.), Plant hormones under challenging environmental factors (pp. 247–266). Springer, Dordrecht. 10.1007/978-94-017-7758-2_10

Li, C., Tan, D. X., Liang, D., Chang, C., Jia, D., & Ma, F. (2015). Melatonin mediates the regulation of ABA metabolism, free-radical scavenging, and stomatal behaviour in two *Malus* species under drought stress. Journal of Experimental Botany, 66, 669–680.

Li, C., Wang, P., Wei, Z., Liang, D., Liu, C., Yin, L., Jia, D., Fu, M., & Ma, F. (2012). The mitigation effects of exogenous melatonin on salinity-induced stress in *Malus hupehensis*. Journal of Pineal Resources, 53, 298–306.

Li, T., Gonzalez, N., Inze, D., & Dubois, M. (2020). Emerging connections between small RNAs and phytohormones. Trends in Plant Science, 25(9), 912–929.

Liang, C., Zheng, G., Li, W., Wang, Y., Hu, B., Wang, H., Wu, H., Qian, Y., Zhu, X. G., & Tan, D. X. (2015). Melatonin delays leaf senescence and enhances salt stress tolerance in rice. Journal of Pineal Research, 59, 91–101.

Liang, D., Gao, F., Ni, Z., Lin, L., Deng, Q., Tang, Y., Wang, X., Luo, X., & Xia, H. (2018). Melatonin improves heat tolerance in kiwifruit seedlings through promoting antioxidant enzymatic activity and glutathione S- transferase transcription. Molecules, 23, 584.

Lichtenthaler, H. K. (1998). The stress concept in plants: an introduction. Annals of the New York Academy of Sciences, 851, 187–198.

Liu, H. & Timko, M. P. (2021). Jasmonic Acid Signaling and Molecular Crosstalk with Other Phytohormones. International Journal of Molecular Science, 22, 2914. 10.3390/ijms22062914

Liu, J., Yang, R., Jian, N., Wei, L., Ye, L., Wang, R., Gao, H., & Zheng, Q. (2020). Putrescine metabolism modulates the biphasic effects of brassinosteroids on canola and *Arabidopsis* salt tolerance. Plant, Cell & Environment, 43 (6), 1348–1359. doi:10.1111/pce.13757

Liu, J., Qiu, G., Liu, C., Li, H., Chen, X., Fu, Q., Lin, Y., & Guo, B. (2022). Salicylic acid, a multifaceted hormone, combats abiotic stresses in plants. Life, 12, 886. 10.3390/life12060886

Liu, Y., Shi, Z., Yao, L., Yue, H., Li, H., & Li, C. (2013). Effect of IAA produced by *Klebsiella oxytoca* Rs-5 on cotton growth under salt stress. The Journal of General and Applied Microbiology, 59(1), 59–65.

Ljung, K. (2013). Auxin metabolism and homeostasis during plant development. Development, 140(5), 943–950.

Long, R., Li, M., Li, X., Gao, Y., Zhang, T., Sun, Y., & Yang, Q. (2017). A novel miRNA sponge form efficiently inhibits the activity of miR393 and enhances the salt tolerance and ABA insensitivity in *Arabidopsis thaliana*. Plant Molecular Biology Reporter, 35, 409–415.

Lu, Y., Li, Y., Zhang, J., Xiao, Y., Yue, Y., & Duan, L. (2013). Overexpression of *Arabidopsis* molybdenum cofactor sulfurase gene confers drought tolerance in maize (*Zea mays* L.). PLoS One , 8, e52126.

Maksymiec, W., Wojcik, M., & Krupa, Z. (2007). Variation in oxidative stress and photochemical activity in *Arabidopsis thaliana* leaves subjected to cadmium and excess copper in the presence or absence of jasmonate and ascorbate. Chemosphere, 66, 421–427.

Marulanda, A., Barea, J. M., & Azcon, R. (2009). Stimulation of plant growth and drought tolerance by native microorganisms (AM fungi and bacteria) from dry environments: mechanisms related to bacterial effectiveness. Journal of Plant Growth Regulation, 28(2), 115–124.

Mitra, D., Rad, K. V., Chaudhary, P., Ruparelia, J., SmruthiSagarika, M., & Boutaj, H. (2021). Involvement of strigolactone hormone in root development, influence and interaction with mycorrhizal fungi in plant: mini-review. Current Research in Microbial Sciences, 2, 100026.

Miura, K., Okamoto, H., Okuma, E., Shiba, H., Kamada, H., Hasegawa, P. M., & Murata, Y. (2013). SIZ1 deficiency causes reduced stomatal aperture and enhanced drought tolerance via controlling salicylic acid-induced accumulation of reactive oxygen species in *A rabidopsis*. The Plant Journal, 73(1), 91–104.

Miura, K. & Tada, Y. (2014). Regulation of water, salinity, and cold stress responses by salicylic acid. Frontiers in Plant Science, 5, 4.

Peres, A. L., Soares, J. S., Tavares, R. G., Righetto, G., Zullo, M. A. T., Mandava, N. B., & Menossi, M. (2019). Brassinosteroids, the sixth class of phytohormones: a molecular view from the discovery to hormonal interactions in plant development and stress adaptation. International Journal of Molecular Science, 20, 331. doi:10.3390/ijms20020331

Qi, J., Song, C. P., Wang, B., Zhou, J., Kangasjarvi, J., Zhu, J. K., & Gong, Z. (2018). Reactive oxygen species signaling and stomatal movement in plant responses to drought stress and pathogen attack. Journal of Integrative Plant Biology, 60, 805–826.

Rhaman, M. S., Imran, S., Rauf, F., Khatun, M., Baskin, C. C., Murata, Y., & Hasanuzzaman, M. (2020). Seed priming with phytohormones: An effective approach for the mitigation of abiotic stress. Plants, 10(1), 37.

Roshchina, V. V. (2010). Evolutionary considerations of neurotransmitters in microbial, plant, and animal cells. Microbial Endocrinology, 1, 17–52.

Rubio-Somoza, I. & Weigel, D. (2013). Coordination of flower maturation by a regulatory circuit of three microRNAs. PLoS Genetics, 9, e1003374.

Ruiz-Lozano, J. M., Aroca, R., Zamarreño, Á.M., Molina, S., Andreo-Jimenez, B., & Porcel, R. (2016). Arbuscular mycorrhizal symbiosis induces strigolactone biosynthesis under drought and improves drought tolerance in lettuce and tomato. Plant Cell Environment, 39, 441–452.

Ruyter-Spira, C., Kohlen, W., Charnikhova, T., van Zeijl, A., van Bezouwen, L., & de Ruijter, N. (2011). Physiological effects of the synthetic strigolactone analog GR24 on root system architecture in *Arabidopsis*: another belowground role for Strigolactones. Plant Physiology, 155, 721–734.

Saeed, W., Naseem, S., & Ali, Z. (2017). Strigolactones biosynthesis and their role in abiotic stress resilience in plants: a critical review. Frontier in Plant Science, 8, 1487.

Sarker, U. & Oba, S. (2020). The response of salinity stress-induced A. tricolor to growth, anatomy, physiology, non-enzymatic and enzymatic antioxidants. Frontiers in Plant Science, 1354. 10.3389/fpls.2020.559876

Sgroy, V., Cassan, F., Masciarelli, O., Del Papa, M. F., Lagares, A., & Luna, V. (2009). Isolation and characterization of endophytic plant growth-promoting (PGPB) or stress homeostasis-regulating (PSHB) bacteria associated to the halophyte *Prosopis strombulifera*. Applied Microbiology and Biotechnology, 85(2), 371–381.

Sharma, J. & Sharma, A. (2022). Role of Brassinosteroids in plants responses to salinity stress: A review. Journal of Applied and Natural Science, 14(2), 582–599.

Sharma, I., Ching, E., Saini, S., Bhardwaj, R., & Pati, P. K. (2013). Exogenous application of brassinosteroid offers tolerance to salinity by altering stress responses in rice variety Pusa Basmati-1. Plant Physiology and Biochemistry, 69, 17–26. DOI: 10.1016/j.plaphy.2013.04.013

Singh, R. P. & Jha, P. N. (2016). A halotolerant bacterium Bacillus licheniformis HSW-16 augments induced systemic tolerance to salt stress in wheat plant (*Triticum aestivum*). Frontiers in Plant Science, 7, 1890.

Skirycz, A. & Inze, D. (2010). More from less: plant growth under limited water. Current Opinion in Biotechnology, 21, 197–203.

Snyman, M. & Cronje, M. (2008). Modulation of heat shock factors accompanies salicylic acid-mediated potentiation of Hsp70 in tomato seedlings. Journal of Experimental Botany, 59, 2125–2132.

Song, J. B., Gao, S., Sun, D., Li, H., Shu, X. X., & Yang, Z. M. (2013). miR394 and LCR are involved in *Arabidopsis* salt and drought stress responses in an abscisic acid-dependent manner. BMC Plant Biology, 13(1), 1–16.

Sorty, A. M., Meena, K. K., Choudhary, K., Bitla, U. M., Minhas, P. S., & Krishnani, K. K. (2016). Effect of plant growth promoting bacteria associated with halophytic weed (*Psoralea corylifolia* L) on germination and seedling growth of wheat under saline conditions. Applied Biochemistry and Biotechnology, 180(5), 872–882.

Soto, M. J., Fernandez-Aparicio, M., Castellanos-Morales, V., García-Garrido, J. M., Ocampo, J. A., Delgado, M. J., & Vierheilig, H. (2010). First indications for the involvement of strigolactones on nodule formation in alfalfa (*Medicago sativa*). Soil Biology and Biochemistry, 42(2), 383–385.

Sun, C., Liu, L., Wang, L., Li, B., Jin, C., & Lin, X. (2021). Melatonin: A master regulator of plant development and stress responses. Journal of Integrative Plant Biology, 63(1), 126–145.

Sun, C., Lv, T., Huang, L., Liu, X., Jin, C., & Lin, X. (2020). Melatonin ameliorates aluminium toxicity through enhancing aluminium exclusion and reestablishing redox homeostasis in roots of wheat. Journal of Pineal Research, 68, e12642.

Sun, H., Tao, J., Hou, M., Huang, S., Chen, S., & Liang, Z. (2015). A strigolactone signal is required for adventitious root formation in rice. Annals of Botany, 115, 1155–1162. doi: 10.1093/aob/mcv052

Sun, H., Tao, J., Liu, S., Huang, S., Chen, S., & Xie, X. (2014). Strigolactones are involved in phosphate-and nitrate-deficiency-induced root development and auxin transport in rice. Journal of Experimental Botany, 65, 6735–6746. doi: 10.1093/jxb/eru029

Sun, H., Xu, F., Guo, X., Wu, D., Zhang, X., & Lou, M. (2019). A Strigolactone signal inhibits secondary lateral root development in rice. Frontiers in Plant Science, 10, 1527. doi: 10.3389/fpls.2019.01527

Sun, J., Miller, J. B., Granqvist, E., Wiley-Kalil, A., Gobbato, E., & Maillet, F. (2015). Activation of symbiosis signaling by arbuscular mycorrhizal fungi in legumes and rice. Plant Cell, 27, 823–838. doi: 10.1105/tpc.114.131326

Tan, M., Li, G., Chen, X., Xing, L., Ma, J., & Zhang, D. (2019). Role of cytokinin, strigolactone, and auxin export on outgrowth of axillary buds in apple. Frontiers in Plant Science, 10, 616. doi: 10.3389/fpls.2019.00616

Vardhini, B. V., Anuradha, S., & Rao, S. S. R. (2006). Brassinosteroids—A Great Potential to Improve Crop Productivity. Indian Journal of Plant Physiology, 11, 1–12.

Vu, L. D., Gevaert, K., & De Smet, I. (2019). Feeling the Heat: Searching for Plant Thermosensors. Trends in Plant Science, 24, 210–219. 10.1016/j.tplants.2018.11.004.

Wang, B., Zhang, J., Xia, X., & Zhang, W. H. (2011). Ameliorative effect of brassinosteroid and ethylene on germination of cucumber seeds in the presence of sodium chloride. Plant Growth Regulation, 65(2), 407–413. doi:10.1007/s10725-011-9595-9

Wang, Y., Reiter, R. J., & Chan, Z. (2018). Phytomelatonin: a universal abiotic stress regulator. Journal of Experimental Botany, 69, 963–974.

Wani, S. H., Kumar, V., Shriram, V., & Sah, S. K. (2016). Phytohormones and their metabolic engineering for abiotic stress tolerance in crop plants. The Crop Journal, 4(3), 162–176. 10.1016/j.cj.2016.01.010

Wei, W., Li, Q. T., Chu, Y. N., Reiter, R. J., Yu, X. M., Zhu, D. H., Zhang, W. K., Ma, B., Lin, Q., & Zhang, J. S. (2015) Melatonin enhances plant growth and abiotic stress tolerance in soybean plants. Journal of Experimental Botany, 66, 695–707.

Wu, F., Gao, Y., Yang, W., Sui, N., & Zhu, J. (2022). Biological functions of Strigolactones and their crosstalk with other phytohormones. Frontiers in Plant Science, 13, 821563. doi: 10.3389/fpls.2022.821563

Xi, L., Wen, C., Fang, S., Chen, X., Nie, J., & Chu, J. (2015). Impacts of strigolactone on shoot branching under phosphate starvation in chrysanthemum (*Dendranthema grandiflorum* cv. Jinba). Frontiers in Plant Science, 6, 694. doi:10.3389/fpls.2015.00694

Xu, L. L., Fan, Z. Y., Dong, Y. J., Kong, J., & Bai, X. Y. (2015). Effects of exogenous salicylic acid and nitric oxide on physiological characteristics of two peanut cultivars under cadmium stress. Plant Biology, 59, 171–182.

Xu, S., Hu, B., He, Z., Ma, F., Feng, J., Shen, W., & Yang, J. (2011). Enhancement of salinity tolerance during rice seed germination by presoaking with hemoglobin. International Journal of Molecular Sciences, 12(4), 2488–2501. 10.3390/ijms12042488

Yamada, Y., Furusawa, S., Nagasaka, S., Shimomura, K., Yamaguchi, S., & Umehara, M. (2014). Strigolactone signaling regulates rice leaf senescence in response to a phosphate deficiency. Planta, 240, 399–408. doi: 10.1007/s00425-014-2096-0

Yan, Z., Chen, J., & Li, X. (2013). Methyl jasmonate as modulator of Cd toxicity in *Capsicum frutescens* var. *fasciculatum* seedlings. Ecotoxicology and Environmental Safety, 98, 203–209.

Zaheer, A., Mirza, B. S., Mclean, J. E., Yasmin, S., Shah, T. M., Malik, K. A., & Mirza, M. S. (2016). Association of plant growth-promoting *Serratia* spp. with the root nodules of chickpea. Research in Microbiology, 167(6), 510–520.

Zhang, H. J., Zhang, N., Yang, R. C., Wang, L., Sun, Q. Q., Li, D. B., Cao, Y. Y., Weeda, S., Zhao, B., & Ren, S. (2012). Melatonin promotes seed germination under high salinity by regulating antioxidant systems, ABA and GA 4 interaction in cucumber (*Cucumis sativus* L.). Journal of Pineal Research, 57, 269–279.

Zhang, N., Zhao, B., Zhang, H. J., Weeda, S., Yang, C., Yang, Z. C., Ren, S., & Guo, Y. D. (2013). Melatonin promotes water-stress tolerance, lateral root formation, and seed germination in cucumber (*Cucumis sativus* L.). Journal of Pineal Research, 54, 15–23.

Zhang, X., Zhang, L., Sun, Y., Zheng, S., Wang, J., & Zhang, T. (2020). Hydrogen peroxide is involved in strigolactone induced low temperature stress tolerance in rape seedlings (*Brassica rapa* L.). Plant Physiology and Biochemistry, 157, 402–415. doi: 10.1016/j.plaphy.2020.11.006

Zhou, X., Zhao, H., Cao, K., Hu, L., Du, T., Baluska, F., & Zou, Z. (2016). Beneficial roles of melatonin on redox regulation of photosynthetic electron transport and synthesis of D1 protein in tomato seedlings under salt stress. Frontiers in Plant Science, 7, 1823.

Zwanenburg, B. & Blanco-Ania, D. (2018). Strigolactones: new plant hormones in the spotlight. Journal of Experimental Botany, 69(9), 2205–2218.

ABBREVIATIONS

IAA	Indole acetic acid
SA	Salicylic acid
SL	Strigolactone
CT	Cytokinin
ET	Ethylene
BR	Brassinosteroid
ABA	Abscisic acid
JA	Jasmonic acid
HM	Heavy metal
KAR	Karrikin

Unit II

Phytohormones in Abiotic Stresses

Section I

Traditional Phytohormones

4 Role of Abscisic Acid in Abiotic Stresses

Diksha Bagal, Ritika Vishnoi, Adeeb Rahman, Vrinda Sharma, and Savita

4.1 INTRODUCTION

Plants, being sessile, cannot move away from stress. They need to respond quickly to prevent the harmful effect of stress. Abiotic stress is the effect of non-living factors on plant growth and development. Drought, flood, salinity, heavy metal and high and low temperature are the various abiotic stresses experienced by plants during their life cycle (Rehman et al., 2021). Abiotic stress can cause biochemical, molecular, morphological and physiological changes in plants, leading to deterioration in cell growth and tissue health, ultimately resulting in reduced crop productivity (Roychoudhury and Banerjee, 2017). Plants survive these extreme conditions through inbuilt mechanisms involving interconnected regulatory pathways (Rahman et al., 2021; Zhang et al., 2022).

Phytohormones are a major player in regulating plants' normal growth and development. Abscisic acid (ABA) is also known as a stress phytohormone involved in seed germination, leaf abscission, inhibition of fruit ripening, etc. It mediates both abiotic and biotic stress responses in plants (Vishwakarma et al., 2017). When a plant experiences any stressful condition, the ABA level increases, leading to signal transduction by binding to its receptors (FCA, ABAR (CHLH or GUN5), GCR2, GTG1/GTG2, PYR1/PYL/RCAR). Plants maintain the homeostasis of the water level and osmotic state by ABA (Bharath et al., 2021). During stressful conditions, ABA acts as a molecular signal and leads to the activation of stress-responsive genes (*NCED3, ABA3/LOS5* and *AAO3*), which helps counter the harmful effect of stress.

Apart from ABA-dependent stress-responsive pathways, ABA-independent pathways are also involved in mitigating stress (Rehman et al., 2021). These two pathways can crosstalk during a plant's response to stressful conditions. ABA-independent pathways activate dehydration-responsive element binding *(DREB)* TFs during drought conditions (Agarwal et al., 2017). They converge with the ABA-dependent pathway by phosphorylating ABA response element binding *(AREB)* TFs (*bZIP* type), which are the master regulators of stress-inducible gene expression. ABA also interacts with other phytohormones like gibberellic acid (GA), jasmonic acid (JA), ethylene (ET) and other chemical compounds like nitric oxide (NO), reactive oxygen species (ROS) etc. during abiotic stress signaling (Roychoudhury et al., 2013). In this chapter, the significance of ABA under abiotic stress, ABA-modulated MAPK signaling and interactions with other signaling molecules is discussed in detail.

4.2 SIGNIFICANCE OF ABSCISIC ACID IN PLANTS UNDER ABIOTIC STRESS

ABA is involved in plants at all stages, from seed germination to senescence (Rehman et al., 2021). The exogenous application of ABA has been found to alleviate the harmful effect of stress (Table 4.1). The stomata closure mediated by ABA is the first line of defense against a diverse set of environmental conditions, like chilling, drought, heavy metal salinity and heat. Calcium (Ca^{2+})-dependent and Ca^{2+}-independent pathways regulate the stomata pore size (Bharath et al., 2021). Various environmental conditions like salinity, heavy metal, drought and high and low temperature

DOI: 10.1201/9781003335788-7

TABLE 4.1
Description of the Protective Effect of ABA During Various Abiotic Stresses in Plants

Plant Species	Stress Type	ABA Concentration	Protective Effects of Exogenous ABA	References
Capsicum	Chilling	0.57 mM	Increase in antioxidant enzymes (APX, GR, GPX, DHAR, etc.) and induces SOD and POD.	Guo et al., 2012
Common bean	Salinity	1 and 10 μM	Improves growth parameters, normal nodule weight and reduction in sodium concentration.	Khadri et al., 2007
Maize	Drought	0.5 mmol L^{-1}	Enhancement in antioxidant system and regulate Ascorbate-glutathione (AsA-GSH) cycle.	Jiang et al., 2022
Rice	Salinity	100 μM	Increase in *OsCam1-1* (the salt-stress sensitive calmodulin) gene expression that aids in proline accumulation, thus salinity stress alleviation.	Sripinyowanich et al., 2013
	Simulated acid rain	0.1, 1, 10 and 100 μM	Improves SAR tolerance by balancing endogenous hormones (ABA, GA, IAA and zeatin) and inducing nutrients uptake.	Liu et al., 2018
	Drought	60 μM	Induction of photosystem II (PSII) by ABA biosynthesis-related genes.	Teng et al., 2014
	Heat	10 μM	Improves heat tolerance by lowering MDA, electrolytic conductivity and expressing cell death-related genes (*OsKOD1, OsCP1* and *OsNAC4*).	Liu et al., 2022
Sugarcane	Cold	100 μM	Increase in the proline amount and decrease in the MDA and GA_3 level.	Huang et al., 2015
Tomato	Cold	20 μM	Promotes plant growth by decreasing the electrolyte leakage and membrane damage.	Kim et al., 2002
Wheat	Salinity and alkali	50 and 100 μmol/L	Increases the shoot growth by regulating the amount of ions and organic solutes on foliar application.	Li et al., 2020
	Drought and salinity	10 μM	Increase in plant biomass, shoot length and proline metabolism.	Kaur and Asthir, 2020
	Heat	100 μM	Increase in antioxidant activities and osmolyte regulation on interaction with NO.	Iqbal et al., 2022
	Osmotic	100 μM	Increases the activity of antioxidant enzymes (APX, GR, MDHAR and DHAR) and the MDA content and electrolyte leakage are reduced.	Shan et al., 2017

limit plants' growth and productivity and are known as abiotic stress (Cramer et al., 2011). Wright and Hiron (1969), for the first time, reported the accumulation of ABA during the drought condition in wheat (Sah et al., 2016). After this finding, ABA accumulation during drought was also reported in other plant species like barley (Thameur et al., 2011), rice (Henson, 1984), soybean (Bensen et al., 1988), sorghum (Kannangara et al., 1983) and maize (Wang et al., 2008). The effect of drought on plants includes reduced stem extension, leaf size, root proliferation, disturbance of plant water relations and water use efficiency (Farooq et al., 2009). Plants tolerate drought conditions mainly via ABA-mediated stomatal closure (Sah et al., 2016). Salinity leads to less water uptake from the soil, and salts enter the transpiration stream in plants, leading to cell injury of transpiring leaves (Parihar et al., 2015). Overlap is found amongst the transcriptome of ABA, drought and salinity stress. Fifty percent of the genes are common to both salinity and drought (Sah et al., 2016). Heavy metals cause chlorosis, low photosynthesis, imbalance in water level and nutrition assimilation, leading to reduced growth and cell death in severe cases (Singh and Prasad, 2015). ABA is accumulated in response to heavy metals. In rice, the exogenous application of ABA leads to resistance against cadmium (Finkelstein and Gibson, 2002). Similarly, high and low temperatures adversely affect plant growth and development. High temperature causes inhibition of seed germination, water loss, alteration in photosynthesis and oxidative stress. Heat shock factors are involved in heat stress response and *HsfAb* is found to connect ABA signaling and ABA-mediated stress response in *Arabidopsis* (Huang et al., 2016). ABA regulates low-temperature stress by activation of diverse stress-responsive genes. Both ABA-dependent and ABA-independent pathways are found to be involved in cold stress response. *C-repeat binding factor/DRE-binding factor* (*CBF/DREB*) transcriptional regulatory cascade is the most extensively studied low-temperature signaling pathway. ABA biosynthesis is not a major player in low-temperature stress response, but it is involved in the activation of low-temperature stress-mediated genes during the later stages. Abiotic stress and treatment of plants with ABA alters the expression of genes involved in ABA biosynthetic pathways. When plants experience abiotic stress, the level of ABA biosynthetic genes, like *NCED3*, *ABA3/LOS5* and *AAO3*, is significantly increased (Chan, 2012). On overexpressing key genes of ABA biosynthesis, the plants obtained were stress-tolerant (Vishal and Kumar, 2018). Hence, the level of endogenous active ABA at the site of abiotic stress signal perception has to be regulated.

4.3 CROSSTALK BETWEEN ABA SIGNALING AND PROTEIN KINASE NETWORKS DURING ABIOTIC STRESS

ABA signaling is usually interconnected with several protein kinases like MAPKs, CPKs, CIPKs, etc. The MAPK signaling pathway is found in all eukaryotes and is triggered in plants when they encounter stress. The pathway consists of three protein kinases: MAPKKK/MAP3K, MAPKK/MEK/MKK and MAPK, which make up the MAPK module (Ichimura et al., 2002). Activation of MAPK is triggered by stimuli such as drought, cold, high temperatures, phytohormones, etc. (Colcombet and Hirt, 2008; Bigeard et al., 2015). The MAPK-ABA module has three essential regulators: pyrabactin resistance/pyrabactin resistance-like/regulatory component of ABA receptor (PYR/PYL/RCAR), protein phosphatase PP2C and SnRK2s. SnRK2 plays a central role in ABA signaling (SNF1-related protein kinases). Especially in drought stress, ABI1 (inhibitor) is sequestered by PYR/PIL/RCAR (ABA receptors), but when ABA is there, it results in autophosphorylation of SnRK2, which then positively regulates AREB/ABF (stress-responsive genes), that affect the amount of MAPKs (MAP3K17 and MAP3K18). Studies showed MAPK belongs to the ABA signaling module or not by creating MAPK mutants and comparing them to ABA-related phenotypes (De Zélicourt et al., 2016). According to some findings, ABA can affect the expression of MAPK cascade components in distinct ways. MPK3, MPK5, MPK7, MPK18, MPK20, MKK9 and MAP3K1/10/14/15/16/17/18/19 are all upregulated in *Arabidopsis* after treatment with ABA (Figure 4.1) (Wang et al., 2011). Some studies have linked ROS to MAPK-ABA signaling in the control of stomatal closure (Desikan et al., 2004). In guard cells, ABA perception activates SnRK2

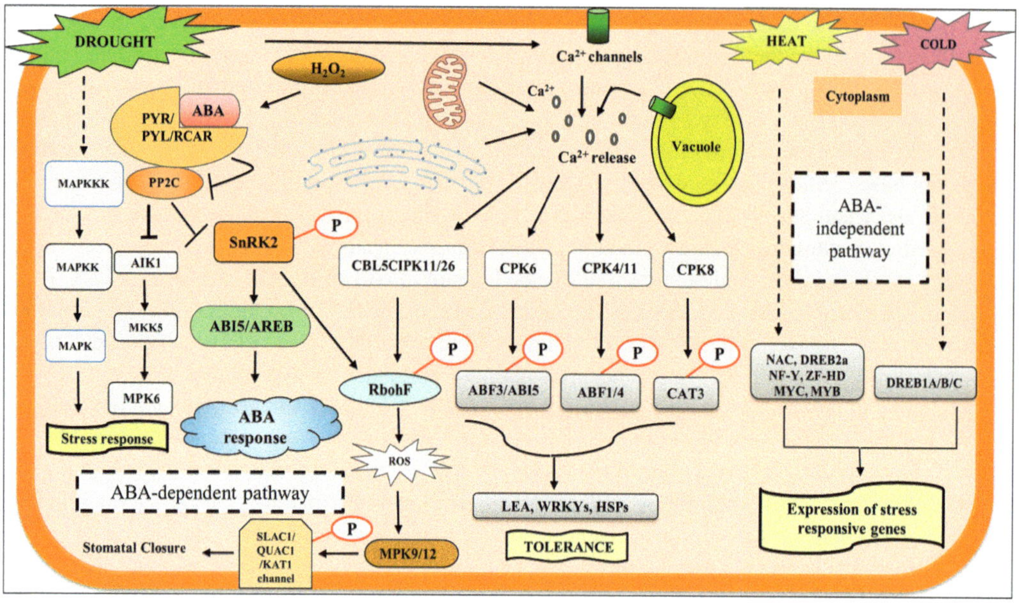

FIGURE 4.1 Interaction networks of ABA signaling and protein kinases in abiotic stress. ABA signaling forms connecting links between abiotic stresses such as drought, heat, cold etc. Depending on the type of stress encountered by the plant, it can activate either an ABA-dependent or ABA-independent pathway. An ABA-dependent pathway usually activates during drought stress, where ABA binds to their receptor PYR/PYL/RCAR and releases the SnRK2 from PP2C and ABI5 inhibition. After the release, SnRK2 autophosphorylates and activates AREB/ARFs genes that are responsible for ABA responses. SnRK2 is also able to phosphorylate RbohD/F, which generates ROS, thereby upregulating MPK9/12. These kinases target the SLAC1 channel and lead to stomatal closure. SnRK2s also phosphorylates QUAC1 and KAT1, which regulate malate and K^+ ions. Other than that, osmotic stress also triggers the activation of various MAPKKKs/MAPKKs/MAPKs that further activate SnRK2s. Not only MAPKs but drought stress also induces Ca^{2+} influx, thereby stimulating Ca^{2+}-dependent protein kinases or calcineurin B (CBL) interacting protein kinases (CIPKs), which are responsible for activating stress-responsive factors such as WRKY, LEA, HSPs etc. Heat and cold stresses activate DREB2a, MYC2, NAC, ZF-HD and DREB1A/B/C via an ABA-independent pathway. ABI - ABA insensitive; AREB/ABFs ABRE-binding factor; MAPKKK - mitogen-activated protein kinase kinase kinase; MAPKK/MKK - mitogen-activated protein kinase kinase; MPK - mitogen-activated protein kinase; PP2C - protein phosphatase 2; PYR/PYL/RCAR - pyrabactin resistance/pyrabactin resistance-like/regulatory component of ABA receptor; RbohF - respiratory burst oxidase homolog F protein; SLAC1 S-type anion channel; KAT1 - potassium channel in *Arabidopsis thaliana* 1; SnRK2 - SNF1-related protein kinase 2, CBL5- calcineurin B-like; CIPK11/26 - CBL interacting protein kinase; CPK6/4/11/8 - Ca^{2+}-dependent protein kinase; CAT - catalase 3; LEA - late embryogenesis abundant; HSPs - heat shock proteins; NAC - NAM-AFA1-CUC2; DREB1a/b/c/2a - dehydration-responsive element binding; NFY - nuclear factor Y; MYC - master regulator of cell cycle entry and proliferative metabolism; MYB - myeloblastosis homolog; ZF-HD - zinc finger homeodomain transcription factors.

kinases, which phosphorylate NADPH oxidase and RbohF produce ROS. ROS promotes MPK9 and MPK12, which target SLAC1(S-type anion channel)/QUACK1/KAT1 and favorably regulate ABA-mediated stomatal closure (Danquah et al., 2014).

Other than MAPKs, there are also other Ca^{2+}-dependent protein kinases (CPKs) that regulate stress responses. Stress, like drought, triggers a Ca^{2+} release, which further activates CBL, interacting with a protein kinase (CIPKs) that is a SnRK3 and CPKs (Chen et al., 2021). These kinases have a potential role in expressing LEA, WRKYs, HSPs, etc. Some kinases, like CBL5-CIPK11/26, also phosphorylates RbohF, which results in oxidative bursts (Han et al., 2019). Modules like *CBL5-CIPK11* can activate SLAC1 (Saito et al., 2018).

It has been observed that the ABA-dependent pathway is activated mostly during drought stress, while the ABA-independent pathway, which doesn't directly depend on ABA for stress response, is activated during heat and cold stresses. In cold stress, there is expression of several *DREB1a/b/c* genes, while heat stress leads to NAC, MYB, MYC, NF-Y, DREB2a and ZF-HD expression (Roychoudhury et al., 2013; Soma et al., 2021).

4.4 INTERACTION OF ABA WITH OTHER SIGNALING MOLECULES DURING ABIOTIC STRESS

4.4.1 ABA and Ethylene Interaction

The plant hormones, ABA and ET, regulate biotic and abiotic stress responses and, according to external and internal impetus, their concentration varies (Müller, 2021). Some studies have reported an antagonistic interaction between ABA and ET, i.e., they affect each other's biosynthetic pathway and signaling pathway, in addition to their antagonistic nature, they can also work in parallel or indicate positive interactions (Müller, 2021). Under abiotic stress, ABA and ethylene are key regulators in stomatal opening and closing and they both result in individual closure of stomata via hydrogen peroxide (H_2O_2) production and ETR1 receptor. ET activates the G_α protein and hence promotes the production of AtrbohF-dependent H_2O_2. The combined outcome of ABA-ET increases antioxidant activity and minimizes H_2O_2 content, resulting in half-open stomata and this is perceived in *Arabidopsis* plants that are exposed to ABA and ET (Müller, 2021). Under water stress, the rate of ET production increases in ABA-deficient seedlings, and it can be prohibited when ABA content is reinstated to normal, i.e., to stop a surfeit of ethylene, high ABA concentration is essential. Therefore, root and shoot growth are sustained, due to ABA accumulation during water stress (Sharp and LeNoble, 2002).

4.4.2 ABA and Gibberellic Acid Interaction

Abiotic stress tolerance of plants is regulated by several hormones, including GA, through the interaction of ABA signaling. Both ABA and GA control seed germination and dormancy (Jogawat, 2019). During unfavorable conditions, seeds exhibit high levels of ABA and low GA, whereas reverse circumstance is seen during favorable conditions. Phytochrome interacting factor3-like5 (PIL5) protein, inhibits seed germination and it activates GA catabolic gene and ABA biosynthetic gene, represses GA biosynthetic gene and ABA catabolic gene (Vishal and Kumar, 2018). In *Arabidopsis*, ABA helps in suppressing GA levels under high-temperature conditions (Toh et al., 2008). GA and ABA act in contrast to each other in response to abiotic stress conditions, and this antagonism between these two hormones is maintained by the DELLA protein. By preserving seed dormancy, this mechanism allows plants to escape extreme abiotic stress conditions. This crosstalk causes delayed germination as a survival mechanism to shield seeds from the damaging effects of abiotic stressors (Skubacz et al., 2016).

4.4.3 ABA and Jasmonic Acid Interaction

Drought or water deficit conditions can increase the accumulation of both ABA and JA (de Ollas and Dodd, 2016). In *Oryza sativa*, seed germination is inhibited synergistically, due to the SAPK10-Bzip72-AOC pathway, when biosynthesis of JA is fostered by ABA (Wang et al., 2020). Drought priming boosted endogenous levels of ABA and JA by inducing biosynthetic genes like *NCED* and *SnRK* (for ABA signaling) and *LOX*, *AOS* and *OPR* (for JA signaling) (Wang et al., 2021). It is possible to track elicitor-induced reprogramming of plant metabolism and development through the interaction of the ABA and JA signaling pathways (Per et al., 2018). The balance between plant development and defensive resistance is coordinated by the interaction between JA and ABA signaling, particularly between PYL (ABA receptor) and JAZ-MYC2 (Jasmonate zim domain protein; master regulator of cell cycle entry and proliferative metabolism) (Yang et al., 2019).

4.4.4 ABA and Nitric Oxide Interaction

Nitric oxide (NO) is a concentration-dependent signaling molecule that has a role in plant development and growth, regulates protein activity through S-nitrosylation and its concentration increases during stress conditions (Fatima et al., 2021). During abiotic stress, ABA and NO crosstalk regulate many processes. In the ABA signaling pathway, NO acts downstream to ABA; the hindrance in NO production leads to the removal of ABA response, but inhibition of ABA production does not affect any response if exogenous NO is supplied (Lozano-Juste and León, 2010). Studies have shown that the production of NO is crucial for ABA-regulated stomatal closure. ABA promotes NO production in the guard cells. So, NO and ABA crosstalk is responsible for stomatal closure and activating the antioxidant machinery during water stress (Hancock et al., 2011). Garcia-Mata et al. (2003) reported that in *Vicia faba*, NO controls the Ca^{2+}-ion channel and elevates the cytosolic Ca^{2+} by stimulating Ca^{2+} release from intracellular storage. NO has a role in regulating Ca^{2+}-independent outward-rectifying K^+ channel directly through S-nitrosylation (Sokolovski and Blatt, 2004) and also the regulation of inward-rectifying K^+ and Cl^- channel by Ca^{2+} release (Sokolovski et al., 2005). As a result, of both Ca^{2+}-dependent and independent modulation of ion channels, a solute is lost from the guard cells, resulting in low turgor pressure and finally stomatal closure.

4.4.5 ABA and Reactive Oxygen Species Interaction

In plant cells, ROS is generated continuously, as a by-product due to aerobic metabolism (Apel and Hirt, 2004). ROS encompasses free radicals such as superoxide anion, singlet oxygen and hydroxyl radical and a non-radical, H_2O_2; all of these result in cell damage and toxicity (Suman et al., 2021). As per recent reports, there is a complex interaction between ROS and some plant hormones, for instance, ABA (Choudhury et al., 2017). There is a very strong link between ABA and ROS for the control of stomatal aperture. Therefore, in the case of stomatal regulation, ABA results in ROS accumulation. In addition to this, it can activate anion channels of the plasma membrane and can hinder inward KAT1 K^+ channels, resulting in reduced size of stomatal aperture. In *Arabidopsis*, under different stresses, redox responsive transcription factor 1 (RRTF1) of family AP2/ERF, is simulated by ABA and ROS and increased *RRTF1* levels result in ROS accumulation (Devireddy et al., 2021). CPK5/CPK6 and CPK4/CPK11 in *Arabidopsis* help in the regulation of ROS production. In *Solanum tuberosum*, ROS production is regulated by CDPK 4 and CDPK 5 (Ca^{2+} dependent protein kinases), which is performed by phosphorylation of NADPH oxidase (Asano et al., 2012).

4.5 CONCLUSION AND FUTURE PERSPECTIVE

Stress can reduce plant growth, development and yield. ABA-mediated stress response pathways and ABAs crosstalk with other signaling molecules and can be used to develop plants that produce more ABA and have a better stress response. Further research is needed on the molecular mechanisms of these pathways. MAPK, as a highly conserved signaling pathway, aids plants in coping with diverse types of stresses. The MAPK signaling pathway serves as a core pathway that interacts with various transcription factors involved in phytohormonal signaling, including ABA. However, the relationship between ABA and the MAPK module is not yet fully understood. Understanding these interactions can help generate stress-tolerant crops in the future.

REFERENCES

Agarwal, P. K., Gupta, K., Lopato, S., & Agarwal, P. (2017). Dehydration responsive element binding transcription factors and their applications for the engineering of stress tolerance. Journal of Experimental Botany, 68(9), 2135–2148.

Apel, K., & Hirt, H. (2004). Reactive oxygen species: metabolism, oxidative stress, and signal transduction. Annual Review of Plant Biology, 55, 373–399.

Asano, T., Hayashi, N., Kikuchi, S., & Ohsugi, R. (2012). CDPK-mediated abiotic stress signaling. Plant Signaling and Behavior, 7(7), 817–821.

Bensen, R. J., Boyer, J. S., & Mullet, J. E. (1988). Water deficit-induced changes in abscisic acid, growth, polysomes, and translatable RNA in soybean hypocotyls. Plant Physiology, 88(2), 289–294.

Bharath, P., Gahir, S., & Raghavendra, A. S. (2021). Abscisic acid-induced stomatal closure: An important component of plant defense against abiotic and biotic stress. Frontiers in Plant Science, 12, 615114.

Bigeard, J., Colcombet, J., & Hirt, H. (2015). Signaling mechanisms in pattern-triggered immunity (PTI). Molecular Plant, 8(4), 521–539.

Chan, Z. (2012). Expression profiling of ABA pathway transcripts indicates crosstalk between abiotic and biotic stress responses in *Arabidopsis*. Genomics, 100(2), 110–115.

Chen, X., Ding, Y., Yang, Y., Song, C., Wang, B., Yang, S., et al. (2021). Protein kinases in plant responses to drought, salt, and cold stress. Journal of Integrative Plant Biology, 63(1), 53–78.

Choudhury, F. K., Rivero, R. M., Blumwald, E., & Mittler, R. (2017). Reactive oxygen species, abiotic stress and stress combination. The Plant Journal, 90(5), 856–867.

Colcombet, J., & Hirt, H. (2008). *Arabidopsis* MAPKs: a complex signalling network involved in multiple biological processes. Biochemical Journal, 413(2), 217–226.

Cramer, G. R., Urano, K., Delrot, S., Pezzotti, M., & Shinozaki, K. (2011). Effects of abiotic stress on plants: a systems biology perspective. BMC Plant Biology, 11(1), 1–14.

Danquah, A., De Zélicourt, A., Colcombet, J., & Hirt, H. (2014). The role of ABA and MAPK signaling pathways in plant abiotic stress responses. Biotechnology Advances, 32(1), 40–52.

de Ollas, C., & Dodd, I. C. (2016). Physiological impacts of ABA–JA interactions under water-limitation. Plant Molecular Biology, 91, 641–650.

De Zélicourt, A., Colcombet, J., & Hirt, H. (2016). The role of MAPK modules and ABA during abiotic stress signaling. Trends in Plant Science, 21(8), 677–685.

Desikan, R., Cheung, M. K., Bright, J., Henson, D., Hancock, J. T., & Neill, S. J. (2004). ABA, hydrogen peroxide and nitric oxide signalling in stomatal guard cells. Journal of Experimental Botany, 55(395), 205–212.

Devireddy, A. R., Zandalinas, S. I., Fichman, Y., & Mittler, R. (2021). Integration of reactive oxygen species and hormone signaling during abiotic stress. The Plant Journal, 105(2), 459–476.

Farooq, M., Wahid, A., Kobayashi, N. S. M. A., Fujita, D. B. S. M. A., & Basra, S. M. A. (2009). Plant drought stress: effects, mechanisms and management. Sustainable Agriculture, 29, 153–188.

Fatima, A., Husain, T., Suhel, M., Prasad, S. M., & Singh, V. P. (2021). Implication of nitric oxide under salinity stress: the possible interaction with other signaling molecules. Journal of Plant Growth Regulation, 41, 1–15.

Finkelstein, R. R., & Gibson, S. I. (2002). ABA and sugar interactions regulating development: cross-talk or voices in a crowd?. Current Opinion in Plant Biology, 5(1), 26–32.

Garcia-Mata, C., Gay, R., Sokolovski, S., Hills, A., Lamattina, L., & Blatt, M. R. (2003). Nitric oxide regulates K+ and Cl-channels in guard cells through a subset of abscisic acid-evoked signaling pathways. Proceedings of the National Academy of Sciences, 100(19), 11116–11121.

Guo, W. L., Chen, R. G., Gong, Z. H., Yin, Y. X., Ahmed, S. S., & He, Y. M. (2012). Exogenous abscisic acid increases antioxidant enzymes and related gene expression in pepper (Capsicum annuum) leaves subjected to chilling stress. Genetics and Molecular Research, 11(4), 4063–4080.

Han, J. P., Köster, P., Drerup, M. M., Scholz, M., Li, S., Edel, K. H., et al. (2019). Fine-tuning of RBOHF activity is achieved by differential phosphorylation and Ca^{2+} binding. New Phytologist, 221(4), 1935–1949.

Hancock, J. T., Neill, S. J., & Wilson, I. D. (2011). Nitric oxide and ABA in the control of plant function. Plant Science, 181(5), 555–559.

Henson, I. E. (1984). Effects of atmospheric humidity on abscisic acid accumulation and water status in leaves of rice (Oryza sativa L.). Annals of Botany, 54(4), 569–582.

Huang, X., Chen, M. H., Yang, L. T., Li, Y. R., & Wu, J. M. (2015). Effects of exogenous abscisic acid on cell membrane and endogenous hormone contents in leaves of sugarcane seedlings under cold stress. Sugar Tech, 17, 59–64.

Huang, Y. C., Niu, C. Y., Yang, C. R., & Jinn, T. L. (2016). The heat stress factor HSFA6b connects ABA signaling and ABA-mediated heat responses. Plant Physiology, 172(2), 1182–1199.

Ichimura, K., Shinozaki, K., Tena, G., Sheen, J., Henry, Y., Champion, A., et al. (2002). Mitogen-activated protein kinase cascades in plants: a new nomenclature. Trends in Plant Science, 7(7), 301–308.

Iqbal, N., Sehar, Z., Fatma, M., Umar, S., Sofo, A., & Khan, N. A. (2022). Nitric oxide and abscisic acid mediate heat stress tolerance through regulation of osmolytes and antioxidants to protect photosynthesis and growth in wheat plants. Antioxidants, 11(2), 372.

Jiang, Z., Zhu, H., Zhu, H., Tao, Y., Liu, C., Liu, J., et al. (2022). Exogenous ABA Enhances the Antioxidant Defense System of Maize by Regulating the AsA-GSH Cycle under Drought Stress. Sustainability, 14(5), 3071.

Jogawat, A. (2019). Crosstalk among phytohormone signaling pathways during abiotic stress. In: A. Roychoudhary & D. Tripathi (eds.), Molecular plant abiotic stress: biology and biotechnology (pp. 209–220). John Wiley & Sons, Ltd.

Kannangara, T., Durley, R. C., Simpson, G. M., & Seetharama, N. (1983). Drought resistance of *Sorghum bicolor*. 6. Changes in endogenous growth regulators of plants grown across an irrigation gradient. Canadian Journal of Plant Science, 63(1), 147–155.

Kaur, G., & Asthir, B. (2020). Impact of exogenously applied ABA on proline metabolism conferring drought and salinity stress tolerance in wheat genotypes. Cereal Research Communications, 48, 309–315.

Khadri, M., Tejera, N. A., & Lluch, C. (2007). Sodium chloride–ABA interaction in two common bean (*Phaseolus vulgaris*) cultivars differing in salinity tolerance. Environmental and Experimental Botany, 60(2), 211–218.

Kim, T. E., Kim, S. K., Han, T. J., Lee, J. S., & Chang, S. C. (2002). ABA and polyamines act independently in primary leaves of cold-stressed tomato (*Lycopersicon esculentum*). Physiologia Plantarum, 115(3), 370–376.

Li, X., Li, S., Wang, J., & Lin, J. (2020). Exogenous abscisic acid alleviates harmful effect of salt and alkali stresses on wheat seedlings. International Journal of Environmental Research and Public Health, 17(11), 3770.

Liu, H., Ren, X., Zhu, J., Wu, X., & Liang, C. (2018). Effect of exogenous abscisic acid on morphology, growth and nutrient uptake of rice (*Oryza sativa*) roots under simulated acid rain stress. Planta, 248, 647–659.

Liu, X., Ji, P., Yang, H., Jiang, C., Liang, Z., Chen, Q., et al. (2022). Priming effect of exogenous ABA on heat stress tolerance in rice seedlings is associated with the upregulation of antioxidative defense capability and heat shock-related genes. Plant Growth Regulation, 98(1), 23–38.

Lozano-Juste, J., & León, J. (2010). Nitric oxide modulates sensitivity to ABA. Plant Signaling and Behavior, 5(3), 314–316.

Müller, M. (2021). Foes or friends: ABA and ethylene interaction under abiotic stress. Plants, 10(3), 448.

Parihar, P., Singh, S., Singh, R., Singh, V. P., & Prasad, S. M. (2015). Effect of salinity stress on plants and its tolerance strategies: a review. Environmental Science and Pollution Research, 22, 4056–4075.

Per, T. S., Khan, M. I. R., Anjum, N. A., Masood, A., Hussain, S. J., & Khan, N. A. (2018). Jasmonates in plants under abiotic stresses: Crosstalk with other phytohormones matters. Environmental and Experimental Botany, 145, 104–120.

Rahman, A., Sinha, K. V., Sopory, S. K., & Sanan-Mishra, N. (2021). Influence of virus–host interactions on plant response to abiotic stress. Plant Cell Reports, 40(11), 2225–2245.

Rehman, A., Azhar, M. T., Hinze, L., Qayyum, A., Li, H., Peng, Z., et al. (2021). Insight into abscisic acid perception and signaling to increase plant tolerance to abiotic stress. Journal of Plant Interactions, 16(1), 222–237.

Roychoudhury, A., & Banerjee, A. (2017). Abscisic acid signaling and involvement of mitogen activated protein kinases and calcium-dependent protein kinases during plant abiotic stress. In: G.K. Pandey (ed.), Mechanism of plant hormone signaling under stress (1, pp. 197–241). John Wiley & Sons, Inc.

Roychoudhury, A., Paul, S., & Basu, S. (2013). Cross-talk between abscisic acid-dependent and abscisic acid-independent pathways during abiotic stress. Plant Cell Reports, 32, 985–1006.

Sah, S. K., Reddy, K. R., & Li, J. (2016). Abscisic acid and abiotic stress tolerance in crop plants. Frontiers in Plant Science, 7, 571.

Saito, S., Hamamoto, S., Moriya, K., Matsuura, A., Sato, Y., Muto, J., et al. (2018). N-myristoylation and S-acylation are common modifications of Ca^{2+}-regulated *Arabidopsis* kinases and are required for activation of the SLAC1 anion channel. New Phytologist, 218(4), 1504–1521.

Shan, C., Zhang, S., & Zhou, Y. (2017). Hydrogen sulfide is involved in the regulation of ascorbate-glutathione cycle by exogenous ABA in wheat seedling leaves under osmotic stress. Cereal Research Communications, 45(3), 411–420.

Sharp, R. E., & LeNoble, M. E. (2002). ABA, ethylene and the control of shoot and root growth under water stress. Journal of Experimental Botany, 53(366), 33–37.

Singh, A., & Prasad, S. M. (2015). Remediation of heavy metal contaminated ecosystem: an overview on technology advancement. International Journal of Environmental Science and Technology, 12, 353–366.

Skubacz, A., Daszkowska-Golec, A., & Szarejko, I. (2016). The role and regulation of ABI5 (ABA-Insensitive 5) in plant development, abiotic stress responses and phytohormone crosstalk. Frontiers in Plant Science, 7, 1884.

Sokolovski, S., & Blatt, M. R. (2004). Nitric oxide block of outward-rectifying K^+ channels indicates direct control by protein nitrosylation in guard cells. Plant Physiology, 136(4), 4275–4284.

Sokolovski, S., Hills, A., Gay, R., Garcia-Mata, C., Lamattina, L., & Blatt, M. R. (2005). Protein phosphorylation is a prerequisite for intracellular Ca^{2+} release and ion channel control by nitric oxide and abscisic acid in guard cells. The Plant Journal, 43(4), 520–529.

Soma, F., Takahashi, F., Yamaguchi-Shinozaki, K., & Shinozaki, K. (2021). Cellular phosphorylation signaling and gene expression in drought stress responses: ABA-dependent and ABA-independent regulatory systems. Plants, 10(4), 756.

Sripinyowanich, S., Klomsakul, P., Boonburapong, B., Bangyeekhun, T., Asami, T., Gu, H., et al. (2013). Exogenous ABA induces salt tolerance in indica rice (*Oryza sativa* L.): the role of OsP5CS1 and OsP5CR gene expression during salt stress. Environmental and Experimental Botany, 86, 94–105.

Suman, S., Bagal, D., Jain, D., Singh, R., Singh, I. K., & Singh, A. (2021). Biotic stresses on plants: Reactive oxygen species generation and antioxidant mechanism. In: T. Aftab & K.R. Hakeem (eds.), Frontiers in plant-soil interaction (pp. 381–411). Academic Press.

Teng, K., Li, J., Liu, L., Han, Y., Du, Y., Zhang, J., et al. (2014). Exogenous ABA induces drought tolerance in upland rice: the role of chloroplast and ABA biosynthesis-related gene expression on photosystem II during PEG stress. Acta Physiologiae Plantarum, 36, 2219–2227.

Thameur, A., Ferchichi, A., & López-Carbonell, M. (2011). Quantification of free and conjugated abscisic acid in five genotypes of barley (*Hordeum vulgare* L.) under water stress conditions. South African Journal of Botany, 77(1), 222–228.

Toh, S., Imamura, A., Watanabe, A., Nakabayashi, K., Okamoto, M., Jikumaru, Y., et al. (2008). High temperature-induced abscisic acid biosynthesis and its role in the inhibition of gibberellin action in *Arabidopsis* seeds. Plant Physiology, 146(3), 1368–1385.

Vishal, B., & Kumar, P. P. (2018). Regulation of seed germination and abiotic stresses by gibberellins and abscisic acid. Frontiers in Plant Science, 9, 838.

Vishwakarma, K., Upadhyay, N., Kumar, N., Yadav, G., Singh, J., Mishra, R. K., et al. (2017). Abscisic acid signaling and abiotic stress tolerance in plants: a review on current knowledge and future prospects. Frontiers in Plant Science, 8, 161.

Wang, C., Yang, A., Yin, H., & Zhang, J. (2008). Influence of water stress on endogenous hormone contents and cell damage of maize seedlings. Journal of Integrative Plant Biology, 50(4), 427–434.

Wang, X., Li, Q., Xie, J., Huang, M., Cai, J., Zhou, Q., et al. (2021). Abscisic acid and jasmonic acid are involved in drought priming-induced tolerance to drought in wheat. The Crop Journal, 9(1), 120–132.

Wang, Y., Hou, Y., Qiu, J., Wang, H., Wang, S., Tang, L., et al. (2020). Abscisic acid promotes jasmonic acid biosynthesis via a 'SAPK10-bZIP72-AOC'pathway to synergistically inhibit seed germination in rice (*Oryza sativa*). New Phytologist, 228(4), 1336–1353.

Wang, Y., Li, L., Ye, T., Zhao, S., Liu, Z., Feng, Y. Q., & Wu, Y. (2011). Cytokinin antagonizes ABA suppression to seed germination of *Arabidopsis* by downregulating ABI5 expression. The Plant Journal, 68(2), 249–261.

Wright, S. T. C., & Hiron, R. W. P. (1969). (+)-Abscisic acid, the growth inhibitor induced in detached wheat leaves by a period of wilting. Nature, 224(5220), 719–720.

Yang, J., Duan, G., Li, C., Liu, L., Han, G., Zhang, Y., & Wang, C. (2019). The crosstalks between jasmonic acid and other plant hormone signaling highlight the involvement of jasmonic acid as a core component in plant response to biotic and abiotic stresses. Frontiers in Plant Science, 10, 1349.

Zhang, H., Zhu, J., Gong, Z., & Zhu, J. K. (2022). Abiotic stress responses in plants. Nature Reviews Genetics, 23(2), 104–119.

5 Auxin in Abiotic Stress

Farheen Islam, Daniya Shahid, Renu Soni, and Neha Singh

5.1 INTRODUCTION

The growth and reproductive capacity of plants are severely affected by unfavorable climatic factors like excessive temperature, salinity, deficiency of water, etc. To overcome these challenges, plants adapt to a complex web of regulatory networks. One key mechanism involves the activation of internal phytohormones like auxin, which triggers a variety of biochemical and physiological processes. These intricate adaptations enable plants to successfully survive and grow despite unfavorable conditions at every stage of their life cycle (Du et al., 2013) (Figure 5.1). Auxin is a phytohormone that is essential during embryogenesis and senescence and regulates several cellular processes. Under optimal conditions, auxin regulates diverse plant developmental processes such as rate of root growth and root system architecture (RSA). Even though the details related to auxin signaling components are now evident, the process of auxin biosynthesis has long been shrouded in uncertainty. In response to changing environmental conditions, the biosynthetic pathway of auxin appears to be differentially regulated (Bao & Li, 2002).

Recently, huge progress has been made in understanding auxin behavior and its functioning mechanism in response to environmentally stressed conditions. Plant roots exhibit remarkable developmental plasticity in response to abiotic stressors such as heat, salinity, cold, nutrient and water

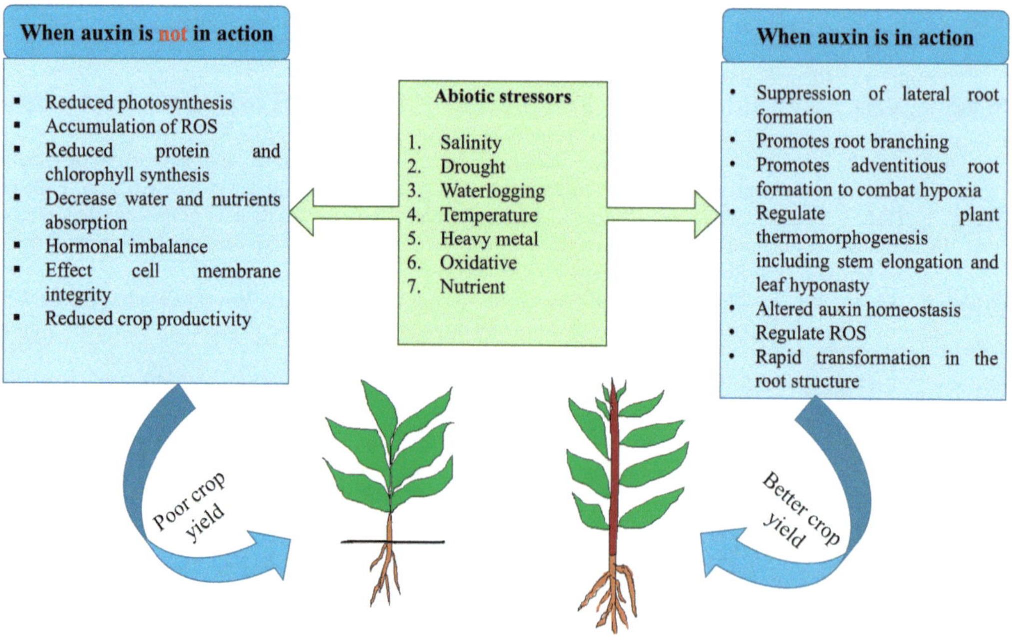

FIGURE 5.1 Role of auxin in combating abiotic stresses. The growth and reproductive capacity of plants are severely affected by unfavorable climatic factors such as drought, salt and excessive temperatures.

DOI: 10.1201/9781003335788-8

deficiencies. This plasticity is made possible through auxin regulation and differential expression of genes involved in numerous pathways related to auxin. Through this mechanism, shoot-derived long-distance auxin signaling is modulated. Research using genomic studies has consistently shown that the expression of critical auxin-responsive genes, including auxin/Indole-3-Acetic Acid (Aux/IAA), Gretchen Hagen3 (GH3), small auxin upregulated RNAs (SAURs) and auxin response factors (ARFs), are altered in response to a wide range of abiotic stress conditions (Sharma et al., 2015). Generally, the amount of free auxin present inside the plant also gets decreased by abiotic stress conditions that may act as a factor in adapting the plant growth under severe environmental conditions. Auxin's function in abiotic stresses has been the subject of recent research. This chapter offers a broad perspective on the significance of auxin signaling in diverse plant responses to abiotic stress.

5.2 ROLE OF AUXIN UNDER ABIOTIC STRESSORS

5.2.1 Salinity Stress

Salt stress results in the accumulation of reactive oxygen species (ROS) and suppression of auxin homeostasis, synthesis and reactions, this could be a mechanism that plants employ to cope with high levels of soil salinity (Fu et al., 2019). Recent understanding of lateral root (LR) development has led to the proof that auxin acts as an essential component that integrates several processes of root morphogenesis (Lavenus et al., 2013), which involves multiple IAAs and auxin response factors (ARFs). During high salinity, a cognate response by roots leads to suppression of LR development in *Arabidopsis thaliana* (L.) Heynh. and *Medicago truncatula* Gaertn. (Galvan-Ampudia & Testerink, 2011). In comparison, low-grade salt exposure causes a decline in the elongation of LR and leads to a rise in the amount of LRs (Zolla et al., 2009). Transcription of auxin synthesis genes (*nitrilase 1* and *2*) can also be induced in *Arabidopsis* during salt stress, indicating that the level of IAA may rise in response to soil saline environment (Bao & Li, 2002). It has also been observed that under salt stress, PIN genes (which has a vital function in regulating asymmetric auxin distribution across diverse developmental pathways) like *PIN1*, *PIN3* and *PIN7* are negatively regulated (Fu et al., 2019) and this negative regulation affects auxin transport, PIN protein abundance and consequently, its signaling (Ribba et al., 2020). Crosstalk between salt stress and auxin also occurs at the germination level and in early-stage seedling development. In a neutral environment, auxin does not majorly influence the germination process of seeds per se, but it shows a negative impact on the germination of seed under high salinity (Jung et al., 2011).

5.2.2 Drought Stress

Auxin plays a crucial function in the reaction of plants toward water-deficient conditions. Members of the GH3 gene family (auxin-responsive gene family), which codes for IAA conjugating enzymes like IAA-amido synthetases, are also involved in responses against drought stress. In rice, *OsGH3.13* enhances *LEA* (late embryogenesis abundant) gene expression, which in turn increases rice seedling growth under low water conditions (Zhang S. et al., 2009). Auxin is known to positively regulate the root and root branching biomass, which may increase the efficiency of water intake and provide resistance against drought (Shi et al., 2014). Auxin transport can be disturbed by ABA to mediate the secretion of proton at the root tip region and regulate the growth of root under a slight water deficit environment (Sharma et al., 2015). In *Arabidopsis*, the activation of genes is related to auxin synthesis; YUCCA7 caused an increased endogenous level of auxin and enhanced resistance to drought (Lee & Luan, 2012).

5.2.3 Temperature Stress

Recent studies show an important relationship between signaling of IAA and high- and low-temperature stress. Increased temperature reduces the auxin levels, mainly the capacity of auxin transport in the reproductive parts of *Capsicum annuum*. Du et al. (2013) observed that in a low-temperature environment of about three days, the amount of IAA increased to 1.6-fold; however, after 6 hours of high temperature, it was enhanced by 1.3-fold compared to the control. The polar transport of auxin is also affected by the cold stress via the selective inhibition of intracellular trafficking of PIN2 and PIN3, carriers related to auxin efflux (Shibasaki et al., 2009). In *Arabidopsis*, a low-temperature condition restricts the gravity response of inflorescence, which is controlled by an uneven auxin distribution. Gravitropism of auxin can also be inhibited by low temperature and, hence, the impact of cold stress on plant growth regulation and development is associated with the response of intracellular IAA (Tian et al., 2022). Heat stress downregulates many *OsGH3* genes in rice, which converts active IAA into an inactive form. Auxin is a crucial factor in the process of heat stress-induced thermos-morphogenesis, which involves leaf hyponasty and the elongation of the stem (Küpers et al., 2020). Elevated temperature reduces the endogenous IAA in developing anthers by lowering the expression of YUCCA and, hence, results in male sterility in *Arabidopsis* and barley. Similarly, several other auxin-responsive genes react to cold stress.

5.2.4 Heavy Metal Stress

While plants require trace amounts of heavy metals for proper growth, excessive concentrations of these metals can have negative effects on the plant's normal metabolic and developmental functions (Gill & Tuteja, 2010; Khan & Khan, 2014, 2017). The reduction in the growth of cell expansion in wheat (Lane et al., 1978) and maize (Hasenstein et al., 1988) caused by heavy metal exposure is linked to changes in auxin transportation and metabolism. Lane et al. (1978) found that high concentrations of lead negatively impact cell elongation and division rates in *Triticum aestivum*, which was regulated by the activity of auxin. Several researchers have investigated the association between auxin signaling and heavy metal stress using various metal ions like aluminum, cadmium and lead, including others (some of which are mentioned in) Table 5.1.

5.2.5 Oxidative Stress

Primary stress results in the formation of reactive oxygen species (ROS). Although ROS are normally present inside the cells, under stressful conditions, the level of ROS surpasses the cell's quenching capability leading to oxidative stress. Plants have their own ROS scavenging mechanisms, which include enzymatic and non-enzymatic antioxidants (Agarwal et al., 2009). Antioxidants keep the ROS level at the baseline and regulate their homeostasis. The crosstalk of hormones and ROS has been exemplified mainly by the function of ABA and H_2O_2 in stomatal closure. However, evidence has also come out for the role of auxin in stress-related hormonal networks, where it can regulate ROS levels. Phytohormones like auxin have been found to play a crucial role in plant tolerance and the alleviation of the oxidative stress that is induced by heavy metals (Bucker-Neto et al., 2017). The functional analysis of genes encoding TIR1/AFB auxin receptors also indicates that they are involved in tolerance to salinity and oxidative stress (Iglesias et al., 2011). Enhanced levels of auxin may trigger the accumulation of H_2O_2, causing inhibition of root cell elongation and root growth. On the contrary, auxin levels may also be increased because of oxidative stress either through tryptophan-dependent auxin biosynthesis or via β-oxidation of precursors leading from lipid peroxidation of membranes (Woodward & Bartel, 2005). It has also been observed that heavy metal stress, like Pb and Cd, increased the content of IAA in the corn coleoptile sections, but only at the concentrations at which elongation growth stimulation was observed. Therefore, it can be concluded that IAA content is necessary for the stimulation of elongation growth by heavy metals accompanied by high levels of H_2O_2 (Malkowski et al., 2020).

Under any environmental stress, both auxin and ROS are affected. Auxin indirectly mediates ROS homeostasis by changing the stability of DELLA proteins or by activating ROS detoxifying enzymes (Fu & Harberd, 2003; Paponov et al., 2008). When auxin is applied exogenously to the plants, the concentration of H_2O_2 is increased (Peer et al., 2013). Both auxin and ROS response are required to adapt to various stresses (De Tullio et al., 2010).

TABLE 5.1
Auxin-Mediated Responses in Plants Under Abiotic Stresses

Abiotic Stress	Auxin-Mediated Responses	Crops	References
Phosphorus deficiency	Phosphorus scarcity in plants leads to the formation of lateral roots via auxin mediation and enhances the sensitivity of plants towards auxin	*Arabidopsis*, *Lupinus albus*	Neumann et al., 2000, Nacry et al., 2005
Nitrogen deficiency	Nitrogen deficiency can lead to altered root architecture, regulated by auxin transport. Several nitrogen uptake transporters are involved in this process and can help plants adjust their response to nitrogen deficiency	*Arabidopsis*	Gojon et al., 2011, Bouguyon et al., 2012
Sulfur deficiency	Plants respond to sulfur deficiency by increasing the accumulation of auxin, which promotes more root growth and enhances access to sulfur	*Arabidopsis*	Frerigmann & Gigolashvili, 2014
Potassium deficiency	Potassium transporters play a crucial role in auxin transport and insufficient potassium levels lead to decreased auxin concentration and transport, resulting in stunted root growth	*Arabidopsis, Oryza sativa, Gossypium herbaceum, Zea mays, Nicotiana tabaccum*	Vicente-Agullo et al., 2004, Zhang Z et al., 2009, Ma et al., 2012, Song et al., 2015
Iron deficiency	Insufficient iron causes an increase in auxin levels in roots and results in the development of branched root hairs. Furthermore, the reaction controlled by auxin involves the interplay of nitric oxide and ethylene in conditions of Fe deficiency	*Arabidopsis, Trifolium pratens, Malus xiaojinensis*	Schmidt et al., 2000, Zheng et al., 2003, Romera et al., 2011, Wu et al., 2012
Aluminum toxicity	Aluminum toxicity causes a reduction in cell expansion and suppresses the growth of roots by interfering with the transportation of auxin	*Arabidopsis, Triticum aestivum, Zea mays*	Kollmeier et al., 2000, Sun et al., 2010, Yang et al., 2011
Cadmium toxicity	Cadmium stress causes an increase in auxin degradation, activity of IAA-oxidase and decrease in the growth of a plant	*Arabidopsis, Pisum sativum*	Chaoui & El Ferjani, 2005, Mei et al., 2009

5.2.6 Flooding Stress

During flooding (or waterlogging), plants face hypoxia, which is sensed by the main roots of the plants. In response to the flooding, plants tend to replace the main root system with adventitious roots to combat the effects of hypoxia (Sauter, 2013). It is found that auxin plays an important role in the formation of adventitious roots by activating proteins like PIN-FORMED and AUXIN RESISTANT1/LIKE AUX1 (da Costa et al., 2020). They are auxin transport proteins and help in regulating the distribution of auxin in response to flooding conditions.

In cucumber waterlogged plants, the levels of auxin, ROS and ethylene were increased and ROS played an important role in the formation of adventitious roots induced by ethylene and auxin (Qi et al., 2019). The soybean plants were able to survive the short-term flooding situation by activating auxin-induced transcriptomic modifications for preserving the energy (Wang et al., 2020). A comparison between two types of soybean plants, one that can tolerate waterlogging and one that is sensitive to it, revealed that the waterlog-tolerant soybean line showed better growth of adventitious roots/aerenchyma cells in the stellar region of the plant stem. Additionally, these plants produced a higher amount of ethylene, which helped them grow well even in waterlogged conditions (Kim et al., 2015). Several studies have shown that ethylene is the primary hormone that helps the plants in adapting to the flooding conditions. Additionally, ethylene modulates other hormones like ABA, GA and auxin to promote the formation of adventitious roots, carbohydrate degradations and internode elongation (Loreti et al., 2016).

5.2.7 Nutrient Deficiency Stress

In nature, plants often face a lack of vital minerals that hinder their growth. To survive and thrive, plants must be able to detect and react to changes in soil nutrient levels. The signaling and transportation of auxin are crucial in governing the developmental responses of plants towards nutrient deficiencies in the soil. According to Kazan (2013), the rapid transformation in the root structure of plants in response to nutrient scarcity is accomplished through the mechanism of auxin signaling. Auxin serves as a mediator for plants to react to nitrogen and other nutrient deficiencies. In *Arabidopsis* seedlings, it was observed that roots grown in low nitrogen conditions have significantly elevated levels of auxin in comparison to those grown in high nitrogen conditions (Kiba et al., 2011). Similarly, additional auxin-mediated responses that occur in response to a lack of various nutrients, such as phosphorus, sulfur and potassium, etc., are highlighted in Table 5.1.

5.3 ABIOTIC STRESS AND THE DIFFERENTIAL REGULATION OF AUXIN-RESPONSIVE GENES

Abiotic stress response alters the expression of numerous auxin-related genes in plants (Yuan et al., 2013). In *Arabidopsis*, low temperature causes a change in the expression of members of the Aux/IAA and ARF TFs. At least 15 of the 31 rice *OsIAA* genes underwent expression studies, showing that eight *OsIAA* genes exhibited enhanced expression during salt stress (Song et al., 2009). After wounding, some of the auxin-related genes like *IAA2*, *IAA3* and *SAUR-ACI* show differential expression, suggesting that there is an interaction between auxin signaling and the wounding pathway. It is interesting to note that auxin positively regulates and wounding negatively regulates all of these genes. Additionally, it was noted that wounding increased the expression of a gene similar to *NPK1*, which has a detrimental function in auxin signaling (Kovtun et al., 1998).

The auxin-responsive genes like *IAA1*, *IAA4*, *At1g29430*, *At4g34760* and *At4g38860* were downregulated when the plants were given ABA, methyl jasmonate and drought treatment (Huang et al., 2008). The tolerance towards salt and oxidative stress increased in the transgenic *Arabidopsis* with an overexpressed auxin-responsive glutathione-S-transferase (*GST*) gene, *OsGSTU4* (Sharma et al., 2015). According to the transcriptome study of Kodaira et al. (2011), during osmotic and ABA treatment many auxin-responsive genes were downregulated. All these results simply show that there is a clear connection between abiotic stress and auxin signaling because auxin-responsive genes show a variable expression under different abiotic stress conditions. In plants, there is a group of transcription factors (B3-type) called ARF genes that play a role in controlling the expression of auxin response genes. These ARF genes are found in various flowering plants like *Arabidopsis*, rice, soybean and maize (Li et al., 2016).

SAUR proteins are an important part of the plant defense system. They help cells grow longer, control the acidity of the cell walls and respond to changes in the environment (Stortenbeker & Bemer, 2019). For instance, when plants are exposed to ABA, SAUR genes become more active (Ren & Gray, 2015; Qiu et al., 2020). However, a recent study by He et al. (2021) discovered that a specific SAUR protein called AtSAUR32 has a special role in helping plants tolerate drought by using the ABA hormone pathway. This finding suggests that AtSAUR32 could be an important factor in understanding how plants respond to ABA.

5.4 FUNCTIONAL GENOMICS OF AUXIN IN ABIOTIC STRESS RESPONSE

Many stress-related genes have been identified at the genome-wide level, thanks to the development of current technologies in functional genomics. According to the genome-wide transcriptome studies, many auxin-related genes showed a variable expression during and after abiotic stress. To understand the connection between differential gene expression under stressed condition, microarray data was used in *Arabidopsis*. Numerous candidate genes have been identified that are involved in signaling pathways for auxin and abiotic stresses. Aux/IAA gene family members have also been discovered to express differentially in *Arabidopsis* during cold acclimation (Sharma et al., 2015).

Similarly, the genome-wide investigation of the GST gene family revealed 31 GST genes that are sensitive to auxin and cytokinin and show variable expression in response to abiotic stress conditions. Rice glutaredoxin genes (GRX) have been differentially expressed under abiotic stress and hormone (auxin) treatment. The differential expressions of various auxin-responsive genes have been identified and their function is clarified using different approaches, like mutants or overexpression analysis, in model plants as well as different crop plants (Sharma et al., 2015).

PIN proteins are auxin export carriers that control intercellular auxin fluxes. Thus, they play an important role in various plant growth and development aspects and also control how plants will adapt to a certain stressful condition. Cardoso et al. (2022) performed an *in-silico* analysis of PIN proteins on *Olea europaea* L. According to their results, upon biotic and abiotic stresses, OePINs are co-expressed under a variety of stress signaling pathways and regulate oxidative stress.

5.5 CONCLUSION

The amount of auxin in plants gets altered by different biotic and abiotic stresses, which disrupt plant growth and development. As auxin is a known major regulator of plant growth and development, therefore under abiotic stress, its importance in providing developmental plasticity to plants needs greater exploration. Thus, regulations of auxin biosynthesis, transport mechanism, metabolism, signal perception, tissue sensitivity and transduction have been the subject of thorough examination. Under abiotic stress conditions, the importance of auxin in mediating LR development is quite critical. Thus, modulation of the auxin concentration in a tissue-specific manner may be a possible strategy for altering root system architecture and enhancing stress tolerance. Several auxin biosynthetic genes, auxin transporters, conjugation molecules, metabolism and signaling components can be potential candidates for improving abiotic stress tolerance in plants. However, many other components of auxin signaling and mechanisms are yet to be discovered for increasing auxin-mediated tolerance in plants. The development of stable auxin-engineered crops is a key challenge that remains to be addressed as these engineered crops perform well under both stressed and unstressed conditions. A better understanding of the crosstalk between stress signaling and auxin would be key in applying biotechnological approaches toward developing tolerant crops against abiotic stresses, especially when climate change is becoming a major global concern.

REFERENCES

Agarwal, P., Agarwal, P. K., Joshi, A. J., Sopory, S. K., & Reddy, M. K. (2009). Overexpression of PgDREB2A transcription factor enhances abiotic stress tolerance and activates downstream stress-responsive genes. Molecular Biology Reports, 37 (2), 1125–1135. 10.1007/s11033-009-9885-8.

Bao, F., & Li, J. Y. (2002). Evidence that the auxin signaling pathway interacts with plant stress response. Journal of Integrative Plant Biology, 44 (5), 532.

Bouguyon, E., Gojon, A., & Nacry, P. (2012). Nitrate sensing and signaling in plants. Seminars in Cell & Developmental Biology, 23 (6), 648–654. 10.1016/j.semcdb.2012.01.004.

Bücker-Neto, L., Paiva, A. L. S., Machado, R. D., Arenhart, R. A., & Margis-Pinheiro, M. (2017). Interactions between plant hormones and heavy metals responses. Genetics and Molecular Biology, 40, 373–386. 10.1590/1678-4685-gmb-2016-0087.

Cardoso, H., Campos, C., Grzebelus, D., Egas, C. & Peixe, A. (2022). Understanding the role of PIN auxin carrier genes under biotic and abiotic stresses in *Olea europaea* L. Biology, 11 (7), 1040. 10.3390/biology11071040.

Chaoui, A., & El Ferjani, E. (2005). Effects of cadmium and copper on antioxidant capacities, lignification and auxin degradation in leaves of pea (*Pisum sativum* L.) Seedlings. Comptes Rendus Biologies, 328 (1), 23–31. 10.1016/j.crvi.2004.10.001.

da Costa C. T., Offringa, R., & Fett-Neto, A. G. (2020). The role of auxin transporters and receptors in adventitious rooting of *Arabidopsis thaliana* pre-etiolated flooded seedlings. Plant Science, 290, 110294. doi: 10.1016/j.plantsci.2019.110294.

De Tullio, M. C., Jiang, K., & Feldman, L. J. (2010). Redox regulation of root apical meristem organization: Connecting root development to its environment. Plant Physiology and Biochemistry, 48 (5), 328–336. 10.1016/j.plaphy.2009.11.005.

Du, H., Liu, H., & Xiong, L. (2013). Endogenous auxin and jasmonic acid levels are differentially modulated by abiotic stresses in rice. Frontiers in Plant Science, 4. 10.3389/fpls.2013.00397.

Frerigmann, H., & Gigolashvili, T. (2014). Update on the role of R2R3-MYBs in the regulation of glucosinolates upon sulfur deficiency. Frontiers in Plant Science, 5. 10.3389/fpls.2014.00626.

Fu, X., & Harberd, N. P. (2003). Auxin promotes *Arabidopsis* root growth by modulating gibberellin response. Nature, 421 (6924), 740–743. 10.1038/nature01387.

Fu, Y., Yang, Y., Chen, S., Ning, N., & Hu, H. (2019). *Arabidopsis* IAR4 modulates primary root growth under salt stress through ROS-mediated modulation of auxin distribution. Frontiers in Plant Science, 10. 10.3389/fpls.2019.00522.

Galvan-Ampudia, C. S., & Testerink, C. (2011). Salt stress signals shape the plant root. Current Opinion in Plant Biology, 14 (3), 296–302. 10.1016/j.pbi.2011.03.019.

Gill, S. S., & Tuteja, N. (2010). Reactive oxygen species and antioxidant machinery in abiotic stress tolerance in crop plants. Plant Physiology and Biochemistry, 48, 909–930. 10.1016/j.plaphy.2010.08.016.

Gojon, A. Krouk, G., Perrine-Walker, F., & Laugier, E. (2011). Nitrate transceptor(s) in plants. Journal of Experimental Botany, 62 (7), 2299–2308. 10.1093/jxb/erq419.

Hasenstein, K. H., Evans, M. L., Stinemetz, C. L., Moore, R., Fondren, W. M., Koon, C., Higby, M. A., & Smucker, A. J. (1988). Comparative effectiveness of metal ions in inducing curvature of primary roots of *Zea mays*. Plant Physiology, 86, 885–889. 10.1104/pp.86.3.885.

He, Y., Liu, Y., Li, M., Lamin-Samu, A. T., Yang, D., Yu, X., Izhar, M., Jan, I., Ali, M., & Lu, G. (2021). The *Arabidopsis* SMALL AUXIN UP RNA32 protein regulates ABA-mediated responses to drought stress. Frontiers in Plant Science, 12, 625493. doi: 10.3389/fpls.2021.625493

Huang, A., Wang, Y., Liu, Y., Wang, G., & She, X. (2019). Reactive oxygen species regulate auxin levels to mediate adventitious root induction in *Arabidopsis* hypocotyl cuttings. Journal of Integrative Plant Biology, 62 (7), 912–926. 10.1111/jipb.12870.

Huang, D., Wu, W., Abrams, S. R., & Cutler, A. J. (2008). The relationship of drought-related gene expression in Arabidopsis thaliana to hormonal and environmental factors. Journal of Experimental Botany, 59, 2991–3007. 10.1093/jxb/ern155.

Iglesias, M. J., Terrile, M. C., & Casalongué, C. A. (2011). Auxin and salicylic acid signaling counteract the regulation of adaptive responses to stress. Plant Signaling & Behavior, 6 (3), 452–454. 10.4161/psb.6.3.14676.

Jung, J. H., & Park, C. M. (2011). Auxin modulation of salt stress signaling in *Arabidopsis* seed germination. Plant Signaling & Behavior, 6 (8), 1198–1200. 10.4161/psb.6.8.15792.

Kazan, K. (2013). Auxin and the integration of environmental signals into plant root development. Annals of Botany (London), 112, 1655–1665. 10.1093/aob/mct229.

Khan, M. I. R., & Khan, N. A. (2014). Ethylene reverses photosynthetic inhibition by nickel and zinc in mustard through changes in PS II activity, photosynthetic-nitrogen use efficiency and antioxidant metabolism. Protoplasma, 251, 1007–1019. 10.1007/s00709-014-0610-7

Khan, M. I. R., & Khan, N. (2017). Reactive Oxygen Species and Antioxidant System in Plants: Role and Regulation Under Abiotic Stress. Springer Nature, Singapore. 10.1007/978-981-10-5254-5

Kiba, T., Kudo, T., Kojima, M., & Sakakibara, H. (2011). Hormonal control of nitrogen acquisition: roles of auxin, abscisic acid, and cytokinin. Journal of Experimental Botany, 62 (4), 1399–1409. 10.1093/jxb/erq410.

Kim, Y. H., Hwang, S. J., Waqas, M., Khan, A. L., Lee, J. H., Lee, J. D., Nguyen, H. T. & Lee, I. J. (2015). Comparative analysis of endogenous hormones level in two soybean (*Glycine max* L.) lines differing in waterlogging tolerance. Frontiers in Plant Science, 6, 714. doi: 10.3389/fpls.2015.00714.

Kodaira, K. S., Qin, F., Tran, L. S. P., Maruyama, K., Kidokoro, S., Fujita, Y., & Yamaguchi-Shinozaki, K. (2011). *Arabidopsis* Cys2/His2 zinc-finger proteins AZF1 and AZF2 negatively regulate abscisic acid-repressive and auxin-inducible genes under abiotic stress conditions. Plant Physiology, 157(2), 742–756. 10.1104/pp.111.182683.

Kollmeier, M., Felle, H. H., & Horst, W. J. (2000). Genotypical differences in aluminum resistance of maize are expressed in the distal part of the transition zone. Is reduced basipetal auxin flow involved in inhibition of root elongation by aluminum? Plant Physiology, 122 (3), 945–956. 10.1104/pp.122.3.945.

Kovtun, Y., L, C. W., Zeng, W., & Sheen, J. (1998). Suppression of auxin signal transduction by a MAPK cascade in higher plants. Nature, 395 (6703), 716–720. 10.1038/27240.

Küpers, J. J., Oskam, L., & Pierik, R. (2020). Photoreceptors regulate plant developmental plasticity through auxin. Plants, 9 (8), 940. 10.3390/plants9080940.

Lane, S. D., Martin, E. S., & Garrod, J. P. (1978). Lead toxicity effect on indole-3-acetic-induced cell elongation. Planta, 144, 79–84. 10.1007/BF00385010.

Lavenus, J. Goh, T., Roberts, I., Guyomarc'h, S., Lucas, M., De Smet, I., Fukaki, H., Beeckman, T., Bennett, M., & Laplaze, L. (2013). Lateral root development in *Arabidopsis*: Fifty shades of auxin. Trends in Plant Science, 18 (8), 450–458. 10.1016/j.tplants.2013.04.006.

Lee, S. C., & Luan, S. (2012). ABA signal transduction at the crossroad of biotic and abiotic stress responses. Plant, Cell & Environment, 35 (1), 53–60. 10.1111/j.1365-3040.2011.02426x.

Loreti, E., van Veen, H., & Perata, P. (2016). Plant responses to flooding stress. Current Opinion in Plant Biology, 33, 64–71. doi: 10.1016/j.pbi.2016.06.005.

Li, S. B., Xie, Z. Z., Hu, C. G., & Zhang, J. Z. (2016). A review of auxin response factors (ARFs) in plants. Frontiers in Plant Science, 7. 10.3389/fpls.2016.00047.

Ma, T. L., Wu, W. H., & Wang, Y. (2012). Transcriptome analysis of rice root responses to potassium deficiency. BMC Plant Biology, 12, 161. 10.1186/1471-2229-12-161.

Malkowski, E., Sitko, K., Szopiński, M., Gieroń, Ż., Pogrzeba, M., Kalaji, H. M., Zieleźnik-& Rusinowska, P. (2020). Hormesis in plants: The role of oxidative stress, auxins and photosynthesis in corn treated with Cd or Pb. International Journal of Molecular Sciences, 21 (6), 2099. 10.3390/ijms21062099.

Mei, H., Cheng, N. H., Zhao, J., Park, S., Escareno, R. A., Pittman, J. K., & Hirschi, K. D. (2009). Root development under metal stress in *Arabidopsis Thaliana* requires the H^+/Cation antiporter CAX4. New Phytologist, 183 (1), 95–105. 10.1111/j.1469-8137.2009.02831x.

Nacry, P., Canivenc, G., Muller, B., Azmi, A., Van Onckelen, H., Rossignol, M., & Doumas, P. (2005). A role for auxin redistribution in the responses of the root system architecture to phosphate starvation in *Arabidopsis*. Plant Physiology, 138 (4), 2061–2074. 10.1104/pp.105.060061.

Neumann, G., et al. (2000). Physiological aspects of cluster root function and development in phosphorus-deficient white lupin (Lupinus albus L.). Annals of Botany, 85, 909–919. 10.1006/anbo.2000.1135.

Paponov, I. A., Paponov, M., Teale, W., Menges, M., Chakrabortee, S., Murray, J. A. H., & Palme, K. (2008). Comprehensive transcriptome analysis of auxin responses in *Arabidopsis*. Molecular Plant, 1 (2), 321–337. 10.1093/mp/ssm021.

Peer, W. A., Cheng, Y., & Murphy, A. S. (2013). Evidence of oxidative attenuation of auxin signalling. Journal of Experimental Botany, 64 (9), 2629–2639. 10.1093/jxb/ert152.

Qi, X., Li, Q., Ma, X., Qian, C., Wang, H., Ren, N., Shen, C., Huang, S., Xu, X., Xu, Q., & Chen, X. (2019). Waterlogging-induced adventitious root formation in cucumber is regulated by ethylene and auxin through reactive oxygen species signalling. Plant Cell Environment, 42, 1458–1470. doi: 10.1111/pce.13504.

Qiu, T., Qi, M., Ding, X., Zheng, Y., Zhou, T., Chen, Y., Han, N., Zhu, M., Bian, H. & Wang, J. (2020). The SAUR41 subfamily of SMALL AUXIN UP RNA gene is abscisic acid inducible to modulate cell expansion and salt tolerance in *Arabidopsis thaliana* seedlings. Annals of Botany, 125, 805–819. doi: 10.1093/aob/mcz160

Ren, H., & Gray, W. M. (2015). SAUR proteins as effectors of hormonal and environmental signals in plant growth. Molecular Plant, 8, 1153–1164. doi: 10.1016/j.molp.2015.05.003

Ribba, T., Garrido-Vargas, F., & O'Brien, J. A. (2020). Auxin-mediated responses under salt stress: From developmental regulation to biotechnological applications. Journal of Experimental Botany, 71 (13), 3843–3853. 10.1093/jxb/eraa241.

Romera, F. J., García, M. J., Alcántara, E., & Pérez-Vicente, R. (2011). Latest findings about the interplay of auxin, ethylene and nitric oxide in the regulation of Fe deficiency responses by strategy I plants. Plant Signaling & Behavior, 5 (12), 167–170. 10.4161/psb.5.12.14111.

Sauter, M. (2013). Root responses to flooding. Current Opinion in Plant Biology, 16 (3), 282–286. 10.1016/j.pbi.2013.03.013.

Schmidt, W., Tittel, J., & Schikora, A. (2000). Role of hormones in the induction of iron deficiency responses in *Arabidopsis* roots. Plant Physiology, 122 (4), 1109–1118. 10.1104/pp.122.4.1109.

Sharma, E., Sharma, R., Borah, P., Jain, M., & Khurana, J. P. (2015). Emerging Roles of Auxin in Abiotic Stress Responses. In: G. Pandey (ed.), Elucidation of Abiotic Stress Signaling in Plants (pp. 299–328), Springer, New York, NY. 10.1007/978-1-4939-2211-6_11

Shi, H., Chen, L., Ye, T., Liu, X., Ding, K., & Chan, Z. (2014). Modulation of auxin content in *Arabidopsis* confers improved drought stress resistance. Plant Physiology and Biochemistry, 82, 209–217. 10.1016/j.plaphy.2014.06.008.

Shibasaki, K., Uemura, M., Tsurumi, S., & Rahman, A. (2009). Auxin response in arabidopsis under cold stress: Underlying molecular mechanisms. The Plant Cell, 21 (12), 3823–3838. 10.1105/tpc.109.069906.

Song, Y., Wang, L., & Xiong, L. (2009). Comprehensive expression profiling analysis of OsIAA gene family I developmental processes and responses to phytohormone and stress treatments. Planta, 229 (3), 577–591. 10.1007/s00425-008-0853-7.

Song, W., Liu, S., Meng, L., Xue, R., Wang, C., Liu, G., Dong, C., Wang, S., Dong, J., & Zhang, Y. (2015). Potassium deficiency inhibits lateral root development in tobacco seedlings by changing auxin distribution. Plant and Soil, 396 (1/2), 163–173. 10.1007/s11104-015-2579-1.

Stortenbeker, N., & Bemer, M. (2019). The SAUR gene family: the plant's toolbox for adaptation of growth and development. Journal of Experimental Botany, 70, 17–27. doi: 10.1093/jxb/ery332

Sun, P., Tian, Q. Y., Chen, J., & Zhang, W. H. (2010). Aluminium-induced inhibition of root elongation in *Arabidopsis* is mediated by ethylene and auxin. Journal of Experimental Botany, 61 (2), 347–356. 10.1093/jxb/erp306.

Tian, J., Ma, Y., Chen, Y., Chen, X., & Wei, A. (2022). Plant hormone response to low-temperature stress in cold-tolerant and cold-sensitive varieties of *Zanthoxylum Bungeanum* Maxim. Frontiers in Plant Science, 13. 10.3389/fpls.2022.847202.

Vicente-Agullo, F., Rigas, S., Desbrosses, G., Dolan, L., Hatzopoulos, P., & Grabov, A. (2004). Potassium carrier TRH1 is required for auxin transport in *Arabidopsis* Roots. The Plant Journal, 40 (4), 523–535. 10.1111/j.1365-313x.2004.02230x.

Wang, X., & Komatsu, S. (2020). Review: Proteomic techniques for the development of flood-tolerant soybean. International Journal of Molecular Sciences, 21, 7497. doi: 10.3390/ijms21207497.

Woodward, A. W., & Bartel, B. (2005). Auxin: Regulation, action, and interaction. Annals of Botany, 95 (5), 707–735. 10.1093/aob/mci083.

Wu, T., Zhang, H. T., Wang, Y., Jia, W. S., Xu, X. F., Zhang, X. Z., & Han, Z. H. (2012). Induction of root Fe(lll) reductase activity and proton extrusion by iron deficiency is mediated by auxin-based systemic signalling in Malus xiaojinensis. Journal of Experimental Botany, 63(2), 859–870. 10.1093/jxb/err314.

Yang, Y., Wang, Q. L., Geng, M. J., Guo, Z. H., & Zhao, Z. (2011). Effect of Indole-3-Acetic acid on aluminum-induced efflux of malic acid from wheat (*Triticum aestivum* L.). Plant and Soil, 346 (1-2), 215–230. 10.1007/s11104-011-0811-1.

Yuan, H., Zhao, K., Lei, H., Shen, X., Liu, Y., Liao, X., & Li, T. (2013). Genome-wide analysis of the GH3 family in apple (*Malus × Domestica*). BMC Genomics, 14, 297. 10.1186/1471-2164-14-297.

Zhang, S. W., Li, C. H., Cao, J., Zhang, Y. C., Zhang, S. Q., Xia, Y. F., Sun, D. Y., & Sun, Y. (2009). Altered architecture and enhanced drought tolerance in rice via the down-regulation of indole-3-acetic acid by TLD1/OsGH3.13 activation. Plant Physiology, 151 (4), 1889–1901. 10.1104/pp.109.146803

Zhang, Z., Yang, F., Li, B., Egrinya Eneji, A., Li, J., Duan, L., Wang, B., Li, Z., & Tian, X. (2009). Coronatine-induced lateral-root formation in cotton (*Gossypium hirsutum*) Seedlings under potassium-sufficient and -deficient conditions about auxin. Journal of Plant Nutrition and Soil Science, 172 (3), 435–444. 10.1002/jpln.200800116.

Zheng, S. J., Tang, C., Arakawa, Y., & Masaoka, Y. (2003). The responses of red clover (*Trifolium pratense* L.) to iron deficiency: A root Fe (III) chelate reductase. Plant Science, 164 (5), 679–687. 10.1016/s0168-9452(02)00422-3.

Zolla, G., Heimer, Y. M., & Barak, S. (2009). Mild salinity stimulates a stress-induced morphogenic response in *Arabidopsis thaliana* roots. Journal of Experimental Botany, 61 (1), 211–224. 10.1093/jxb/erp290

6 Regulatory Role of Cytokinin in Abiotic Stress Tolerance of Plants

Shivani Nagar, Shashi Meena, Sourabh Karwa, Rajkumar Dhakar, Dhandapani Raju, Sudhir Kumar, and Deepika Kumar Umesh

6.1 INTRODUCTION

Abiotic stress is one of the primary causes of crop loss worldwide, reducing average crop yields by more than 50%. Abiotic stress causes morphological, physiological, biochemical and molecular changes that adversely affect plant growth and productivity. To overcome the negative effect of abiotic stress, plants develop different stress avoidance or tolerance mechanisms. Many research reports suggest that plant growth regulators play a crucial role in mitigating the adverse effects of abiotic stress on plants (Shabbir et al., 2022; Zhang et al., 2023). Bahrun et al. (2002) observed that plants respond to most stresses by altering hormonal homeostasis, stability, content, biosynthesis and compartmentalization in the plant at an individual level. Plant hormones act as stress signaling molecules, response determiners and regulators in response to changing environmental conditions (Wolters & Jurgens, 2009). The role of cytokinin is studied abundantly in understanding its responses during stress. Cytokinins promote cell division and act in synergy and antagonism with other plant hormones, influencing various events during plant growth. Mainly cytokinins are produced in meristematic regions in the root system and transported via the xylem to the shoot. These cytokinins control plants' developmental and physiological processes from tissue to the whole plant level (Miller et al., 1961a,b; Cortleven et al., 2019). Cytokinins promote leaf expansion and seed germination (Miller et al., 1961a,b; Cortleven et al., 2019), biosynthesis and accumulation of chlorophyll and conversion of etioplasts into chloroplasts (Blum et al., 1981), cambial activity in stem and root (Matsumoto-Kitano et al., 2008) and delay leaf senescence (Richmond & Lang, 1957; Wybouw & De Rybel, 2019). Currently, a comprehensive understanding of the effects of cytokinin on plant tolerance to stress is lacking. The effects of cytokinin on physiological and molecular responses under abiotic stress are covered below.

6.2 CYTOKININ BIOSYNTHESIS AND SIGNALING UNDER ABIOTIC STRESS

Cytokinins are natural adenine derivatives with isoprenoid or aromatic side chains at the N6 position of the adenine ring. Based on the chemical structure, natural cytokinins are divided into isoprenoid and aromatic cytokinins. Isoprenoid cytokinins are primarily composed of isopentenyl adenine, cis-zeatin, trans-zeatin and dihydro-zeatin. Among them, isopentenyl adenine and trans-zeatin are considered the main active cytokinins (Binn, 1994). The aromatic cytokinins include orthotopolin, mesotopolin and their derivatives, benzyladenine and others. Synthetic cytokinins, such as kinetin, 6-benzyl amino purine, benzyladenine and trans-zeatin riboside, also exhibit cytokinin activity and affect plant growth on external application. The terpenoid biosynthesis pathway synthesizes cytokinins. The key enzymes involved in cytokinin biosynthesis are

DOI: 10.1201/9781003335788-9

isopentenyl transferase (IPT) (adenosine phosphate-isopentenyl transferases and tRNA-isopentenyl transferases (tRNA-IPTs) and LONELY GUY (LOG). The adenosine phosphate-isopentenyl transferases initiate cytokinin biosynthesis by catalyzing dimethylallyl diphosphate (DMAPP) to form isopentenyl ribotide. Then isopentenyl ribotides are catalyzed by cytochrome P450 monooxygenase CYP735As into trans-zeatin ribotides. This isopentenyl adenine and trans-zeatin-ribotides are the precursors of most adenine-type and trans-zeatin-type cytokinins, generating isopentenyl adenine and trans-zeatin through ribosides (Mok & Mok, 2001).

The biosynthesis of cis-zeatin is initiated by tRNA-IPTs, which use DMAPP to catalyze the prenylation of tRNA and then further generate cis-zeatin-ribotides. LOGs are a new cytokinin-activating enzyme with a conserved lysine decarboxylase-binding domain. LOGs directly convert the inactive cytokinin ribotides into a biologically active free base form. However, the level of active cytokinin is also regulated by binding to sugars (most commonly glucose) or by irreversible cleavage of cytokinin oxidases. Glucosyl conjugates are unable to bind histidine kinase (HK) cytokinin receptors and are inactive in bioassays. Cytokinin oxidase catalyzes irreversible degradation and makes it non-functional (Binns, 1994; Huang et al., 2012).

In response to stress, a decline in endogenous cytokinin levels has been observed by many researchers (Bano et al., 1994; Shashidhar et al., 1996). Drought stress reduces isopentenyl adenine, isopentenyl adenosine, trans-zeatin and zeatin riboside (Alvarez et al., 2008). In *Arabidopsis*, varied endogenous cytokinin levels were created by either loss of expression of IPT genes or overexpression of CKX-encoding genes. The *Arabidopsis* CKX overexpressor lines (35 S: CKX1, 35 S: CKX2, 35 S:CKX3, 35 S: CKX4) and of ipt1, ipt3, ipt5 and ipt7 mutants were evaluated for stress tolerance. The results revealed that reducing cytokinin levels improved drought and salt stress tolerance (Werner et al., 2003; Nishiyama et al., 2011).

Many HK proteins act as cytokinin receptors including *Arabidopsis* histidine kinase 2 (AHK2), *Arabidopsis* histidine kinase 3 (AHK3) and cytokinin response 1 (CRE1)/wooden leg1 (WOL1)/ *Arabidopsis* histidine kinase 4 (AHK4), which are primarily located in the plasma membrane (Huang et al., 2002; Hwang & Sheen, 2001). The structure of HKs consists of a kinase domain-containing histidine residue, a receptor (Rec) domain, an N terminal extracellular signal input domain and several transmembrane domains (Hwang et al., 2012). The cytokinin binding to the HKs signal input domain provides a signal to initiate autophosphorylation of the conserved histidine residue in the kinase domain. AHK receptors have different ligand binding affinities and expression patterns in *Arabidopsis* and maize, potentially contributing to their functional specificity. The phosphate group is transferred to downstream AHPs and ARRs to form a regulatory circuit, leading to the related gene expression (Mock & Mock, 2001). This multi-step phosphorylation system plays a vital role in transmission of stress signals via cytokinins.

Nevertheless, the expression of *Arabidopsis* histidine phosphor transfer proteins (AHPs) is reduced under drought stress. The triple mutant of AHPs i.e., ahp2-ahp3-ahp5 triple mutants, showed a drought-tolerance phenotype similar to receptor mutants (Nishiyama et al., 2013). Downstream targets of cytokinin signaling *Arabidopsis* A-type RRs (ARRs) ARR5 and ARR15 and the C-type RR ARR22 are behaving independently of the cytokinin receptors under drought stress. This suggests that these downstream components are regulated by other additional signaling pathways (Kang et al., 2012). In contrast, The B-type RR double mutant arr1arr12 demonstrates increased salt tolerance (Mason et al., 2010). This suggests the negative regulation of salt responses mediated by cytokinin signaling. However, despite greater cytokinin sensitivity, arr3-arr4-arr5-arr6 quadruple RRA mutants are also more salt-tolerant (Mason et al., 2010). This further supports the notion of RRAs interacting with other signaling systems independent of cytokinin. Under stress, ABA and cytokinin signaling pathways interact with each other and show crosstalk. The interaction between cytokinin and ABA signaling pathways is more evident for processes such as seed germination, later root formation and greening of cotyledons. A new mechanism, cytokinin–ABA interaction, was reported where cytokinin itself modulates the ABA response through nitric oxide. Under drought, ABA-mediated stomatal closure involves the production of nitric oxide. Cytokinin inhibits this process through direct interaction with and reduction of the NO molecule, acting as scavenger of this signaling molecule (Liu et al., 2013). The present knowledge on the interactions between components of the cytokinin signaling pathway and stress responses suggests that the response of the cytokinin signaling component is variable and may be condition-specific.

6.3 ROLE OF CYTOKININ IN PHYSIOLOGICAL METABOLISM UNDER STRESS

Cytokinins are known to regulate several aspects of plant growth and development in response to abiotic stress (Hare et al., 1997; Rivero et al., 2007; O'Brien & Benkova, 2013). Cytokinin homeostasis and signaling are rapidly altered under various abiotic stress conditions. For example, studies have shown that drought and/or salinity reduce cytokinin contents and transport in various plant species. The decrease in cytokinin levels in shoots under the stressed conditions may be due to IPT1,3,5 repression and/or CKX1,3,6 activation in *Arabidopsis* and decreased transport of cytokinins from the roots to the xylem under abiotic stress, the ratio between abscisic acid (ABA) and cytokinin is modulated in the xylem sap. Cytokinins regulate stomatal behavior (Haisel et al., 2008), the formation and protection of cellular structures and the induction and activation of protein synthesis (Chernyad'ev, 2005). Cytokinins maintain open stomata and thus increase stomatal conductance (gs) and transpiration.

6.4 ROLE OF CYTOKININ IN ABIOTIC STRESS TOLERANCE

6.4.1 Water Deficit Stress

Cytokinin plays an essential role in many aspects of plant growth and development, such as cell division, photosynthesis, senescence, chloroplast development and assimilates partitioning under water stress (Binns, 1994). Cytokinin homeostasis is essential for the development of stress-tolerant crop plants. It can diminish the damage caused by stress-induced negative changes in plant metabolism. Cytokinins play essential roles in regulating plant responses to water stress. Water stress inhibits the synthesis and causes degradation of cytokinins (Rivero et al., 2007, 2009). Decreased content of cytokinins was found in alfalfa under drought (Goicoechea et al., 1997). The decrease in cytokinin accumulation during drought stress is associated with an increase in ABA concentrations during stress (Davies & Zhang, 1991), leading to stomatal closure and an increase in stomatal resistance (Goicoechea et al., 1997). Both synthetic and naturally occurring cytokinins increase the transpiration rate of excised leaves and increase the stomatal aperture in the isolated epidermis (Incoll & Jewel, 1987). The transpiration rate of tobacco plants grown in vitro increased with increasing benzyl amino purine concentration but decreased at significantly higher cytokinin concentration (Pospisilová et al., 2000). Similarly, stomatal aperture of *Digitalis* was greater when grown *in vitro* in a medium with kinetin or benzyl amino purine, and the effect was concentration-dependent. Water stress usually accelerates leaf senescence and, in contrast, cytokinins delay leaf senescence (Rivero et al., 2007; Zwack et al., 2013; Joshi et al., 2019).

Increased endogenous cytokinin content through exogenous application promoted survival under water deficit stress conditions by delaying the leaf senescence. Suppression of drought-induced leaf senescence in transgenic tobacco plants, which accumulate cytokinins under stress, has been linked to enhanced expression of dehydrins and heat shock proteins and increased drought tolerance (Rivero et al., 2010). Enhanced expression of genes coding for proteases is a joint event both in senescence and under various environmental stresses (Martínez-Ballesta et al., 2010), which is necessary for the reorganization of plant metabolism, remodeling of cell protein components, degradation of damaged or unnecessary proteins and nutrient remobilization (Feller, 2004). Cytokinins also regulate physiological processes and maintain protein levels under drought stress in SAG12-ipt creeping bentgrass (Merewitz et al., 2010). Isopentenyl transferase overexpressing plants with increased endogenous cytokinin content under stress conditions has improved stress tolerance in various plant species, including monocot species such as rice. The suppression of drought-induced leaf senescence results in outstanding drought tolerance of the transgenic plants and minimal yield loss when the plants rewatered with only 30% of the amount of water used under control conditions. Under water stress, cytokinins can delay leaf senescence because they directly affect photosynthetic parameters, e.g., chlorophyll and photosynthetic protein synthesis and degradation, chloroplast composition and ultrastructure, electron transport and enzyme activities (Synkova et al., 1997). Cytokinin is known to stimulate photosynthetic enzymes like Rubisco and, more generally, the development of functional chloroplasts. The production of drought-tolerant crops grown under restricted water regimes without yield losses would considerably minimize drought-related losses and ensure food production in water-limited lands.

6.4.2 Heat Stress

The physiological effect of high temperatures on plants includes protein denaturation and aggregation, decreased fluidity of membrane lipids, degradation of chlorophyll, inactivation of enzymes in chloroplast and mitochondria, inhibition of protein synthesis, reduced rate of photosynthesis and protein degradation. In plant cells, membrane-based processes such as photosynthesis and respiration are essential. Heat stress results in the misfolding of newly synthesized proteins and the denaturation of existing proteins. Protein thermostability is believed to be provided in part by chaperones, the heat tolerance of plants is a complex trait, most probably controlled by multiple genes. The analysis of hormones, proteome and transcriptome also confirms that cytokinin plays an essential role in plant resistance to heat stress. The study reported that cytokinin upregulated the expression of many heat shock (HS) response proteins. The accumulation of endogenous cytokinins can maintain average plant growth under high-temperature stress and positively impact plants under heat stress. In creeping bentgrass (*Agrostis stolonifera*), the overexpression of *IPT* under *SAG12* maintains the formation and elongation of roots, reduces the loss of chlorophyll and delays the senescence of leaves, enhancing plant heat resistance under heat stress (Xu et al., 2009).

Further, the overexpression of *IPT* induced by two different promoters (*SAG12:ipt*, *HSP18:ipt*) leads to a significant increase in heat stress proteins in plants, increasing heat tolerance (Merewitz et al., 2010). Reduction in cytokinin degradation under stress can also help in attaining the optimum level of endogenous cytokinin. The optimum cytokinin levels can be achieved by downregulation of cytokinin oxidase/dehydrogenase (a negative regulator of cytokinin synthesis). Under heat stress, the cytokinin oxidase activity of heat-sensitive rice varieties increases significantly, resulting in low cytokinin levels and rice yield. In comparison, heat-resistant rice keeps a low level of cytokinin oxidase, resulting in increased heat resistance (Wu et al., 2017). Thus, the research report on the role of cytokinin under heat stress suggests that cytokinin plays an essential role in the plant's heat stress adaptation responses. Nevertheless, the precise mechanism of cytokinin functions under heat stress remains to be elucidated completely.

6.4.3 Salinity Stress

Salinity stress is a significant and prominent issue worldwide in agricultural salinity, restraining crop productivity and quality (Yamaguchi & Blumwald, 2005; Shabala & Cuin, 2008). Presently, it has been reported that nearly 20% of the world's cultivated area, nearly 33% of the irrigated agricultural land and half of the world's irrigated lands are affected by salinity by 2050 (Nachshon, 2018). It is more rigorously expanding, posing a more significant threat to the sustainable development of agriculture and its productivity. Various studies suggest the role of cytokinins in mitigating salinity stress. In salinity stress, the level of cytokinin increased in IPT overexpressing lines of *Arabidopsis* for the reprogramming transcript expression, which are related to the growth and development of plants. Also, it masked the harmful effects of stress (Golan et al., 2016).

Javid et al. (2011) reported a rise in grain yield of rice under salinity stress. Cytokinins are considered antagonistic to ABA and auxin synergists/antagonists in different plants for different functions (Pospisilová et al., 2000). It was presumed that under salinity stress, wheat plant cytokinin interacts with various hormones like auxin and ABA to cope with the stress symptoms (Iqbal et al., 2006). Zwack and Rashotte (2013) suggested that cytokinins have anti-senescence activity, which was regulated by osmotic turgor pressure in leaves, because of which it maintains the rate of photosynthesis by reconstituting chloroplast machinery under stress conditions. This suggested that cytokinins help in maintaining the electron donor capacity of the PSII system (Shao et al., 2010). It has been reported that mutation in *OsCKX2* genes leads to a higher accumulation of cytokinin in inflorescence meristems, which was responsible for promoting grain per panicle (Ashikari et al., 2005; Li et al., 2013; Mrízová et al., 2013). Similar results were reported by Joshi et al. (2018), which suggested that by mutating *OsCKX2* genes via controlling cytokinin levels, the panicle branching to some extent and promote grain yield in rice under salinity stress. Exogenous cytokinin applications alleviated leaf senescence induced by salt stress in perennial ryegrass. It also regulates antioxidant enzyme activity by suppressed Na^+ accumulation (Ma et al., 2016). As the Na^+ and K^+ equilibrium is an essential factor under salinity stress, cytokinin upregulated a high-affinity K^+ transporter (HKT) and helps in maintaining balance for plants to cope with salinity stress (Deinlein et al., 2014; Ma et al., 2016). In wheat, grain yield was also increased by 25% by overexpression of the HKT gene compared with the control when grown in the salt soil (James et al., 2011; James et al., 2012; Munns et al., 2012).

6.4.4 Cold Stress

Another significant detrimental stress in agriculture is cold (Mboup et al., 2012). Cold stress, classified as low temperature, means freezing or slightly above the freezing point. Under cold stress, plants usually lower down the growth and respiration rate, but simultaneously increase the accumulation of osmolytes, sucrose and other essential macromolecules that are used to maintain fluidity in lipid bilayers (Crowe et al., 2001; Rekarte-Cowie et al., 2008; Knight & Knight, 2012). In stress conditions, sucrose accumulation triggers signaling cascades that trigger various gene expressions, also helps to maintain the cellular water potential under stress conditions (Rekarte-Cowie et al., 2008). The role of cytokinin in response to cold temperatures is not entirely clear. In cold stress, cytokinin content and signaling are downregulated, and because of that, growth retards (Maruyama et al., 2014). In contrast, exogenous application of cytokinin may increase the tolerance in *Arabidopsis* (Jeon et al., 2010). During cold stress, a large number of transcripts get upregulated (Fowler & Thomashow 2002; Kreps et al., 2002; Lee et al., 2005; Le et al., 2008, 2014), which include CBRs (c-repeating binding factor (CBF1, CBF2 and CBF3), *CORs* (cold-responsive genes) and inducers of CBF expression 1 (ICF1) (Zwack et al., 2016). Other genes that are also regulated in cold stress were *HOS1*, *SIZ1*, *OST1* and *CAMTA3*. CRF (cytokinin response factor) is downregulated in cold stress. Cytokinins and receptors play a vital function, but their aspect, molecular action and duration required more clarification in cold stress.

6.5 CONCLUSION

Plants respond by bringing desirable physiological and morphological changes to adapt to adverse situations and enable optimal growth. Phytohormones play critical roles in helping plants to adapt to adverse environmental conditions. Research findings have helped to understand the role of cytokinin in abiotic stress. Attention has focused on cytokinins that are critical regulators of plant senescence. Cytokinins have a potent effect on plant physiology and environmental responses and are intimately involved in regulating cell division, apical dominance, chloroplast development, anthocyanin production and maintenance of the source-sink relationship. Cytokinin-mediated changes in metabolic processes that ultimately result in altered growth patterns are suitable for withstanding environmental stress. Their possible involvement in responses to adverse environmental conditions is also suggested. The more recent overexpression studies prove the role of cytokinin in abiotic stress tolerance. Thus, manipulating cytokinin levels at the proper developmental stages in an appropriate organ or tissue will be helpful in the improvement of plant stress tolerance.

REFERENCES

Alvarez, S., Marsh, E. L., Schroeder, S. G., & Schachtman, D. P. (2008). Metabolomic and proteomic changes in the xylem sap of maize under drought. Plant, Cell & Environment, 31(3), 325–340.

Ashikari, M., Sakakibara, H., Lin, S., Yamamoto, T., Takashi, T., Nishimura, A., ... Matsuoka, M. (2005). Cytokinin oxidase regulates rice grain production. Science, 309, 741– 745.

Bahrun, A., Jensen, C. R., Asch, F., & Mongensen, V. O. (2002). Drought induced changes in xylem pH, ionic composition, and ABA concentration act as early signals in field-grown maize (*Zea mays* L.). Journal of Experimental Botany, 53, 251–263.

Bano, A., Hansen, H., Dörffling, K., & Hahn, H. (1994). Changes in the contents of free and conjugated abscisic acid, phaseic acid and cytokinins in xylem sap of drought stressed sunflower plants. Phytochemistry, 37(2), 345–347.

Binns, A. N. (1994). Cytokinin accumulation and action: biochemical, genetic and molecular approaches. Annual Review in Plant Physiology and Plant Molecular Biology (Annual Review in Plant Physiology), 45, 173–196.

Blum, A., & Ebercon, A. (1981). Cell membrane stability as a measure of drought and heat tolerance in wheat. Crop Science, 21, 43–47.

Chernyad'ev, I. I. (2005). Effect of water stress on the photosynthetic apparatus of plants and the protective role of cytokinins: a review. Applied Biochemistry and Microbiology, 41, 115–128.

Cortleven, A., Leuendorf, J. E., Frank, M., Pezzetta, D., Bolt, S., & Schmülling, T. (2019). Cytokinin action in response to abiotic and biotic stresses in plants. Plant, Cell & Environment, 42, 998–1018. 10.1111/pce.13494.

Crowe, J. H., Crowe, L. M., Oliver, A. E., Tsvetkova, N., Wolkers, W., & Tablin, F. (2001). The trehalose myth revisited: introduction to a symposium on stabilization of cells in the dry state. Cryobiology, 43, 89–105. 10.1006/cryo.2001.2353.

Davies, W. J., & Zhang, J. (1991) Root signals and the regulation of growth and development of plants in drying soil. Annual Review in Plant Physiology and Plant Molecular Biology (Annual Review in Plant Physiology), 42, 55–76.

Deinlein, U., Stephan, A. B., Horie, T., Luo, W., Xu, G., & Schroeder, J. I. (2014). Plant salt-tolerance mechanisms. Trends in Plant Science, 19(6), 371–379.

Feller, U. (2004). Proteolysis. In: L. D. Noodén (ed.), Plant Cell Death Processes (pp. 107–123). Elsevier Inc.

Fowler, S., & Thomashow, M. F. (2002). *Arabidopsis* transcriptome profiling indicates that multiple regulatory pathways activated during cold acclimation in addition to the CBF cold response pathway. Plant Cell, 14, 1675–1690.

Goicoechea, N., Antolin, M. C., & Sanchez-Diaz, M. (1997). Gas exchange is related to the hormonal balance in mycorrhizal or nitrogen-fixing alfalfa subjected to drought. Physiologia Plantarum, 100, 989–997.

Golan, Y., Shirron, N., Avni, A., Shmoish, M., & Gepstein, S. (2016). Cytokinins induce transcriptional reprograming and improve *Arabidopsis* plant performance under drought and salt stress conditions. Frontiers in Environmental Science 4, 63. doi: 10.3389/fenvs.2016.00063\

Haisel, D., Vankova, R., Synkova, H., & Pospisilova, J. (2008). The impact of transzeatinO-glucosyltransferase gene over-expression in tobacco on pigment content and gas exchange. Biologia Plantarum, 52, 49–58.

Hare, P. D., Cress, W. A., & Van Staden, J. (1997). The involvement of cytokinins in plant responses to environmental stress. Plant Growth Regulation, 23, 79–103.

Huang, S., Cerny, R. E., Qi, Y., Bhat, D., Aydt, C. M., Hanson, D. D., Malloy, K. P., & Ness, L. A. (2002). Transgenic studies on the involvement of cytokinin and gibberellin in male development. Plant Physiology, 131, 1270–1282. 10.1104/pp.102.018598.

Hwang, I., & Sheen, J. (2001). Two-component circuitry in Arabidopsis cytokinin signal transduction. Nature, 413, 383–389. 10.1038/35096500.

Hwang, I., Sheen, J., & Müller, B. (2012). Cytokinin signaling networks. Annual Review of Plant Biology, 63, 353–380.

Incoll, L. D. , & Jewer, P. C. (1987). Cytokinins and stomata. In:E. Zeiger et al (eds), Stomatal function (pp. 281–292). Stanford: Stanford University Press.
Iqbal, M., Ashraf, M., Jamil, A., & Ur-Rehman, S. (2006). Does seed priming induce changes in the levels of some endogenous plant hormones in hexaploid wheat plants under salt stress? Journal of Integrative Plant Biology, 48, 81–189.
James, R. A., Blake, C., Byrt, C. S., & Munns, R. (2011). Major genes for Na^+ exclusion, *Nax1* and *Nax2* (wheat HKT1; 4 and HKT1; 5), decrease Na^+ accumulation in bread wheat leaves under saline and waterlogged conditions. Journal of Experimental Botany, 62(8), 2939–2947.
James, R. A., Blake, C., Zwart, A. B., Hare, R. A., Rathjen, A. J., & Munns, R. (2012). Impact of ancestral wheat sodium exclusion genes *Nax1* and *Nax2* on grain yield of durum wheat on saline soils. Functional Plant Biology, 39(7), 609–618.
Javid, M. G., Sorooshzadeh, A., Moradi, F., Sanavy, S. A. M. M., & Allahdadi, I. (2011). The role of phytohormones in alleviating salt stress in crop plants. Australian Journal of Crop Science, 5, 726.
Jeon, J., Kim, N. Y., Kim, S., Kang, N. Y., Novák, O., Ku, S. J., Cho, C., Lee, D. J., Lee, E. J., Strnad, M., & Kim, J. (2010). A subset of cytokinin two-component signaling system plays a role in cold temperature stress response in *Arabidopsis*. Journal of Biological Chemistry, 285(30), 23371–23386.
Joshi, R., Sahoo, K. K., Tripathi, A. K., Kumar, R., Gupta, B. K., Pareek, A., & Singla-Pareek, S. L. (2018). Knockdown of an inflorescence meristem-specific cytokinin oxidase–*OsCKX2* in rice reduces yield penalty under salinity stress condition. Plant Cell & Environment, 41(5), 936–946.
Joshi, S., Choukimath, A., Isenegger, D., Panozzo, J., Spangenberg, G. & Kant, S. (2019). Improved wheat growth and yield by delayed leaf senescence using developmentally regulated expression of a cytokinin biosynthesis gene. Frontiers in Plant Science, 10, 1285.
Kang, N. Y., Cho, C., Kim, N. Y., & Kim, J. (2012). Cytokinin receptor-dependent and receptor-independent pathways in the dehydration response of *Arabidopsis thaliana*. Journal of Plant Physiology, 169(14), 1382–1391.
Knight, M. R., & Knight, H. (2012). Low-temperature perception leading to gene expression and cold tolerance in higher plants. New Phytologist, 195, 737–751. 10.1111/j.1469-8137.2012.04239.x.
Kreps, J. A., Wu, Y., Chang, H. S., Zhu, T., Wang, X., & Harper, J. F. (2002). Transcriptome changes for *Arabidopsis* in response to salt, osmotic, and cold stress. Plant Physiology, 130, 2129–2141.
Le, M. Q., Engelsberger, W. R., & Hincha, D. K. (2008). Natural genetic variation in acclimation capacity at sub-zero temperatures after cold acclimation at 4 °C in different Arabidopsis thaliana accessions. Cryobiology, 57, 104–112. 10.1016/j.cryobiol.2008.06.004.
Le, M. Q., Pagter, M., & Hincha, D. K. (2014). Global changes in gene expression, assayed by microarray hybridization and quantitative RT-PCR, during acclimation of three *Arabidopsis thaliana* accessions to sub-zero temperatures after cold acclimation. Plant Molecular Biology, 87, 1–15.
Lee, B. H., Henderson, D. A., & Zhu, J. K. (2005). The *Arabidopsis* coldresponsive transcriptome and its regulation by ICE1. Plant Cell, 17, 3155–3175.
Li, S., Zhao, B., Yuan, D., Duan, M., Qian, Q., Tang, L., et al. (2013). Rice zinc finger protein DST enhances grain production through controlling *Gn1a/OsCKX2* expression. Proceedings of the National Academy of Sciences of the United States of America, 110, 3167–3172.
Liu, W. Z., Kong, D. D., Gu, X. X., Gao, H. B., Wang, J. Z., Xia, M., Gao, Q., Tian, L. L., Xu, Z. H., Bao, F., & Hu, Y. (2013). Cytokinins can act as suppressors of nitric oxide in *Arabidopsis*. Proceedings of the National Academy of Sciences, 110(4), 1548–1553.
Ma, X., Zhang, J., & Huang, B. (2016). Cytokinin-mitigation of salt-induced leaf senescence in perennial ryegrass involving the activation of antioxidant systems and ionic balance. Environmental and Experimental Botany, 125, 1–11.
Martínez-Ballesta, M. C., Alcaraz-López, C., Muries, B., Mota-Cadenas, C., & Carvajal, M. (2010). Physiological aspects of rootstock–scion interactions. Scientia Horticulturae, 127, 112–118. 10.1016/j.scienta.2010.08.002.
Maruyama, K., et al. (2014). Integrated analysis of the effects of cold and dehydration on rice metabolites, phytohormones, and gene transcripts. Plant Physiology, 164, 1759–1771. 10.1104/pp.113.231720.
Mason, M. G., Jha, D., Salt, D. E., Tester, M., Hill, K., Kieber, J. J., & Eric Schaller, G. (2010). Type-B response regulators *ARR1* and *ARR12* regulate expression of *AtHKT1*; 1 and accumulation of sodium in *Arabidopsis* shoots. The Plant Journal, 64(5), 753–763.
Matsumoto-Kitano, M., Kusumoto, T., Tarkowski, P., Kinoshita-Tsujimura, K., Václavíková, K., Miyawaki, K., & Kakimoto, T. (2008). Cytokinins are central regulators of cambial activity. Proceedings of the National Academy of Sciences, 105(50), 20027–20031.

Mboup, M., Fischer, I., Lainer, H., & Stephan, W. (2012). Trans-species polymorphism and allele-specific expression in the cbf gene family of wild tomatoes. Molecular Biology and Evolution, 29, 3641–3652.
Merewitz, E., Gianfagna, T., & Huang, B. (2010). Effects of SAG12-ipt andHSP18.2-ipt. expression on cytokinin production, root growth and leaf senescence in creeping bentgrass exposed to drought stress. Journal of the American Society for Horticultural Science, 135, 230–239.
Miller, C. O. (1961a). A kinetin-like compound in maize. Proceedings of the National Academy of Sciences of the United States of America, 47(2), 170.
Miller, C. O. (1961b). Kinetin and related compounds in plant growth. Annual Review of Plant Physiology, 12(1), 395–408.
Mok, D. W., & Mok, M. C. (2001). Cytokinin metabolism and action. *Annu. Rev*. Plant Physiol. Plant Molecular Biology, 52, 89–118.
Mrízová K., Jiskrová E., Vyroubalová Š., Novák O., Ohnoutková L., Pospíšilová H., et al. (2013). Overexpression of cytokinin dehydrogenase genes in barley (Hordeum vulgare cv. Golden Promise) fundamentally affects morphology and fertility. PLoS One, 8, e79029.
Munns, R., James, R. A., Xu, B., Athman, A., Conn, S. J., Jordans, C., ... & Gilliham, M. (2012). Wheat grain yield on saline soils is improved by an ancestral Na^+ transporter gene. Nature Biotechnology, 30(4), 360–364.
Nachshon, Uri. (2018). Cropland soil salinization and associated hydrology: Trends, processes and examples. Water, 10, 1030. 10.3390/w10081030.
Nishiyama, R., Watanabe, Y., Fujita, Y., Le, D. T., Kojima, M., Werner, T., Vankova, R., Yamaguchi-Shinozaki, K., Shinozaki, K., Kakimoto, T., & Sakakibara, H. (2011). Analysis of cytokinin mutants and regulation of cytokinin metabolic genes reveals important regulatory roles of cytokinins in drought, salt and abscisic acid responses, and abscisic acid biosynthesis. The Plant Cell, 23(6), 2169–2183.
Nishiyama, R., Watanabe, Y., Leyva-Gonzalez, M. A., Van Ha, C., Fujita, Y., Tanaka, M., Seki, M., Yamaguchi-Shinozaki, K., Shinozaki, K., Herrera-Estrella, L., & Tran, L. S. P. (2013). *Arabidopsis* AHP2, AHP3, and AHP5 histidine phosphotransfer proteins function as redundant negative regulators of drought stress response. Proceedings of the National Academy of Sciences, 110(12), 4840–4845.
O'Brien, J. A., & Benková, E., (2013). Cytokinin cross-talking during biotic and abiotic stress responses. Frontiers in Plant Science, 4, 451.
Pospíšilová, J., Synková, H., & Rulcová, J. (2000). Cytokinins and water stress. Biologia Plantarum, 43, 321–328. 10.1023/a:1026754404857.
Rekarte-Cowie, I., Ebshish, O. S., Mohamed, K. S., & Pearce, R. S. (2008). Sucrose helps regulate cold acclimation of *Arabidopsis* thaliana. Journal of Experimental Botany, 59(15), 4205–4217.
Richmond, A. E., & Lang, A. (1957). Effect of kinetin on protein content and survival of detached Xanthium leaves. Science, 125, 650–651.
Rivero, R. M., Kojima, M., Gepstein, A., Sakakibara, H., Mittler, R., Gepstein, S., & Blumwald, E. (2007). Delayed leaf senescence induces extreme drought tolerance in a flowering plant. Proceedings of the National Academy of Sciences, 104(49), 19631–19636.
Rivero, R. M., Shulaev, V., & Blumwald, E. (2009). Cytokinin-dependent photorespiration and the protection of photosynthesis during water deficit. Plant Physiology, 150(3), 1530–1540.
Rivero, R. M., Gimeno, J., Van Deynze, A., Walia, H., & Blumwald, E. (2010). Enhanced cytokinin synthesis in tobacco plants expressing PSARK:: IPT prevents the degradation of photosynthetic protein complexes during drought. Plant and Cell Physiology, 51(11), 1929–1941.
Shabala, S., & Cuin, T. A. (2008). Potassium transport and plant salt tolerance. Physiologia Plantarum, 133, 651–669.
Shabbir, R., et al. (2022). Combined abiotic stresses: challenges and potential for crop improvement. Agronomy, 12, 2795. 10.3390/agronomy12112795.
Shao, R., Wang, K., & Shangguan, Z. (2010). Cytokinin-induced photosynthetic adaptability of *Zea mays* L. to drought stress associated with nitric oxide signal: Probed by ESR spectroscopy and fast OJIP fluorescence rise. Journal of Plant Physiology, 167, 472–479. doi: 10.1016/j.jplph.2009.10.020
Shashidhar, V. R., Prasad, T. G., & Sudharshan, L. (1996). Hormone signals from roots to shoots of sunflower (*Helianthus annuus* L.). Moderate soil drying increases delivery of abscisic acid and depresses delivery of cytokinins in xylem sap. Annals of Botany, 78(2), 151–155.
Synkova, H., Wilhelmova, N., Sestak, Z., & Pospisilova, J. (1997). Photosynthesis in transgenic plants with elevated cytokinin contents. In: M. Pessarakli (ed.), Handbook of Photosynthesis (pp. 541–552). Marcel Dekker.
Werner, T., Motyka, V., Laucou, V., Smets, R., Van Onckelen, H., & Schmülling, T. (2003). Cytokinin-deficient transgenic *Arabidopsis* plants show multiple developmental alterations indicating opposite functions of cytokinins in the regulation of shoot and root meristem activity. The Plant Cell, 15(11), 2532–2550.

Wolters, H., & Jurgens, G. (2009). Survival of the fl exible: hormonal growth control and adaptation in plant development. Nature Genetics, 10, 305–317.

Wu, C., Cui, K., Wang, W., Li, Q., Fahad, S., Hu, Q., Huang, J., Nie, L., Mohapatra, P. K., & Peng, S. (2017). Heat-induced cytokinin transportation and degradation are associated with reduced panicle cytokinin expression and fewer spikelets per panicle in rice. Frontiers in Plant Science, 8, 371.

Wybouw, B., & De Rybel, B. (2019). Cytokinin – A developing story. Trends in Plant Science, 24, 177–185. 10.1016/j.tplants.2018.10.012.

Xu, Y., Tian, J., Gianfagna, T., & Huang, B. (2009). Effects of SAG12-ipt expression on cytokinin production, growth and senescence of creeping bentgrass (*Agrostis stolonifera* L.) under heat stress. Plant Growth Regulation, 57, 281–291.

Yamaguchi, T., & Blumwald, E. (2005). Developing salt-tolerant crop plants: challenges and opportunities. Trends in Plant Science, 10, 615–620.

Zhang, Y., Xu, J., Li, R., Ge, Y, Li, Y., & Li, R. (2023). Plants' response to abiotic stress: Mechanisms and strategies. International Journal of Molecular Sciences, 24, 10915. 10.3390/ijms241310915.

Zwack, P. J., & Rashotte, A. M. (2013). Cytokinin inhibition of leaf senescence. Plant Signaling & Behavior, 8(7), e24737.

Zwack, P. J., Compton, M. A., Adams, C. I., & Rashotte, A. M. (2016). Cytokinin response factor 4 (CRF4) is induced by cold and involved in freezing tolerance. Plant Cell Reports, 35(3), 573–584.

7 Ethylene in Abiotic Stress

*Rohit Kumar Mahto**, *Shikha Tripathi, Devendra Pratap Singh, Kishori Lal, Deepesh Kumar, Rahul Kumar, Ayyagari Ramlal, Shubham Kumar Singh, and Shiv Shankar Sharma*

7.1 INTRODUCTION

Plants being sessile are exposed to various types of environmental stresses such as cold, heat, heavy metal, drought, flooding, nutrient deficiency, etc. that affect the growth and development of plants which ultimately leads to less productivity (Sharma et al., 2019; Chen et al., 2021). However, to overcome these stresses, many plants have evolved different mechanisms for the detection of the environmental changes and adaptability to these changes (Husain et al., 2020). The role of various plant hormones and their signaling is well established in abiotic stress tolerance (Verma et al., 2016). Among different hormones that influence the plant growth and development, the phytohormone ethylene plays an important role (Sharma et al., 2019). Ethylene is a gaseous hormone synthesized in plants by two key enzymes aminocyclopropane-1-carboxylic acid (ACC) synthases (ACSs) and ACC oxidases (ACOs) (Bleecker & Kende, 2000). Ethylene not only regulates the growth and development of plants by modulating the various development and physiological processes but also plays a role in stress responses (Chen et al., 2021). In recent years, there have been major developments in our understanding of molecular mechanisms that regulate ethylene production. Ethylene plays an important role along with other hormones, combining different signals and allowing for the beginning of a situation that favors the growth, development and reproductive success of plants. Ethylene controls the growth of leaves, flowers and fruits as well as the processes of senescence. Crop growth and development under a variety of environmental conditions determine agricultural production. Growth and flexibility of plant organs can affect plant production by altering photosynthesis, nutrient efficiency and yield index. It promotes or inhibits growth and senescence processes depending on its concentration, timing of application and the species. The application of ethephon, an ethylene-releasing compound enhanced ethylene evolution and increased leaf area of mustard at a lower concentration, while inhibited at higher concentration (Khan, 2005; Khan et al., 2008). Ethylene governs the development of leaves, flowers and fruits. It may also promote, inhibit or induce senescence depending upon the optimal or sub-optimal ethylene levels (Konings & Jackson, 1979; Khan, 2005; Pierik et al., 2006). It appears quite interesting to examine how the same hormone influences the two contradictory processes of growth and senescence. This chapter describes the roles of ethylene in various abiotic stresses.

7.2 ETHYLENE SIGNALING UNDER VARIOUS ABIOTIC STRESS CONDITIONS

Plants are sessile and directly affected by the different environmental factors, which are responsible for modifications in their morphological, physiological anatomical, biochemical and

* Contributed equally as first author.

DOI: 10.1201/9781003335788-10

molecular aspects (Abeles et al., 2012; Arraes et al., 2015; Tripathi et al., 2017). Different abiotic stresses such as drought, flood, heat, cold, heavy metals, etc. led to nearly 50% loss in crops (Bray, 2000). To cope with these different abiotic stresses, plants have developed certain defense mechanisms for their long survival. Under various abiotic stress conditions, phytohormones and their signaling play an essential role in mitigating stress (Khan et al., 2013). Among various phytohormones in plants, ethylene is considered a multifunctional phytohormone and acts as a stress hormone that regulates both growth and senescence. In plants, ethylene synthesis is triggered when they are exposed to different abiotic stresses (Wang et al., 2018) also seen in streptophytes (de Poel & de Vries, 2023).

7.2.1 Heavy Metal Stress

During abiotic stresses, ethylene plays an important role in the regulation of various growth and developmental processes in plants. Studies suggest that ethylene production in plants under metal stress depends upon the type of metals and their concentration (Thao et al., 2015). Among various metals, cadmium is one of the most phytotoxic heavy metals that stimulate ethylene production in plants (Arteca et al., 2007). A study on mustard exhibited that ethylene plays a key role in cadmium stress tolerance, acting by increasing the ROS accumulation resulting regulate the defense system (Masood et al., 2012). In normal and stressful conditions, ethylene is the vital regulator of plant growth and photosynthesis in the plant cells, and its synthesis increased under metal stress like a study on *Arabidopsis* under cadmium stress. A study on *Arabidopsis thaliana* plants revealed that when plants were exposed to cadmium stress it resulted in an increase in the biosynthesis of ACC and ethylene via ACS2 and ACS6 mediated higher expression. On the exposure to heavy metals, ethylene production has increased but it remains unknown how the increased ethylene levels happened on a molecular basis. Cadmium-induced the biosynthesis of ACC and ethylene in *Arabidopsis thaliana* plants mainly via the increased expression of *ACS2* and *ACS6*. This was confirmed in the *acs2-1acs6-1* double knockout mutants, which showed a decreased ethylene production, positively affecting leaf biomass and resulting in a delayed induction of ethylene-responsive gene expressions without significant differences in Cd contents between wild-type and mutant plants. Furthermore, Cu is also reported in *Nicotiana glutinosa* to induce the gene expression of *ACO1* and *ACO3* gene and it is supposed that ethylene production is also increased by upregulation of the *ACO* genes (Kim et al., 1998; Rudus̈ et al., 2013; Keunen et al., 2016). Under different abiotic stress conditions, ethylene also contributes to the synthesis and regulation of ROSs. Steffens (2014) reported the interaction of ethylene and ROS under metal stress in paddy crops and showed that ROS levels increased under various metal stresses. Ethylene-responsive transcription factor (AP2/ERF) gene family is primarily responsible to regulate ROS synthesis and signaling under several abiotic stresses (Wang et al., 2014). ERF1 decreases the expression of superoxide dismutase (SOD) and peroxidase (POD), which results in increased ROS level in the plants (Wang et al., 2014).

7.2.2 Drought Stress

Drought stress is considered the most limiting factor in plant growth, development and crop productivity. Plants under water-deficit conditions face alteration in an array of physiological, biological, developmental and molecular processes, leading to inhibition of growth and a significant reduction in the rate of photosynthesis (Bray, 1997; Chaves et al., 2003; Hirayama & Shinozaki, 2010). It has been observed that ethylene is involved in both the opening and closing of the stomatal aperture via crosstalk with ABA. Various studies have been conducted to reveal the crosstalk between abscisic acid (ABA) and ethylene signaling. Tanaka et al. (2005), working on *Arabidopsis thaliana,* created an eto1 mutant (ethylene-overproducing) and two ethylene-insensitive mutants etr1-1 and ein3-1. The drought-stressed eto1 (ethylene overproducer 1) mutant exhibited less sensitivity to ABA and showed delayed stomata closure compared to the drought-treated wild type. Furthermore, the application of ethylene or its precursor 1-aminocyclopropane-1-carboxylic acid (ACC) leads to inhibition of the ABA-dependent stomatal closure. During concomitant ABA and ACC application to etr1-1 and ein3-1 mutants, they were able to close the stomata, while wild-type plants were not that potent in it. Desikan et al. (2006) revealed the involvement of H_2O_2 in ethylene-mediated stomatal closure. H_2O_2 is generated by NADPH oxidase and is one of the most important molecules that play an essential role in ABA-induced stomatal closure.

7.2.3 Flood Stress

Under flood conditions, ethylene is a known ruler of flood tolerance to the different rice varieties that grow under slightly submerged conditions (Bailey-Serres et al., 2012). Due to their sessile nature, plants acquire new adaptations to cope with changing environments (Sharma et al., 2020). Rice is grown under submerged conditions. They evolved and adapted to flooding stress by elongating their hollow internodes, which helped gas exchange with the atmosphere. Ethylene synthesis and accumulation increased during water lodging conditions. Xu et al. (2006) reported that submersed conditions of paddy crops trigger flood response by encoding ethylene response factors, which are actively involved in ethylene signaling. Under submerged conditions, ethylene accumulates in the plant and induces the expression of these two genes. There are several ERFs and genes which help rice plants to survive in submergence conditions. In *Arabidopsis*, there are five groups of VII ethylene response factors, characterized by the N-terminal consensus sensitive to oxygen and most of the group members, particularly group-1 proteins RAP2.2 and RAP2.12 can silence and attenuate the hypoxic response (Bailey-Serres et al., 2012). The genes involved in the molecular mechanism of ethylene signaling during deep-water response such as SNORKEL1 (SK1) and SNORKEL2 (SK1), encode ethylene response factors (VII AP2). Expression of these two SNORKELs (SK1 and SK2) genes helps the elongation of the stem in a broad water paddy via regulating the gibberellin action important for internode elongation. Similarly, SUBIA a transcription factor also encodes the ERF/AP2 transcription factor, which is also an ethylene-producible gene, that enhances the rice tolerance against dehydration and oxidative stress by increasing the expression of genes involved in dehydration readjustment and ROS refining, these indicate that these genes and ERFs are supporting to survive the rice plants under submerged conditions (Xu et al., 2006; Hattori et al., 2009).

7.2.4 Heat Stress

Heat stress survival and response are complicated phenomena in plants. A potential essential regulator in biotic and abiotic stress situations is the plant hormone ET. When ET interacts with one of its receptor complexes, constitutive triple response two kinases are rendered inactive. This causes ET-insensitive 2 to first be dephosphorylated, followed by the beginning of cleavage on the C-terminal of EIN2 and finally its translocation to the nucleus, where it activates EIN3/EIL1 (Schauberger et al., 2017). Although the classical HSPs (chaperones) through the HSP network are crucial, even this response is complex and requires multiple HSPs. The HSR is further controlled by additional transcription factors and numerous signaling pathways, all of which help cells survive high-temperature stress by controlling a variety of effector components (Koguchi et al., 2017). In rice seedlings under heat stress with ethylene precursor treatment as opposed to rice seedlings under heat stress alone, quantitative reverse transcriptase-polymerase chain reaction revealed higher expression levels of heat shock factors like HSFA1a and HSFA2a, c, d, e and f, and ethylene-signaling-related genes like ethylene insensitive 2, ethylene insensitive-like 1 and ethylene insensitive 2 (Schauberger et al., 2017). The significant fold activation of ERF021 under heat stress conditions points to the critical role of ET in soybean heat stress resistance (Poór et al., 2021). In the case of the pea plant, a gene named PsACO2 ethylene biosynthesis, the gene expression pattern had the strongest correlation with the evolution of ethylene in the majority of tissues. When PsACO2 transcript increased, tissue ethylene evolution also increased (in pedicels and pericarps from non-pollinated flowers, non-pollinated ovaries exposed to heat stress and stigma/style tissue from pollinated or non-pollinated flowers at 0 DAA) (Savada et al., 2017).

7.2.5 Salt Stresses

Salinity stress poses negative effects on plant growth, including osmotic stress, oxidative stress, ion toxicity and nutrient deficiency (Chinnusamy et al., 2006; Zhu, 2007). The high salinity condition leads to inhibition of growth, decreases crop yield and even death under prolonged high salinity conditions. Under salinity stress, generally presumed that ethylene functions by improving the response of plants against the stress (Arraes et al., 2015; Tao et al., 2015). In *Arabidopsis,* as well as in several other crops, including maize, tomato and grapevines, ethylene has emerged as one of the essential phytohormones and positive mediators for stress tolerance against salinity (Munns et al., 2008; Siddikee et al., 2012; Freitas et al., 2018; Xu et al., 2019). It was reported that ethylene controls stress responses and plant growth and development by stabilizing ROSs. In *Arabidopsis*, ethylene overproducer 1 (ETO1) acts as a positive mediator of salt response by increasing ROS formation as well as maintaining Na^+/K^+ homeostasis (Yang et al., 2017). In cotton and many other plants, it has been reported that long exposure to salinity stress leads to up regulations of various sets of genes of ethylene involved in salinity stress regulation such as (i) *ACS1, ACS12, ACO1* and *ACO3* homologs genes involved in ethylene biosynthesis; (ii) *ERF1, ERF2, EIN3* and *MEKK1-MKK2-MPK4/6* kinases genes, which are involved in ethylene signaling pathways; (iii) *ETR1, ETR2* and *EIN4* receptors of ethylene genes and the genes, which are actively involved in feedback mechanism like *CONSTITUTIVE TRIPLE RESPONSE 1* (CTR1) genes (Peng et al., 2014). Wang et al. (2017) reported that *ETR1* expression is suppressed during salt stress in *Arabidopsis* and also reported that the *etr* mutants (loss-of-function) exhibited enhanced tolerance while gain of function in *etr-1* mutants was more sensitive to salinity stresses (Zhou et al., 2006; Cao et al., 2007; Peng et al., 2014; Wang et al., 2017).

7.3 CONCLUSION AND FUTURE PROSPECTS

Ethylene controls many aspects in a plant's life including seed germination, root formation, flower growth, fruit ripening, senescence and also involved in regulating biotic and abiotic stress responses. Ethylene is involved in different types of abiotic stresses and across plants and other kingdoms (streptophytes) similarly, environmental stimuli also have also promoted the synthesis of ethylene. In the years to come, delineating molecular mechanisms and crosstalks of hormones with each other is required for further understanding and improving abiotic stress tolerance in crops for better survival and production.

REFERENCES

Abeles, F. B., Morgan, P. W., & Saltveit Jr, M. E. (2012). Ethylene in plant biology. Academic Press.

Arraes, F. B. M., Beneventi, M. A., Lisei de Sa, M. E., Paixao, J. F. R., Albuquerque, E. V. S., Marin, S. R. R., … & Grossi-de-Sa, M. F. (2015). Implications of ethylene biosynthesis and signaling in soybean drought stress tolerance. BMC Plant Biology, 15, 1–20.

Arteca, R. N., & Arteca, J. M. (2007). Heavy-metal-induced ethylene production in *Arabidopsis* thaliana. Journal of Plant Physiology, 164(11), 1480–1488.

Bailey-Serres, J., Fukao, T., Gibbs, D. J., Holdsworth, M. J., Lee, S. C., Licausi, F., … & van Dongen, J. T. (2012). Making sense of low oxygen sensing. Trends in Plant Science, 17(3), 129–138.

Bleecker, A. B., & Kende, H. (2000). Ethylene: a gaseous signal molecule in plants. Annual Review of Cell and Developmental Biology, 16(1), 1–18.

Bray, E. A. (1997). Plant responses to water deficit. Trends in Plant Science, 2(2), 48–54.

Bray, E. A. (2000). Responses to abiotic stresses. Biochemistry and Molecular Biology of Plants, 1158–1203, American Society of Plant Physiologists, Rockville.

Cao, W. H., Liu, J., He, X. J., Mu, R. L., Zhou, H. L., Chen, S. Y., & Zhang, J. S. (2007). Modulation of ethylene responses affects plant salt-stress responses. Plant Physiology, 143(2), 707–719.

Chaves, M. M., Maroco, J. P., & Pereira, J. S. (2003). Understanding plant responses to drought—from genes to the whole plant. Functional Plant Biology, 30(3), 239–264.

Chen, H., Bullock Jr, D. A., Alonso, J. M., & Stepanova, A. N. (2021). To fight or to grow: The balancing role of ethylene in plant abiotic stress responses. Plants, 11(1), 33.

Chinnusamy, V., Zhu, J., & Zhu, J. K. (2006). Salt stress signaling and mechanisms of plant salt tolerance. In: J. K. Setlow (ed.), Genetic engineering: Principles and methods (27, pp. 141–177). Springer. 10.1007/0-387-25856-6_9.

de Poel, B. V., & de Vries, J. (2023). Evolution of ethylene as an abiotic stress hormone in streptophytes. Environmental and Experimental Botany, 214, 105456.

Desikan, R., Last, K., Harrett-Williams, R., Tagliavia, C., Harter, K., Hooley, R., ... & Neill, S. J. (2006). Ethylene-induced stomatal closure in *Arabidopsis* occurs via AtrbohF-mediated hydrogen peroxide synthesis. The Plant Journal, 47(6), 907–916.

Freitas, V. S., de Souza Miranda, R., Costa, J. H., de Oliveira, D. F., de Oliveira Paula, S., de Castro Miguel, E., ... & Gomes-Filho, E. (2018). Ethylene triggers salt tolerance in maize genotypes by modulating polyamine catabolism enzymes associated with H2O2 production. Environmental and Experimental Botany, 145, 75–86.

Hattori, Y., Nagai, K., Furukawa, S., Song, X. J., Kawano, R., Sakakibara, H., ... & Ashikari, M. (2009). The ethylene response factors SNORKEL1 and SNORKEL2 allow rice to adapt to deep water. Nature, 460(7258), 1026–1030.

Hirayama, T., & Shinozaki, K. (2010). Research on plant abiotic stress responses in the post-genome era: past, present and future. The Plant Journal, 61(6), 1041–1052.

Husain, T., Fatima, A., Suhel, M., Singh, S., Sharma, A., Prasad, S. M., & Singh, V. P. (2020). A brief appraisal of ethylene signaling under abiotic stress in plants. Plant Signaling & Behavior, 15(9), 1782051.

Keunen, E., Schellingen, K., Vangronsveld, J., & Cuypers, A. (2016). Ethylene and metal stress: small molecule, big impact. Frontiers in Plant Science, 7, 23.

Khan, N. A. (2005). The influence of exogenous ethylene on growth and photosynthesis of mustard (*Brassica juncea*) following defoliation. Scientia Horticulturae, 105(4), 499–505. doi: 10.1016/j.scienta.2005.02.004.

Khan, N. A., Mir, M. R., Nazar, R., & Singh, S. (2008). The application of ethephon (an ethylene releaser) increases growth, photosynthesis and nitrogen accumulation in mustard (*Brassica juncea* L.) under high nitrogen levels. Plant Biology, 10(5), 534–538.

Khan, M. I. R., Iqbal, N., Masood, A., Per, T. S., & Khan, N. A. (2013). Salicylic acid alleviates adverse effects of heat stress on photosynthesis through changes in proline production and ethylene formation. Plant Signaling & Behavior, 8(11), e26374.

Kim, Y. S., Choi, D., Lee, M. M., Lee, S. H., & Kim, W. T. (1998). Biotic and abiotic stress-related expression of 1-aminocyclopropane-l-carboxylate oxidase gene family in *Nicotiana glutinosa* L. Plant and Cell Physiology, 39(6), 565–573.

Koguchi, M., Yamasaki, K., Hirano, T., & Sato, M. H. (2017). Vascular plant one-zinc-finger protein 2 is localized both to the nucleus and stress granules under heat stress in *Arabidopsis*. Plant Signaling & Behavior, 12(3), e1295907.

Konings, H., & Jackson, M. B. (1979). A relationship between rates of ethylene production by roots and the promoting or inhibiting effects of exogenous ethylene and water on root elongation. Zeitschrift für Pflanzenphysiologie, 92, 385–397. doi: 10.1016/S0044-328X(79)80184-1.

Masood, A., Iqbal, N., & Khan, N. A. (2012). Role of ethylene in alleviation of cadmium-induced photosynthetic capacity inhibition by sulphur in mustard. Plant, Cell & Environment, 35, 524–533. doi: 10.1111/j.1365-3040.2011.02432.x.

Munns, R., & Tester, M. (2008). Mechanisms of salinity tolerance. Annual Reviews in Plant Biology, 59, 651–681.

Peng, Z., He, S., Gong, W., Sun, J., Pan, Z., Xu, F., et al. (2014). Comprehensive analysis of differentially expressed genes and transcriptional regulation induced by salt stress in two contrasting cotton genotypes. BMC Genomics, 15, 1–28.

Pierik, R., Tholen, D., Poorter, H., Visser, E. J., & Voesenek, L. A. (2006). The Janus face of ethylene: growth inhibition and stimulation. Trends in Plant Science, 11(4), 176–183.

Poór, P., Nawaz, K., Gupta, R., Ashfaque, F., & Khan, M. I. R. (2021). Ethylene involvement in the regulation of heat stress tolerance in plants. Plant Cell Reports, 41, 1–24.

Ruduś, I., Sasiak, M., & Kępczyński, J. (2012). Regulation of ethylene biosynthesis at the level of 1-aminocyclopropane-1-carboxylate oxidase (ACO) gene. Acta Physiologiae Plantarum, 35, 295–307. 10.1007/s11738-012-1096-6.

Savada, R. P., Ozga, J. A., Jayasinghege, C. P., Waduthanthri, K. D., & Reinecke, D. M. (2017). Heat stress differentially modifies ethylene biosynthesis and signaling in pea floral and fruit tissues. Plant Molecular Biology, 95, 313–331.

Schauberger, B., Archontoulis, S., Arneth, A., Balkovic, J., Ciais, P., Deryng, D., et al. (2017). Consistent negative response of US crops to high temperatures in observations and crop models. Nature Communications, 8(1), 13931.

Sharma, A., Kumar, V., Sidhu, G. P. S., Kumar, R., Kohli, S. K., Yadav, P., et al. (2019). Abiotic stress management in plants: Role of ethylene. In: A. Roychoudhary & D. Tripathi (eds.), Molecular plant abiotic stress: Biology and biotechnology (pp. 185–208). John Wiley and Sons Inc.

Sharma, A., Kumar, V., Shahzad, B., Ramakrishnan, M., Singh Sidhu, G. P., Bali, A. S., et al. (2020). Photosynthetic response of plants under different abiotic stresses: a review. Journal of Plant Growth Regulation, 39, 509–531.

Siddikee, M. A., Chauhan, P. S., & Sa, T. (2012). Regulation of ethylene biosynthesis under salt stress in red pepper (*Capsicum annuum* L.) by 1-aminocyclopropane-1-carboxylic acid (ACC) deaminase-producing halotolerant bacteria. Journal of Plant Growth Regulation, 31, 265–272.

Steffens, B. (2014). The role of ethylene and ROS in salinity, heavy metal, and flooding responses in rice. Frontiers in Plant Science, 5, 685.

Tanaka, Y., Sano, T., Tamaoki, M., Nakajima, N., Kondo, N., & Hasezawa, S. (2005). Ethylene inhibits abscisic acid-induced stomatal closure in *Arabidopsis*. Plant Physiology, 138(4), 2337–2343.

Tao, J. J., Chen, H. W., Ma, B., Zhang, W. K., Chen, S. Y., & Zhang, J. S. (2015). The role of ethylene in plants under salinity stress. Frontiers in Plant Science, 6, 1059.

Thao, N. P., Khan, M. I. R., Thu, N. B. A., Hoang, X. L. T., Asgher, M., Khan, N. A., & Tran, L. S. P. (2015). Role of ethylene and its cross talk with other signaling molecules in plant responses to heavy metal stress. Plant Physiology, 169(1), 73–84.

Tripathi, D. K., Shweta, Singh, S., Yadav, V., Arif, N., Singh, S., et al. (2017). Silicon: a potential element to combat adverse impact of UV-B in plants. In: V. Singh (eds), UV-B radiation: from environmental stressor to regulator of plant growth, pp. 175–195. UK: John Wiley & Sons Ltd.

Verma, V., Ravindran, P., & Kumar, P. P. (2016). Plant hormone-mediated regulation of stress responses. BMC Plant Biology, 16, 1–10.

Wang, B., Sun, Y. F., Song, N., Wei, J. P., Wang, X. J., Feng, H., et al. (2014). MicroRNAs involving in cold, wounding and salt stresses in Triticum aestivum L. Plant Physiology and Biochemistry, 80, 90–96.

Wang, F., Wang, L., Qiao, L., Chen, J., Pappa, M. B., Pei, H., ... & Dong, C. H. (2017). *Arabidopsis* CPR5 regulates ethylene signaling via molecular association with the ETR1 receptor. Journal of Integrative Plant Biology, 59, 810–824. 10.1111/jipb.12570.

Wang, Y., Zou, W., Xiao, Y., Cheng, L., Liu, Y., Gao, S., et al. (2018). MicroRNA1917 targets CTR4 splice variants to regulate ethylene responses in tomato. Journal of Experimental Botany, 69(5), 1011–1025.

Xu, K., Xu, X., Fukao, T., Canlas, P., Maghirang-Rodriguez, R., Heuer, S., et al. (2006). Sub1A is an ethylene-response-factor-like gene that confers submergence tolerance to rice. Nature, 442(7103), 705–708.

Xu, L., Xiang, G., Sun, Q., Ni, Y., Jin, Z., Gao, S., & Yao, Y. (2019). Melatonin enhances salt tolerance by promoting MYB108A-mediated ethylene biosynthesis in grapevines. Horticulture Research, 6, 1–14.

Yang, C., Li, W., Cao, J., Meng, F., Yu, Y., Huang, J., et al. (2017). Activation of ethylene signaling pathways enhances disease resistance by regulating ROS and phytoalexin production in rice. The Plant Journal, 89(2), 338–353.

Zhou, H. L., Cao, W. H., Cao, Y. R., Liu, J., Hao, Y. J., Zhang, J. S., & Chen, S. Y. (2006). Roles of ethylene receptor NTHK1 domains in plant growth, stress response and protein phosphorylation. FEBS Letters, 580(5), 1239–1250.

Zhu, J-K. (2007). Plant salt stress. Encyclopedia of Life Sciences, 1–3. doi: 10.1002/9780470015902.a0001300.pub2

8 Roles of Gibberellic Acid in Mitigating Abiotic Stresses

Rohit Babar, Pravin Mane, and Sangram B Chavan

8.1 INTRODUCTION

Gibberellins (GAs) are recognized as one of the quintessential classical hormones, constituting a group that includes auxins, cytokinins, abscisic acid, and ethylene. Each category of these hormones is intricately linked to specific plant traits and physiological responses (Castro-Camba et al., 2022). Historically, GAs have been predominantly linked to the regulation of plant stature and seed dormancy. Nevertheless, recent research suggests that this characterization might be overly simplistic, as GAs, alongside other phytohormones, exert direct or indirect influences on numerous plant traits. GAs played a pivotal role in the Green Revolution of the latter half of the 20th century, and the enhanced agronomic features observed in many plant varieties, characterized by dwarf phenotypes and increased biomass, were found to be associated with GA activity and signalling pathways. The investigation into GA commenced in Japan, where rice crops were afflicted by a fungal infection manifesting as excessive growth and a lack of grain production. This phenomenon was attributed to the secretion of the fungus *Gibberella fusikorai*, later reclassified as *Fusarium fusikorai.* The nomenclature "gibberellin" originated from the genus name *Gibberella* (Castro-Camba et al., 2022). Through subsequent chemical analysis, the active constituent, gibberellin, demonstrated significant efficacy in promoting plant growth and inducing flowering, a fact substantiated by its chemical characteristics (Hedden and Sponsel, 2015; Sponsel, 2016). Following purification, raw gibberellic acid appears as a white to pale yellow substance, characterized by the chemical formula $C_{19}H_{22}O_6$. Notably, a total of 126 GAs have been identified, including GA_1, GA_3, GA_4, GA_7, among others (Camara et al., 2018).

Agriculture depends on climatic conditions, however, climate change always remain as a major constraint in agriculture (Gornall et al., 2010). Global warming is primarily responsible for climate change, which leads to rise in temperature thereby affecting yield and production. The IPCC (Intergovernmental Panel on Climate Change; https://www.ipcc.ch/) in its sixth report forecasted a global temperature rise of 1.5°C in the next decade if the GHG (greenhouse gas) emissions not controlling the temperature increase by 1.5 °C or 2°C will be beyond the extent. The increasing temperature disturbs the ecosystem; it enhances abiotic stresses. Abiotic stresses include drought, waterlogging, hailstorms, salinity, nutrient toxicity, cyclones, coldness, etc. These are adversely affected by agriculture and the losses are about 51–82% (Oshunsanya et al., 2019).

8.2 MECHANISM OF GAs

It is challenging to identify the precise tissue or organ in which GAs are generated and located since the biosynthesis of active GAs is a complicated, multistep process involving a variety of intermediates. A significant advance in our comprehension of the signaling mechanism for this hormone was made in 2005 with the identification of the GA receptor using the gibberellin insensitive dwarf1 (*gid1*) mutant of rice (Ueguchi-Tanaka et al., 2005; Sun, 2011). *Arabidopsis* has three isoforms of the *gid1* gene, namely *gid1A*, *gid1B* and *gid1C,* which play an important role in the extreme elimination of GA reactions (Willige et al., 2007; Schwechheimer, 2012) and

DOI: 10.1201/9781003335788-11

controlling various development processes (Suzuki et al., 2009). The presence of GAs in phloem sap and their capacity to move through this medium were confirmed by the first findings on GA mobility in higher plants, which were revealed over 50 years ago (Hoad & Bowen, 1968). Findings showed that *gai* mRNA may travel from partner cells to sieve tubes and that the translated product stays at the locations to which it is delivered, resulting in a reduction in the expression of several genes and attenuating GA responses (Xu et al., 2013; Binenbaum et al., 2018). By stimulating the *26 S proteasome*-dependent breakdown of the DELLA proteins, GAs encourage growth by releasing the growth restriction imposed by DELLA (Fu et al., 2002; Xu et al., 2014).

GA signaling is controlled in many ways. GAs have been regulating their production by promoting the transcription of GA2oxs, catabolic enzymes that deactivate GAs and by suppressing the transcription of the genes encoding GA_3 OXIDASES (GA_3ox) and GA20ox, the rate-limiting enzymes in the synthesis of active GAs. GAs encourage both DELLA protein gene transcription and degradation at the same time. The GA receptor and *SLY1* F-box protein genes' transcription is suppressed by GAs. By doing this (and assuming that transcription levels match protein levels), GA-induced DELLA protein breakdown is readjusted by decreased GA production, increased DELLA protein synthesis and decreased GA sensitivity since there are fewer GA receptors and F-box proteins available. This regulatory system is dependent on the degradation of the DELLA protein, and GA-insensitive mutants exhibit dysregulated expression of the GA biosynthesis and signaling genes (Bouquin et al., 2001; Cao et al., 2006; Willige et al., 2007; Liang et al., 2014).

8.3 ROLE OF GIBBERELLIC ACID IN ABIOTIC STRESS MANAGEMENT

Global warming is threatening the world by causing frequent occurrences of numerous natural calamities. These conditions are directly or indirectly impacting the climate and subsequently agriculture. A continuously changing climate is forcing agriculture and its allied sectors into a great forfeiture (Vermeulen et al., 2012; Chauhan et al., 2014; Watts et al., 2018). This climate change gives rise to so many abrupt conditions due to which agricultural production is greatly affected (Morton, 2007; Raza et al., 2019). These abrupt conditions are usually comprised of long dry spells or drought, sudden heavy rains, hailstorms, extreme weather conditions like high or low temperatures and many other aspects that significantly reduce food production (Pereira, 2016; Pathak et al., 2012). This scenario is hampering food security all across the globe (Schmidhuber & Tubiello 2007; Tirado et al., 2010).

Nowadays, it is elementary to identify the key remedial measures to cope with various abiotic stresses and ultimately head towards sustainable food security. By keeping this as their main motive, many researchers are exploring a wide range of methodologies, which may include advanced agronomical practices, the use of stress-tolerant varieties, the application of IoT, ICT, implicating beneficial microbes, etc. There is a scope to elucidate the impact of gibberellins, especially GA_3 in stress mitigation at the physiological and molecular level. This chapter provides insight into the significance of GA_3 and its various physiological and biochemical aspects with different abiotic stresses.

8.3.1 Drought Stress Management

It is an important abiotic factor that is associated with moisture, the long dry period and the loss of available moisture from the atmosphere and soil that adversely affects crops. Drought significantly affected morphological traits of crops like chlorophyll content, height, stem diameter, stomata conductance, etc. (Akter et al., 2014). Zawaski and Busov (2014) have reported that a key mechanism for controlling development and physiological adaptation in response to recent or impending adverse situations is GA catabolism and repressive signaling. Foliar application of GAs might be increasing the drought tolerance in sunflowers in a pot; however, GA_3 significantly increased the biosynthesis of antioxidant compounds in sunflowers compared to those untreated

(Jan et al., 2019). Kaur et al. (2000) found that according to the findings of the current study, GA_3 and kinetin promote improved seedling growth under stressful circumstances, and it is due to the high acid invertase activity in a shoot. By strengthening essential plant components and reducing ion loss through ion leakage by managing cell permeability, gibberellic acid and glycine betaine were successful in enhancing cowpea plants' tolerance to drought stress (Miri et al., 2021). Al Mahmud et al. (2019) reported that exogenous GAs helped the seedlings survive by upregulating antioxidant defense mechanisms and the glyoxalase system, whereas the severity of drought stress in wheat relies on the growth stage and it increases with an increase in the duration of stress.

8.3.2 Heat Stress Management

The vital physiological and biochemical processes in plants usually get adversely affected by heat stress, thus making it a major yield-limiting factor in various crop plants across the globe (Bita & Gerats, 2013; Chavez-Arias et al., 2018). A few attempts have been briefly discussed below, showing the leveraging impact of GAs in mitigating heat stress. Guo et al. (2022) tested the responses of two tomato genotypes, Ahmar and Roma, by spraying GA_3 at concentrations of 25, 50, 75 and 100 mg L^{-1}, whereas untreated plants were kept as controls for heat stress. An experiment was carried out in controlled temperature conditions, such as 25°C i.e., normal growth condition and 45°C, i.e., heat-stressed condition. Root, shoot morphological and physiological parameters were recorded to study the ameliorative effect of GA_3. From the recorded data, it is revealed that non-sprayed plants have shown the least root growth compared to plants treated with GA_3 at a rate of 75 mg L^{-1} across the genotypes under heat stress. A similar trend was also observed in shoot length. GA_3 improved the proline, nitrogen, phosphorus and potassium levels in the leaves of both genotypes under elevated temperatures. Si and GA_3 applications together promote plant growth and metabolic control in date palms under heat stress (Khan et al., 2020).

8.3.3 Waterlogging Stress Management

Waterlogging is one of the major abiotic stresses in a plant's restriction of aerobic respiration during waterlogging affects growth and a variety of cognitive stages, including seed germination, vegetative growth and subsequent reproductive growth. It also reduces energy metabolism. The balance of several hormones is the cornerstone for ensuring that plants experience appropriate physiological metabolism, growth and development (Miransari & Smith, 2014; Pan et al., 2021). Therefore, phytohormones are directly associated with the control of the entire life process of plants (Fahad et al., 2015). The exogenous application of GAs mitigates the flooding or waterlogging stress in plants; many scientists reported evidence of the role of GAs in managing waterlogging. Islam et al. (2021) studied the application of different levels of GAs to manage waterlogging stress in mungbeans. The obtained results were significant in plant heights and number of leaves. Similar results reported that the application of GAs showed that soybeans' waterlogging tolerance line (WTL) exhibited a preference for high amounts of gibberellic acid (GA), particularly the bioactive GA_4 levels (Kim et al., 2015). Foliar applications of GA_3 (200 ppm) and salicylic acid (150 ppm) each increased the number of pods and seeds for the Burangrang soybean variety under waterlogging conditions compared to normal conditions (Damanik & Siregar, 2019). Islam et al. (2021) suggested that GA_3 might be applied to treat mungbean damage brought on by waterlogging, as well as potential damage to other cash crops.

8.3.4 Salinity Stress Management

The main factor limiting plant growth and development is soil salinity. GA is one of the plant growth regulators that increases salt tolerance and lessens the effects of salt stress on plants. With the decrease in growth due to salinity and the gradual effects of exogenous GA_3 application on various morphological, physiological and biochemical activities, it is possible to conclude that applying GA_3 helps reduce salinity stress and that its effectiveness is higher in salt-tolerant cultivars (Misratia et al., 2015). Shahzad et al. (2021) found that $GA_3P + GA_3FS$ could be a useful strategy for enhancing maize development and growth and lowering oxidative stress in salt-contaminated soils. Two wheat cultivars were subjected to GA_3 treatments (100 ppm), which enhanced growth standards, photosynthetic pigments and subsequently crop output (Shaddad et al., 2013). Ali et al. (2021) revealed that GA_3 and nitrogen could effectively prevent salt damage to early seedling growth. According to the findings, $GA_3P + GA_3FS$ could be a useful strategy for enhancing maize growth and development and lowering oxidative stress in salt-contaminated soils (Shahzad et al., 2021).

8.3.5 Chilling or Cold Temperature Stress Management

It is a serious environmental component that influences plant development, growth, productivity and distribution in cold stress. The metabolic pathways of anthers are altered by cold stress to cause pollen sterility, which means reduced grain yield (Sharma & Nayyar, 2016); this is a complex phenomenon of ABA, GAs and carbohydrates. To continue plant metabolism during cold stress the need is to optimize the level of GAs. Fruit's modulation of the cold response during storage appears to be significantly influenced by GAs (Ding et al., 2015).

8.3.6 Heavy Metal Stress Management

As it challenges the sustainability of agricultural growth worldwide, heavy metal stress in plants has drawn significant interest from throughout the world. Heavy metals include cadmium (Cd), mercury (Hg), lead (Pb), arsenic (As), chromium (Cr), copper (Cu), nickel, zinc (Zn), iron (Fe), etc. The plant accustomed to them in high concentrations could adversely affect morphological characters, nutrition, metabolic activities and yield attributes of a crop or plants (Hossain et al., 2012). Plants can maintain their capacity for growth and developmental plasticity owing to phytohormones, which function as chemical messengers that increase their resilience to heavy metal stress. Zhu et al. (2012) reported that gibberellic acid (5 μM) has been shown to lessen Cd toxicity in *Arabidopsis* by lowering Cd absorption and lipid peroxidation. In addition to controlling Fe transport and translocation, GAs hindered iron (Fe) translocation by reducing *OsYSL2* gene expression (Wang et al., 2017). Emamverdian et al. (2020) demonstrated in sunflowers under Cu stress where GA treatment increased energy trapping in PSII reaction centers and increased LHCII complex stability, PN and *Fv/Fm*. Additionally, under Cd stress, GAs are said to accelerate the growth rate, chlorophyll content and net CO_2 assimilation rate of soybeans. By controlling shoot and root development, enhancing plant metabolism generally, increasing photosynthetic pigments and improving shoot and root morphometry and transcript abundance of *VrPCS1*, *VrIRT1*, *VrIRT2* and *VrCD29*, GA_3 treatment mitigated Cd stress. As a result, it is suggested to use GA_3 to regulate Cd-induced phytotoxicity in mungbean plants effectively (Hakla et al., 2021).

8.4 CONCLUSION

Nowadays, abiotic stresses are found to be more frequent and recurrent in occurrence, so there is only one way to cope with them to mitigate or manage them by adopting novel strategies and technologies. The development of stress-tolerant varieties is merely a time-consuming process. So it is advisable to identify a quick remedial solution that possesses both crop growth promoting ability and stress tolerance imparting ability. PGRs are observed to be a potent solution for gaining sustainable yield, even under a variety of abiotic stresses. Among the described classes of PGRs, GAs have reported better results in context with a healthier crop stand and maintaining better crop physiology that ultimately leads to a sustainable yield under stressful scenarios too. GAs have the key ability to control oxidative stress progressions and undesirable movements of antioxidant enzymes; consequently, they support the crop plant in conquering the harmful impacts of abiotic stresses. Hence, it could be concluded that GAs play a pivotal role in regulating plant growth and development from germination to death, and that modulation of GAs is an attractive approach for conferring the varied abiotic stress management and mitigation as well.

8.5 FUTURE PROSPECTIVE

The present book chapter highlights the importance of GAs for the management of abiotic stresses. Several studies have reported that GAs enhance the growth and production of crops as well as enhance biotic and abiotic stress tolerance in plants. There are many roles of GAs in plant systems, but many things are yet to be understood. Along with this, GAs are complex compounds and it is essential to discover the interactive phenomenon of GAs and major stress-signaling hormones.

REFERENCES

Akter, N., Rafiqul Islam, M., Abdul Karim, M., & Hossain, T. (2014). Alleviation of drought stress in maize by exogenous application of gibberellic acid and cytokinin. Journal of Crop Science and Biotechnology, 17(1), 41–48.

Al Mahmud, J., Biswas, P. K., Nahar, K., Fujita, M., & Hasanuzzaman, M. (2019). Exogenous application of gibberellic acid mitigates drought-induced damage in spring wheat. Acta Agrobotanica, 72(2), 1–18.

Ali, A. Y. A., Ibrahim, M. E. H., Zhou, G., Nimir, N. E. A., Elsiddig, A. M. I., Jiao, X., et al. (2021). Gibberellic acid and nitrogen efficiently protect early seedlings growth stage from salt stress damage in Sorghum. Scientific Reports, 11(1), 1–11.

Binenbaum, J., Weinstain, R., & Shani, E. (2018). Gibberellin localization and transport in plants. Trends in Plant Science, 23(5), 410–421.

Bita, C. E., & Gerats, T. (2013). Plant tolerance to high temperature in a changing environment: scientific fundamentals and production of heat stress-tolerant crops. Frontiers in Plant Science, 4, 273.

Bouquin, T., Meier, C., Foster, R., Nielsen, M. E., & Mundy, J. (2001). Control of specific gene expression by gibberellin and brassinosteroid. Plant Physiology, 127(2), 450–458.

Camara, M. C., Vandenberghe, L. P., Rodrigues, C., de Oliveira, J., Faulds, C., Bertrand, E., & Soccol, C. R. (2018). Current advances in gibberellic acid (GA3) production, patented technologies and potential applications. Planta, 248, 1049–1062. 10.1007/s00425-018-2959-x.

Cao, D., Cheng, H., Wu, W., Soo, H. M., & Peng, J. (2006). Gibberellin mobilizes distinct DELLA-dependent transcriptomes to regulate seed germination and floral development in *Arabidopsis*. Plant Physiology, 142(2), 509–525.

Castro-Camba, R., Sánchez, C., Vidal, N., & Vielba, J. M. (2022). Plant development and crop yield: The role of gibberellins. Plants, 11(19), 1–27. 10.3390/plants11192650.

Chauhan, B. S., Mahajan, G., Randhawa, R. K., Singh, H., & Kang, M. S. (2014). Global warming and its possible impact on agriculture in India. Advances in Agronomy, 123, 65–121.

Chavez-Arias, C. C., Ligarreto-Moreno, G. A., & Restrepo-Díaz, H. (2018). Evaluation of heat stress period duration and the interaction of daytime temperature and cultivar on common bean. Environmental and Experimental Botany, 155, 600–608.

Damanik, R. I., & Siregar, L. A. M. (2019). Growth and production of soybean (*Glycine max* L. Merril) varieties in response to waterlogging at vegetative (V_5) growth phase by application of gibberellic acid and salicylic acid. In IOP Conference Series: Earth and Environmental Science, vol. 260, no. 1, p. 012143, IOP Publishing.

Ding, Y., Sheng, J., Li, S., Nie, Y., Zhao, J., Zhu, Z., et al. (2015). The role of gibberellins in the mitigation of chilling injury in cherry tomato (*Solanum lycopersicum* L.) fruit. Postharvest Biology and Technology, 101, 88–95.

Emamverdian, A., Ding, Y., & Mokhberdoran, F. (2020). The role of salicylic acid and gibberellin signaling in plant responses to abiotic stress with an emphasis on heavy metals. Plant Signalling and Behaviour, 15(7), 1777372.

Fahad, S., Hussain, S., Bano, A., Saud, S., Hassan, S., Shan, D., & Huang, J. (2015). Potential role of phytohormones and plant growth-promoting rhizobacteria in abiotic stresses: consequences for changing environment. Environmental Science and Pollution Research, 22, 4907–4921.

Fu, X., Richards, D. E., Ait-Ali, T., Hynes, L. W., Ougham, H., Peng, J., & Harberd, N. P. (2002). Gibberellin-mediated proteasome-dependent degradation of the barley DELLA protein SLN1 repressor. The Plant Cell, 14(12), 3191–3200.

Gornall, J., Betts, R., Burke, E., Clark, R., Camp, J., Willett, K., & Wiltshire, A. (2010). Implications of climate change for agricultural productivity in the early twenty-first century. Philosophical Transactions of the Royal Society B: Biological Sciences, 365(1554), 2973–2989.

Grennan, A. K. (2006). Gibberellin metabolism enzymes in rice. Plant Physiology, 141(2), 524–526.

Guo, T., Gull, S., Ali, M. M., Yousef, A. F., Ercisli, S., Kalaji, H. M., & Ghareeb, R. Y. (2022). Heat stress mitigation in tomato (*Solanum lycopersicum* L.) through foliar application of gibberellic acid. Scientific Reports, 12(1), 1–13.

Hakla, H. R., Sharma, S., Urfan, M., Yadav, N. S., Rajput, P., Kotwal, D., et al. (2021). Gibberellins target shoot-root growth, morpho-physiological and molecular pathways to induce cadmium tolerance in *Vigna radiata* L. Agronomy, 11(5), 896.

Hedden, P., & Sponsel, V. (2015). A century of gibberellin research. Journal of Plant Growth Regulation, 34(4), 740–760.

Hoad, G. V., & Bowen, M. R. (1968). Evidence for gibberellin-like substances in phloem exudate of higher plants. Planta, 82(1), 22–32.

Hossain, M. A., Piyatida, P., da Silva, J. A. T., & Fujita, M. (2012). Molecular mechanism of heavy metal toxicity and tolerance in plants: central role of glutathione in detoxification of reactive oxygen species and methylglyoxal and in heavy metal chelation. Journal of Botany, 2012, 872875.

Islam, M. R., Hasan, M., Akter, N., and Akhtar, S. (2021). Cytokinin and gibberellic acid alleviate the effect of waterlogging in mungbean (*Vigna radiata* L. wilczek). Journal Clean WAS (JCleanWAS), 5(1), 21–26.

Jan, A. U., Hadi, F., Akbar, F., & Shah, A. (2019). Role of potassium, zinc and gibberellic acid in increasing drought stress tolerance in sunflower (*Helianthus annuus* l.). Pakistan Journal of Botany, 51(3), 809–815.

Kaur, S., Gupta, A. K., & Kaur, N. (2000). Effect of GA3, kinetin and indole acetic acid on carbohydrate metabolism in chickpea seedlings germinating under water stress. Plant Growth Regulation, 30(1), 61–70.

Khan, A., Bilal, S., Khan, A. L., Imran, M., Shahzad, R., Al-Harrasi, A., & Lee, I. J. (2020). Silicon and gibberellins: synergistic function in harnessing ABA signaling and heat stress tolerance in date palm (*Phoenix dactylifera* L.). Plants, 9(5), 620.

Kim, Y. H., Hwang, S. J., Waqas, M., Khan, A. L., Lee, J. H., Lee, J. D., & Lee, I. J. (2015). Comparative analysis of endogenous hormones level in two soybean (*Glycine max* L.) lines differing in waterlogging tolerance. Frontiers in Plant Science, 6, 714.

Liang, Y. C., Reid, M. S., & Jiang, C. Z. (2014). Controlling plant architecture by manipulation of gibberellic acid signalling in petunia. Horticulture Research, 1, 1–16.

MacMillan, J. (2001). Occurrence of gibberellins in vascular plants, fungi, and bacteria. Journal of Plant Growth Regulation, 20(4), 387–442.

Miransari, M., & Smith, D. L. (2014). Plant hormones and seed germination. Environmental and Experimental Botany, 99, 110–121.

Miri, M., Ghooshchi, F., Tohidi-Moghadam, H. R., Larijani, H. R., & Kasraie, P. (2021). Ameliorative effects of foliar spray of glycine betaine and gibberellic acid on cowpea (*Vigna unguiculata* L. Walp.) yield affected by drought stress. Arabian Journal of Geosciences, 14(10), 1–9.

Misratia, K. M., Islam, M. R., Ismail, M. R., Oad, F. C., Hanafi, M. M., & Puteh, A. (2015). Interactive effects of gibberellic acid (GA_3) and salt stress on growth, biochemical parameters and ion accumulation of two rice (*Oryza sativa* L.) varieties differing in salt tolerance. Journal of Food, Agriculture and Environment, 13(1), 66–70.

Morton, J. F. (2007). The impact of climate change on smallholder and subsistence agriculture. Proceedings of the National Academy of Sciences, 104(50), 19680–19685.

Oshunsanya, S. O., Nwosu, N. J., & Li, Y. (2019). Abiotic stress in agricultural crops under climatic conditions. In: M. Jhariya, A. Banerjee, R. Meena, & D. Yadav (eds.), Sustainable agriculture, forest and environmental management (pp. 71–100). Springer, Singapore. 10.1007/978-981-13-6830-1_3.

Pan, J., Sharif, R., Xu, X., & Chen, X. (2021). Mechanisms of waterlogging tolerance in plants: Research progress and prospects. Frontiers in Plant Science, 11, 627331.

Pathak, H., Aggarwal, P. K., & Singh, S. D. (2012). Climate change impact, adaptation and mitigation in agriculture: methodology for assessment and applications. Indian Agricultural Research Institute, New Delhi, 302.

Pereira, A. (2016). Plant abiotic stress challenges from the changing environment. Frontiers in Plant Science, 7, 1123.

Raza, A., Razzaq, A., Mehmood, S. S., Zou, X., Zhang, X., Lv, Y., & Xu, J. (2019). Impact of climate change on crops adaptation and strategies to tackle its outcome: A review. Plants, 8(2), 34.

Schmidhuber, J., & Tubiello, F. N. (2007). Global food security under climate change. Proceedings of the National Academy of Sciences, 104(50), 19703–19708.

Schwechheimer, C. (2012). Gibberellin signaling in plants–the extended version. Frontiers in Plant Science, 2, 107.

Shaddad, M. A. K., Abd, E. S. H., & Mostafa, D. (2013). Role of gibberellic acid (GA3) in improving salt stress tolerance of two wheat cultivars. International Journal of Plant Physiology and Biochemistry, 5(4), 50–57.

Shahzad, K., Hussain, S., Arfan, M., Hussain, S., Waraich, E. A., Zamir, S., et al. (2021). Exogenously applied gibberellic acid enhances growth and salinity stress tolerance of maize through modulating the morpho-physiological, biochemical and molecular attributes. Biomolecules, 11(7), 1005.

Sharma, K. D., & Nayyar, H. (2016). Regulatory networks in pollen development under cold stress. Frontiers in Plant Science, 7, 402.

Sponsel, V. M. (2016). Signal achievements in gibberellin research: the second half-century. In: P. Hedden and S. G. Thomas (eds), Annual Plant Reviews, vol. 49. UK: John Wiley & Sons Ltd, pp. 1–36.

Sun, T. P. (2011). The molecular mechanism and evolution of the GA–GID1–DELLA signalling module in plants. Current Biology, 21(9), R338–R345.

Suzuki, H., Park, S. H., Okubo, K., Kitamura, J., Ueguchi-Tanaka, M., Iuchi, S., & Nakajima, M. (2009). Differential expression and affinities of *Arabidopsis* gibberellin receptors can explain variation in phenotypes of multiple knock-out mutants. The Plant Journal, 60(1), 48–55.

Tirado, M. C., Clarke, R., Jaykus, L. A., McQuatters-Gollop, A., & Frank, J. M. (2010). Climate change and food safety: A review. Food Research International, 43(7), 1745–1765.

Ueguchi-Tanaka, M., Ashikari, M., Nakajima, M., Itoh, H., Katoh, E., Kobayashi, M., et al. (2005). GIBBERELLIN INSENSITIVE DWARF1 encodes a soluble receptor for gibberellin. Nature, 437(7059), 693–698.

Vermeulen, S. J., Campbell, B. M., & Ingram, J. S. (2012). Climate change and food systems. Annual Review of Environment and Resources, 37(1), 195–222.

Wang, B., Wei, H., Xue, Z., & Zhang, W. H. (2017). Gibberellins regulate iron deficiency-response by influencing iron transport and translocation in rice seedlings (*Oryza sativa*). Annals of Botany, 119(6), 945–956.

Watts, N., Amann, M., Arnell, N., Ayeb-Karlsson, S., Belesova, K., Berry, H., & Costello, A. (2018). The 2018 report of the Lancet Countdown on health and climate change: shaping the health of nations for centuries to come. The Lancet, 392(10163), 2479–2514.

Willige, B. C., Ghosh, S., Nill, C., Zourelidou, M., Dohmann, E. M., Maier, A., & Schwechheimer, C. (2007). The DELLA domain of GA INSENSITIVE mediates the interaction with the GA INSENSITIVE DWARF1A gibberellin receptor of *Arabidopsis*. The Plant Cell, 19(4), 1209–1220.

Xu, H., Iwashiro, R., Li, T., & Harada, T. (2013). Long-distance transport of Gibberellic Acid Insensitive mRNA in *Nicotiana benthamiana*. BMC Plant Biology, 13(1), 1–9.

Xu, H., Liu, Q., Yao, T., & Fu, X. (2014). Shedding light on integrative GA signaling. Current Opinion in Plant Biology, 21, 89–95.

Zawaski, C., & Busov, V. B. (2014). Roles of gibberellin catabolism and signalling in growth and physiological response to drought and short-day photoperiods in Populus trees. PLoS One, 9(1), e86217.

Zhu, X. F., Jiang, T., Wang, Z. W., Lei, G. J., Shi, Y. Z., Li, G. X., & Zheng, S. J. (2012). Gibberellic acid alleviates cadmium toxicity by reducing nitric oxide accumulation and expression of IRT1 in *Arabidopsis* thaliana. Journal of Hazardous Materials, 239, 302–307.

Section II

Non-Traditional Phytohormones

9 Roles of Brassinosteroids in Abiotic Stresses

Shashi Meena, Sheel Yadav, Sukumar Taria, Sudhir Kumar, and Shivani Nagar

9.1 INTRODUCTION

Plants inevitably face various types of abiotic stress factors, including temperature (heat, chilling and freezing), water (drought, waterlogging), salinity, nutrient toxicity and deficiency, light (intense and weak) and radiation (UV-A/B), among others. These stressors substantially impact agriculture globally (Bhandari & Nailwal, 2020; Umar et al., 2021), affecting plant growth, development, acclimation and stress defense mechanisms. Under extremely stressful conditions, the accumulation of reactive oxygen species (ROS) disrupts the balance between ROS and antioxidants in plant cells, leading to oxidative stress and severe damage to vital plant metabolites such as proteins, lipids, carbohydrates and DNA (Anjum et al., 2015) (Figure 9.1). Plants employ complex metabolic processes, including ROS-scavenging systems and antioxidative pathways, to cope with abiotic stresses. These defense mechanisms enhance their tolerance and survival in harsh environments. Brassinosteroids (BRs), a group of plant steroid hormones, are essential for plant growth, development, stress management and pheromone production (Vardhini & Anjum, 2015). Recent studies have shown the ability to synthesize BRs in vitro (Bartwal et al., 2013), and their interaction with other phytohormones regulates various physiological and developmental processes (Saini et al., 2015). This review emphasizes the significant role of BRs in mitigating abiotic stresses.

9.2 BRASSINOSTEROIDS

Brassinosteroids are a group of endogenous steroidal phytohormones found in the pollen grains of various species. They share structural similarities with animal steroid hormones and play a role in plant growth and development (Clouse & Sasse, 1998). The discovery of brassinosteroids occurred in the 1970s when their growth-promoting effects on stem elongation were observed (Mitchell et al., 1970). Brassinolide, the first isolated polyhydroxylated steroidal plant hormone, was obtained from rape pollen (Grove et al., 1979). Since then, numerous brassinosteroid-related compounds have been identified from different plant organs (Kothari & Lachoweic, 2021). Castasterone was identified as the second brassinosteroid (Yokota et al., 1982). Brassinosteroids are categorized based on the number of carbons in their structure (Vardhini et al., 2019), and the three biologically active ones in plants are brassinolide, 28-homobrassinolide and 24-epibrassinolide. Brassinosteroids play a crucial role in regulating physiological and developmental processes (Vardhini et al., 2006). They affect plant growth, differentiation and tropisms by influencing auxin transport (Xu, 2006). Their levels vary across different plant tissues, with higher concentrations in growing tissues like pollen and immature seeds (Clouse & Sasse, 1998). Brassinosteroids are essential for cell elongation, differentiation and embryo growth in seeds (Leubner-Metzger, 2001). Interestingly, changes in BR activity may contribute to altered stature and increased yields in certain cereal varieties associated with the Green Revolution (Castorina & Consonni, 2020).

DOI: 10.1201/9781003335788-13

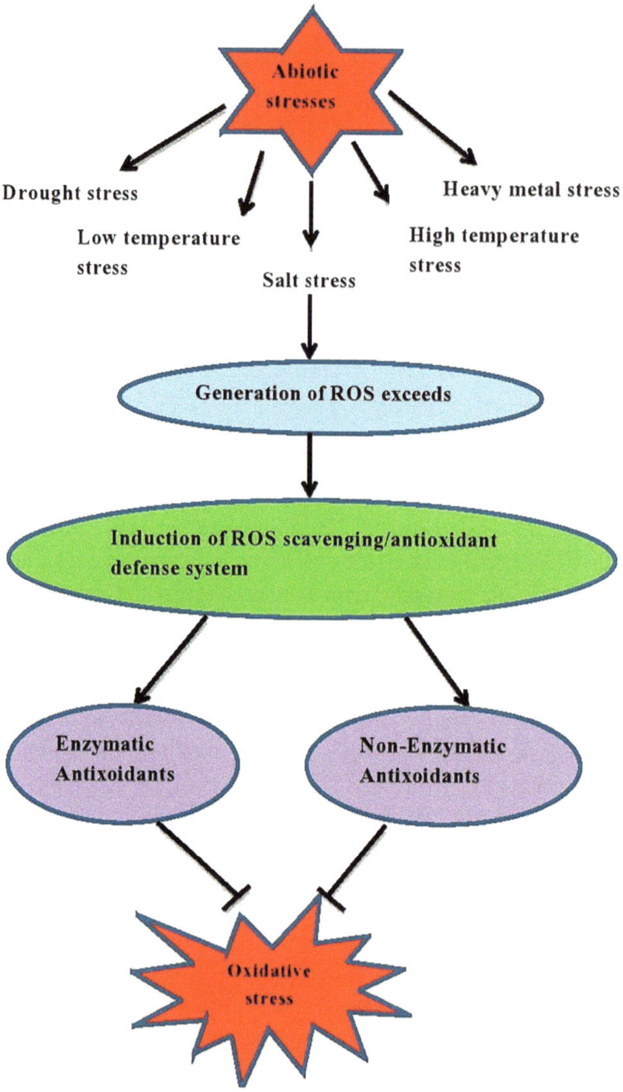

FIGURE 9.1 The diagram depicts the impact of major abiotic stresses on plants and their antioxidant defense system's role in regulating oxidative stress.

9.3 THE MECHANISM INVOLVED IN BRs-INDUCED STRESS-TOLERANCE

BRs signal transduction involves the BRI1 receptor, a plasma membrane-bound leucine-rich-repeat receptor-like kinase. BRs are perceived by BRI1, activating its kinase activity and regulating transcription through BES1 and BZR1 transcription factors, leading to stress tolerance. Exogenous BRs bind to BRI1, recruiting BAK1 as a co-receptor kinase and releasing BKI1. This binding activates BRI1 through transphosphorylation and promotes downstream phosphorylation events. The dephosphorylation of BIN2 by activated BSU1 allows the accumulation of unphosphorylated BZR1 and BZR2/BES1, which regulate BR-targeted genes for stress tolerance. The biosynthesis and inactivation of BRs are tightly regulated to maintain optimal levels. Exogenous BRs enhance stress tolerance, despite not significantly changing endogenous levels. Concentration-dependent responses are observed with exogenous BR application.

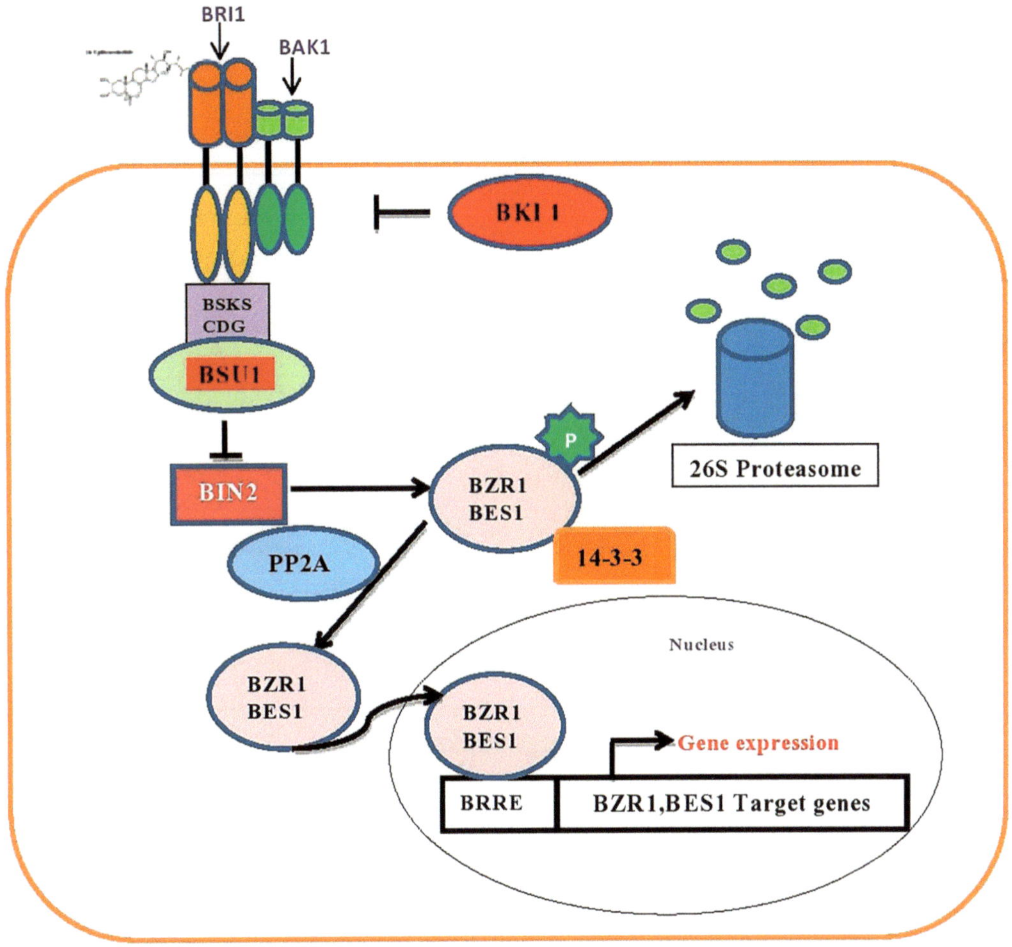

FIGURE 9.2 Brassinosteroid signaling model in *Arabidopsis*. Whereas BRI1: BRASSINOSTEROID INSENSITIVE 1 (receptor kinase), BKI1: BRI1 KINASE INHIBITOR (negative regulator), BZR1: BRASSINAZOLE RESISTANT 1 (transcription factor), BES1: BRI1-EMS SUPPRESSOR 1 (transcription factor), BSU1: BRI1 SUPPRESSOR 1 (phosphatase), BIN2: BRASSINOSTEROID INSENSITIVE 2 (kinase), BSK1: BR-SIGNALING KINASE 1 (positive regulator), BRRE: BRASSINOSTEROID RESPONSE ELEMENT (DNA sequence).

In the absence or low levels of BR, the kinase BIN2 phosphorylates the transcription factors BZR1 and BZR2/BES1, inhibiting their DNA-binding activity and retaining them in the cytoplasm through interaction with 14-3-3 proteins. When BR levels are high, BSU1 inactivates BIN2, leading to its degradation. PP2A is involved in the dephosphorylation of BZR1 and BZR2. Unphosphorylated BZR1 and BZR2 can then enter the nucleus and bind to target gene promoters, resulting in gene activation or repression (Wang et al., 2012) (Figure 9.2).

9.4 ROLE OF BRASSINOSTEROIDS IN VARIOUS ABIOTIC STRESSES

Brassinosteroids are a class of plant hormones that have been found to play a crucial role in mitigating the harmful effects of abiotic stresses on plants. Under stress conditions, plants experience reduced photosynthetic activity and an increase in the production of reactive oxygen species (ROS), which can cause oxidative damage. Brassinosteroids protect plants by modulating the activities of antioxidative enzymes and antioxidants (Hayat et al., 2007). They promote the synthesis of

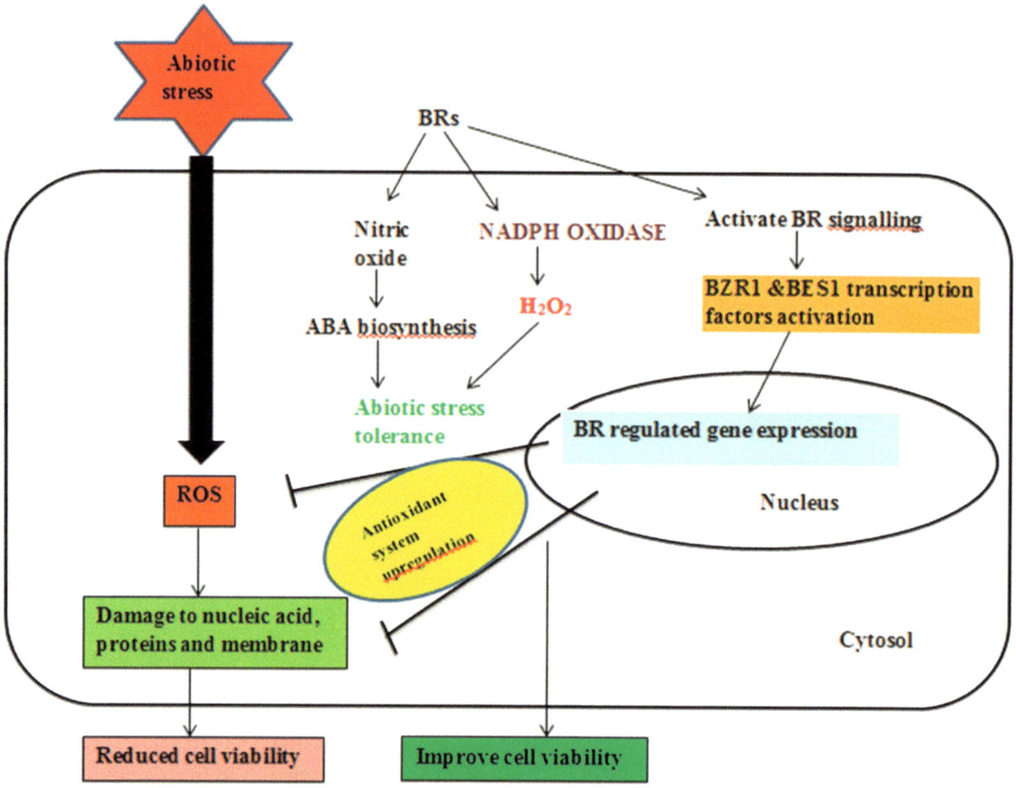

FIGURE 9.3 Model of BR-induced abiotic responses in *Arabidopsis*.

osmoprotectants, which help maintain cellular homeostasis and detoxify ROS (Rontein et al., 2002). Additionally, brassinosteroids enhance the antioxidant defense system in plants, reducing lipid peroxidation and preventing oxidative damage (Mittler, 2002). Exogenous application of brassinosteroids has been shown to improve plant growth and development under stress conditions. They work in conjunction with other plant hormones to regulate plant responses to oxidative stress (Cao et al., 2005). Overall, brassinosteroids play a vital role in stress mitigation by enhancing plant resilience and promoting growth in adverse environmental conditions (Figure 9.3).

9.4.1 Salt Stress

Salt stress induces oxidative stress in plants, negatively impacting various processes and reducing agricultural productivity (Anwar et al., 2018). Brassinosteroids (BRs) have been found to alleviate the inhibitory effects of salt stress by enhancing chlorophyll content, photosynthesis and antioxidative enzyme activities while reducing reactive oxygen species (ROS) and malondialdehyde (MDA) levels (Wu et al., 2017). BRs also promote groundnut seedling growth and germination under salt-stress conditions. Exogenous application of 24-epibrassinolide and 28-homobrassinolide improves seed germination rates in groundnut (Vardhini & Rao, 1998) and *Eucalyptus camaldulensis* (Sasse et al., 1995), respectively, in saline environments.

24-epibrassinolide and 28-homobrassinolide alleviate salinity-induced effects in rice, such as germination rate inhibition, seedling growth inhibition and photosynthetic pigment loss (Anuradha & Ram Rao, 2003). Pretreating rice seeds with low concentrations of brassinosteroids enhances plant growth under saline conditions (Rao et al., 2002). Exogenous application of 24-epibrassinolide improves salt-sensitive rice seedling growth (Özdemir et al., 2004) and promotes seed germination in *Brassica napus* under salinity stress (Kagale et al., 2007). 28-homoBL mitigates salt stress in *Zea mays* by enhancing antioxidant activity and reducing lipid peroxidation (Arora et al., 2008). Similar effects are observed in BR-treated canola seedlings (Talaat, 2013). The application of brassinolide alleviates salt stress in *Malus hupehensis* Rehd. by maintaining photosynthetic capacity, enhancing antioxidant activity, regulating gene expression and promoting osmotic balance (Su et al., 2020).

9.4.2 Low-Temperature Stress

Plants are prone to low-temperature stress during winter or autumn. Plants are more vulnerable to low-temperature stress during winter and autumn, which includes chilling and frost stress (Saltveit, 2001). Chilling stress negatively affects the photosynthetic machinery, chlorophyll contents, sugar accumulation, lipid peroxidation, water balance and flower bud development, leading to significant yield losses (Allen & Ort, 2001; Anwar et al., 2018). Injection of EBL into rape seedlings reduces cold-induced plasma membrane permeability. EBL treatment in rape seedlings exposed to 2°C for 3 days reduces ion leakage and increases photosynthetic pigments (Janeczko et al., 2007). Cold-related gene expression is enhanced in *Brassica napus* and *Arabidopsis thaliana* treated with EBL and subjected to cold stress (Kagale et al., 2007). EBL plays a critical role in alleviating chilling stress effects in cucumber seedlings by enhancing RUBISCO activation and photosynthetic gene expression. 24-epibrassinolide application increases proline, ATP and superoxide dismutase activity, improving chilling stress tolerance in rice (Rao et al., 2002). Brassinosteroids help maintain chlorophyll content and enhance fruit-setting in maize (Katsumi, 1991) and tomato plants (Kamuro & Takatsuto, 1991) under cold conditions. Treatment with 24-epibrassinolide increases heat shock protein concentrations in *Brassica napus* seedlings, enhancing their cold stress tolerance (Dhaubhadel et al., 2002). Exogenous application of BRs increases resistance to low-temperature stress in tomato (Shu et al., 2016), cucumber (Wei et al., 2015) and pepper plants (Li et al., 2015). BR-treated pepper seedlings exhibit upregulated gene expression and enhanced levels of salicylic acid, jasmonic acid and ethylene biosynthesis under chilling stress (Wei et al., 2015; Li et al., 2016). Foliar treatment with BRs improves rapeseed performance under freezing stress by promoting growth and photosynthesis (Ma et al., 2009).

Low-temperature stress during winter affects plants, including chilling and frost stress (Saltveit, 2001). Chilling stress leads to yield losses by negatively impacting photosynthesis, chlorophyll content, sugar accumulation, lipid peroxidation, water balance and flower bud development (Anwar et al., 2018). EBL injection reduces cold-induced membrane permeability and increases photosynthetic pigments in rape seedlings. EBL treatment enhances cold-related gene expression in *Brassica napus* and under cold stress (Janeczko et al., 2007; Kagale et al., 2007). EBL helps alleviate chilling stress effects in cucumber (Wei et al., 2015) by enhancing RUBISCO activation and photosynthetic gene expression. 24-epibrassinolide application improves chilling stress tolerance in rice (Rao et al., 2002) by increasing proline, ATP and superoxide dismutase activity. Brassinosteroids maintain chlorophyll content and enhance fruit-setting in maize (Katsumi, 1991) and tomato plants (Kamuro & Takatsuto, 1991) under cold conditions. Treatment with 24-epibrassinolide enhances cold stress tolerance in *Brassica napus* by increasing heat shock protein concentrations (Dhaubhadel et al., 2002). BR-treated pepper seedlings show upregulated gene expression and enhanced levels of salicylic acid, jasmonic acid and ethylene biosynthesis under chilling stress (Li et al., 2016).

Gallo et al. (2017) found that foliar application of BRs enhances growth and resilience in *Fagus sylvatica* L. plantations, reducing the negative impacts of frost stress. Chen et al. (2019) demonstrated that foliar treatment with 24-epiBL mitigates cold-induced oxidative stress by regulating the ascorbate-glutathione cycle. Cucumber seedlings treated with BRs under PEG and cold stress show improved physiological parameters, enzymatic antioxidant activity, reduced ROS levels and increased ethylene levels via enhanced transcription of ethylene signaling biosynthesis genes (Zhu et al., 2016). Exogenous application of brassinosteroid and silicon in tomatoes under chilling stress effectively reduces the negative impacts on growth, photosynthetic rate, pigments, antioxidative activity and osmolyte accumulation (Bashir & John, 2023).

9.4.3 High-Temperature Stress

High-temperature stress threatens global crop production, affecting leaf water potential, photosynthesis and water availability (Zhang et al., 2005). Brassinosteroid analogue (BB6) treatment induces heat shock protein production, protecting mRNA in tomato leaf discs exposed to high temperatures (Sam et al., 2001). 24-epibrassinolide and MH5 influence the activity of antioxidative enzymes and reduce cell damage caused by high-temperature stress in tomato leaf discs (Mazorra et al., 2002). Foliar spraying of 24-epibrassinolide induces the expression of mitochondrial small heat shock proteins, providing tolerance against high temperatures in tomato plants (Singh & Shano, 2005). Exogenous application of EBL protects Rubisco, Calvin cycle enzymes and RuBP regeneration in tomato plants under high temperatures (Ogweno et al., 2008). 24-epibrassinolide increases transcripts of cold-responsive structural genes in *Arabidopsis* seedlings compared to untreated seedlings (Kagale et al., 2007).

Studies have demonstrated that treating *Solanum melongena* (eggplant) with 24-epibrassinolide (EBR) under high-temperature conditions enhances plant growth and mitigates the negative impact on photosynthetic processes. The application of EBR at concentrations of 0.05–0.2 mM increases the activities of antioxidant enzymes in eggplant exposed to high temperatures. It also leads to higher contents of ascorbic acid, proline, soluble sugars and proteins, while decreasing glutathione levels (Wu et al., 2014).

BRs are plant hormones that enhance tolerance to high-temperature stress in various plant species. BR treatment has been shown to improve tolerance in bananas (Nassar, 2004), *Ficus concinna* (Jin et al., 2015), *Brassica* and *Arabidopsis* (Kagale et al., 2007) by maintaining the physiological and antioxidant defense system. BRs increase enzymatic and non-enzymatic antioxidant defenses, enhance glyoxalase system activity (Jin et al., 2015) and promote the production of heat shock proteins (HSPs) to protect proteins from denaturation (Chen et al., 2021). BRs also help maintain the balance between reduced and oxidative radicals and protect plant photosynthesis by improving pigment contents, photosystem activity and Rubisco function (Zhao et al., 2017). The effects of BRs on growth and development under high-temperature stress can vary depending on plant species, developmental stage and environmental conditions (Yang et al., 2021). Foliar application of BRs has the potential to increase tolerance, regulate growth and fertility and maintain yields in cereal crops under heat stress (Kothari & Lachowiec, 2021). In wheat, BR application alleviates the negative effects of combined heat stress and drought by accelerating photosynthesis and gene expression related to Rubisco activase (Zhao et al., 2017).

9.4.4 Water Stress

BRs have shown the potential to mitigate the negative effects of water stress on plant growth and crop production. They help maintain tissue-water status, stimulate gene expressions related to stress response and enhance membrane stability (Farooq et al., 2009). BR application has been found to increase grain yield and improve water content and soluble protein levels in drought-sensitive wheat varieties (Desoky et al., 2021). In sugar beet, synthetic BRs alleviate mild drought stress (Schilling et al., 1991). Additionally, BR treatment enhances drought tolerance in *Arabidopsis* and *Brassica* seedlings by regulating drought-responsive genes (Kagale et al., 2007). BRs also promote the accumulation of osmoprotectants like soluble sugars and proline in soybean, aiding in stress protection (Fariduddin et al., 2009).

Application of 24-epibrassinolide has shown positive effects on plant growth and drought stress alleviation in various crops. In spring wheat, it increased water content in leaves (Prusakova et al., 2000). French beans treated with brassinosteroids during flowering exhibited improved root nodulation, cytokinin content and nitrogenase enzyme activity under drought stress (Upreti & Murti, 2004). Brassinosteroid analogs enhanced antioxidant enzyme activities and reduced the activities of IAA oxidase and PPO in sorghum seedlings subjected to water stress (Vardhini et al., 2011). Chickpea plants treated with brassinosteroids showed enhanced plant parameters under water stress conditions (Singh, 1993). In radish seedlings, brassinosteroid application increased the activity of antioxidative enzymes, mitigating the harmful effects of drought stress (Mahesh et al., 2013).

Foliar application of brassinosteroids in apple seedlings stress (Kumari & Thakur, 2019) and 2-deoxybrassinosteroid analogues in *Arabidopsis* (Díaz et al., 2021) has demonstrated their potential in maintaining growth, physiological processes and enhancing drought tolerance. These studies suggest that foliar application of brassinosteroid analogues could be a promising strategy for improving crop production under drought-stress conditions. Brassinosteroids positively impact plant growth, water status, antioxidant activities and gene expression, thereby mitigating the adverse effects of drought stress and enhancing plant tolerance.

9.4.5 Heavy Metal Stress

Heavy metal contamination in soil can be harmful to plant growth and metabolism, leading to toxicity symptoms. Brassinosteroids, such as 24-epibrassinolide, have shown potential in mitigating heavy metal toxicity in plants. They regulate cell membrane permeability and ion transport, reducing heavy metal uptake and accumulation (Vardhini et al., 2016). Studies have demonstrated that 24-epibrassinolide effectively reduces heavy metal absorption in various plant species (Khripach et al., 1996). For example, it has been shown to improve plant height, chlorophyll content, photosynthetic rate and enzymatic activities in plants exposed to cadmium (Bajguz, 2010). Additionally, 24-epibrassinolide application has been found to decrease zinc and cadmium levels in tomato fruits and reduce the accumulation of caesium and strontium ions in barley grains grown in heavy metal-contaminated soil. Overall, brassinosteroids can enhance plant tolerance to heavy metals and regulate physiological processes, mitigating the negative impacts of heavy metal toxicity.

In *Chlorella vulgaris*, 24-epibrassinolide completely blocks copper, lead, cadmium and zinc accumulation (Bajguz, 2000). A 10^{-8} M solution of 24-epibrassinolide effectively prevents heavy metal buildup in algal cells (Bajguz, 2002). Brassinosteroids alleviate aluminum (Al) toxicity in mungbean and protect photosystem II (PS2) in winter rape plants under cadmium (Cd) stress (Abdullahi et al., 2002; Janeczko et al., 2005). Exogenous brassinosteroid application improves shoot emergence, biomass production and reduces heavy metal uptake in *Brassica juncea* seedlings. Brassinosteroids enhance the activity of antioxidative enzymes in maize seedlings, reducing the harmful effects of nickel (Ni) (Bhardwaj et al., 2007). Treatment with brassinosteroids in *Brassica juncea* plants under cadmium stress improves various parameters related to growth, defense enzymes and nitrogen assimilation (Hayat et al., 2007). Brassinosteroids mitigate pesticide toxicity in *Cucumis sativus* seedlings by increasing CO_2 assimilation and antioxidant activity. Combining brassinosteroids and nitric oxide (NO) alleviates chromium-induced phytotoxicity in soybean by modulating antioxidative defense and glyoxalase systems (Basit et al., 2023). 24-epibrassinolide and gibberellic acid (GA) alleviate zinc toxicity in *Medicago sativa* by promoting antioxidant defense and regulating heavy metal detoxification (Ren et al., 2023).

9.5 CONCLUSION

Brassinosteroids are plant growth regulators that have been found to have significant effects on plant growth and development, including cell division, cell elongation and reproductive development. They also confer tolerance to various environmental stresses such as heat, cold, drought, heavy metals, salt and pathogens. Brassinosteroids act as stress-protective molecules by inducing complex biochemical reactions, stimulating antioxidative enzyme production and enhancing the plant's defense mechanisms. They are considered environmentally friendly and non-toxic, making them suitable for use in stress conditions to improve crop yield and mitigate the negative impacts of environmental pressures on agriculture.

REFERENCES

Abdullahi, B. A., Gu, X. G., Gan, Q. L., & Yang, Y. H. (2002). Brassinolide amelioration of aluminium toxicity in mungbean seedling growth. Journal of Plant Nutrition, 26(9), 1725–1734.

Allen, D. J., & Ort, D. R. (2001). Impacts of chilling temperatures on photosynthesis in warm-climate plants. Trends in Plant Science, 6, 36–42. 10.1016/s1360-1385(00)01808-2.

Anjum, N. A., Sofo, A., Scopa, A., Roychoudhury, A., Gill, S. S., Iqbal, M., et al. (2015). Lipids and proteins—major targets of oxidative modifications in abiotic stressed plants. Environmental Science and Pollution Research, 22, 4099–4121.

Anuradha, S., & Ram Rao, S. S. (2003). Application of brassinosteroids to rice seeds (*Oryza sativa* L.) reduced the impact of salt stress on growth, prevented photosynthetic pigment loss and increased nitrate reductase activity. Plant Growth Regulation, 40, 29–32.

Anwar, A., Bai, L., Miao, L., Liu, Y., Li, S., Yu, X., & Li, Y. (2018). 24-Epibrassinolide ameliorates endogenous hormone levels to enhance low-temperature stress tolerance in cucumber seedlings. International Journal of Molecular Sciences, 19(9), 2497.

Arora, N., Bhardwaj, R., Sharma, P., & Arora, H. K. (2008). Effects of 28-homobrassinolide on growth, lipid peroxidation and antioxidative enzyme activities in seedlings of *Zea mays* L. under salinity stress. Acta Physiologiae Plantarum, 30, 833–839.

Bajguz, A. (2000). Blockade of heavy metals accumulation in Chlorella vulgaris cells by 24-epibrassinolide. Plant Physiology and Biochemistry, 38(10), 797–801.

Bajguz, A. (2002). Brassinosteroids and lead as stimulators of phytochelatins synthesis in Chlorella vulgaris. Journal of Plant Physiology, 159(3), 321–324.

Bajguz, A. (2010). An enhancing effect of exogenous brassinolide on the growth and antioxidant activity in Chlorella vulgaris cultures under heavy metals stress. Environmental and Experimental Botany, 68(2), 175–179.

Bartwal, A., Mall, R., Lohani, P., Guru, S. K., & Arora, S. (2013). Role of secondary metabolites and brassinosteroids in plant defense against environmental stresses. Journal of Plant Growth Regulation, 32, 216–232.

Bashir, S., & John, R. (2023). Alleviation of chilling stress by supplementation of brassinosteroid and silicon in *Solanumlycopersicum* L. Plant and Soil, 486, 165–181. 10.1007/s11104-022-05866-8.

Basit, F., Tao, J., An, J., Song, X., Sheteiwy, M. S., Holford, P., et al. (2023). Nitric oxide and brassinosteroids enhance chromium stress tolerance in *Glycine max* L. (Merr.) by modulating antioxidative defense and glyoxalase systems. Environmental Science and Pollution Research, 30(18), 51638–51653.

Bhandari, S., & Nailwal, T. K. (2020). Role of brassinosteroids in mitigating abiotic stresses in plants. Biologia, 75(12), 2203–2230.

Bhardwaj, R., Arora, N., Sharma, P., & Arora, H. K. (2007). Effects of 28-homobrassinolide on seedling growth, lipid peroxidation and antioxidative enzyme activities under nickel stress in seedlings of *Zea mays* L. Asian Journal of Plant Science, 6(5), 765–772.

Cao, S., Xu, Q., Cao, Y., Qian, K., An, K., Zhu, Y., et al. (2005). Loss-of-function mutations in DET2 gene lead to an enhanced resistance to oxidative stress in *Arabidopsis*. Physiologia Plantarum, 123(1), 57–66.

Castorina, G., & Consonni, G. (2020). The role of brassinosteroids in controlling plant height in Poaceae: a genetic perspective. International Journal of Molecular Sciences, 21(4), 1191.

Chen, Z. Y., Wang, Y. T., Pan, X. B., & Xi, Z. M. (2019). Amelioration of cold-induced oxidative stress by exogenous 24-epibrassinolide treatment in grapevine seedlings: Toward regulating the ascorbate--glutathione cycle. Scientia Horticulturae, 244, 379–387.

Chen, J., Fei, K., Zhang, W., Wang, Z., Zhang, J., & Yang, J. (2021). Brassinosteroids mediate the effect of high temperature during anthesis on the pistil activity of photo-thermosensitive genetic male-sterile rice lines. The Crop Journal, 9(1), 109–119.

Clouse, S. D., & Sasse, J. M. (1998). Brassinosteroids: essential regulators of plant growth and development. Annual Review of Plant Biology, 49(1), 427–451.

Desoky, E. S. M., Mansour, E., Ali, M. M., Yasin, M. A., Abdul-Hamid, M. I., Rady, M. M., & Ali, E. F. (2021). Exogenously used 24-epibrassinolide promotes drought tolerance in maize hybrids by improving plant and water productivity in an arid environment. Plants, 10(2), 354.

Dhaubhadel, S., Browning, K. S., Gallie, D. R., & Krishna, P. (2002). Brassinosteroid functions to protect the translational machinery and heat-shock protein synthesis following thermal stress. The Plant Journal, 29(6), 681–691.

Díaz, K., Espinoza, L., Carvajal, R., Silva-Moreno, E., Olea, A. F., & Rubio, J. (2021). Exogenous application of brassinosteroid 24-norcholane 22 (S)-23-dihydroxy type analogs to enhance water deficit stress tolerance in *Arabidopsis thaliana*. International Journal of Molecular Sciences, 22(3), 1158.

Fariduddin, Q., Yusuf, M., Hayat, S., & Ahmad, A. (2009). Effect of 28-homobrassinolide on antioxidant capacity and photosynthesis in *Brassica juncea* plants exposed to different levels of copper. Environmental and Experimental Botany, 66(3), 418–424.

Farooq, M., Wahid, A., Kobayashi, N. S. M. A., Fujita, D. B. S. M. A., & Basra, S. M. A. (2009). Plant drought stress: effects, mechanisms and management. In: E. Lichtfouse (eds),Sustainable Agriculture, 153–188. https://doi.org/10.1007/978-90-481-2666-8_12

Gallo, J., Baláš, M., Linda, R., & Kuneš, I. (2017). Growth performance and resistance to ground late frosts of *Fagus sylvatica* L. plantation treated with a brassinosteroid compound. Journal of Forest Science, 63(3), 117–125.

Grove, M. D., Spencer, G. F., Rohwedder, W. K., Mandava, N., Worley, J. F., Warthen Jr, J. D., et al. (1979). Brassinolide, a plant growth-promoting steroid isolated from *Brassica napus* pollen. Nature, 281(5728), 216–217.

Hayat, S., Ali, B., Hasan, S. A., & Ahmad, A. (2007). Brassinosteroid enhanced the level of antioxidants under cadmium stress in *Brassica juncea*. Environmental and Experimental Botany, 60(1), 33–41.

Janeczko, A., Gullner, G., Skoczowski, A., Dubert, F., & Barna, B. (2007). Effects of brassinosteroid infiltration prior to cold treatment on ion leakage and pigment contents in rape leaves. Biologia Plantarum, 51, 355–358.

Janeczko, A., Koscielniak, J., Pilipowicz, M., Szarek-Lukaszewska, G., & Skoczowski, A. (2005). Protection of winter rape photosystem 2 by 24-epi brassinolide under cadmium stress. Photosynthetica, 43, 293–298.

Jin, S. H., Li, X. Q., Wang, G. G., & Zhu, X. T. (2015). Brassinosteroids alleviate high-temperature injury in *Ficus concinna* seedlings via maintaining higher antioxidant defense and glyoxalase systems. AoB Plants, 7, 1–12.

Kagale, S., Divi, U. K., Krochko, J. E., Keller, W. A., & Krishna, P. (2007). Brassinosteroid confers tolerance in *Arabidopsis thaliana* and Brassica napus to a range of abiotic stresses. Planta, 225, 353–364.

Kamuro, Y., & Takatsuto, S. (1991). Capability for and problems of practical uses of brassinosteroids. In: H. G. Cutler, T. Yokota, & G. Adam (eds.), Brassinosteroids (pp. 280–291), ACS Publication.

Katsumi, M. (1991). Physiological modes of brassinolide action in cucumber hypocotyl growth. In: H. G. Cutler, T. Yokota, & G. Adam (eds.), Brassinosteroids (pp. 246–254), ACS Publication.

Khripach, V. A., Voronina, L. V., & Malevannaya, N. N. (1996). Preparation for the diminishing of heavy metals accumulation of agricultural plants. Pat. Appl. RU, 95(101,850).

Kothari, A., & Lachowiec, J. (2021). Roles of brassinosteroids in mitigating heat stress damage in cereal crops. International Journal of Molecular Sciences, 22(5), 2706.

Kumari, S., & Thakur, A. (2019). The Effects of Water Stress and Brassinosteroid on Apple Varieties. International Journal of Economic Plants, 6(1), 1–6.

Leubner-Metzger, G. (2001). Brassinosteroids and gibberellins promote tobacco seed germination by distinct pathways. Planta, 213, 758–763.

Li, J., Yang, P., Gan, Y., Yu, J., & Xie, J. (2015). Brassinosteroid alleviates chilling-induced oxidative stress in pepper by enhancing antioxidation systems and maintenance of photosystem II. Acta Physiologiae Plantarum, 37. 10.1007/s11738-015-1966-9.

Li, J., Yang, P., Kang, J., Gan, Y., Yu, J., Calderón-Urrea, A., et al. (2016). Transcriptome analysis of pepper (Capsicum annuum) revealed a role of 24-epibrassinolide in response to chilling. Frontiers in Plant Science, 7, 1281.

Ma, N., Liu, D., Zhang, C., Li, J., & Li, G. (2009). Regulation effects of exogenous hormones on growth and photosynthesis and yield of rapeseed (*Brassica napus* L.) after Frozen. Acta Agronomica Sinica, 35(7), 1336–1343.

Mahesh, K., Balaraju, P., Ramakrishna, B., & Rao, S. S. R. (2013). Effect of brassinosteroids on germination and seedling growth of radish (*Raphanus sativus* L.) under PEG-6000 induced water stress. American Journal of Plant Sciences, 2013, 2305–2313.

Mazorra, L. M., Nunez, M., Hechavarria, M., Coll, F., & Sánchez-Blanco, M. J. (2002). Influence of brassinosteroids on antioxidant enzymes activity in tomato under different temperatures. Biologia Plantarum, 45, 593–596.

Mitchell, J. W., Mandava, N., Worley, J. F., Plimmer, J. R., & Smith, M. V. (1970). Brassins—a new family of plant hormones from rape pollen. Nature, 225, 1065–1066.

Mittler, R. (2002). Oxidative stress, antioxidants and stress tolerance. Trends in Plant Science, 7(9), 405–410.

Nassar, A. H. (2004). Effect of homobrassinolide on in vitro growth of apical meristems and heat tolerance of banana shoots. International Journal of Agriculture and Biology, 6(5), 771–775.

Ogweno, J. O., et al. (2022). Brassinosteroidsalleviate heat-induced inhibition ofphotosynthesis by increasing carboxylation efficiency and enhancing antioxidant systems in Lycopersiconesculentum. Journal of Plant Growth Regulation, 27, 49–57.

Özdemir, F., Bor, M., Demiral, T., & Türkan, İ. (2004). Effects of 24-epibrassinolide on seed germination, seedling growth, lipid peroxidation, proline content and antioxidative system of rice (*Oryza sativa* L.) under salinity stress. Plant Growth Regulation, 42, 203–211.

Prusakova, L. D., Chizhova, S. I., Ageeva, L. F., Golantseva, E. N., & Yakovlev, A. F. (2000). Effects of Epibrassinolide and Ekost on the drought resistance and productivity of spring wheat. Agrokhimiya, 3(3), 50–54.

Rao, S. S. R., Vardhini, B. V., Sujatha, E., & Anuradha, S. (2002). Brassinosteroids–a new class of phytohormones. Current Science, 82(10), 1239–1245.

Ren, Y., Li, X., Liang, J., Wang, S., Wang, Z., Chen, H., & Tang, M. (2023). Brassinosteroids and gibberellic acid actively regulate the zinc detoxification mechanism of *Medicago sativa* L. seedlings. BMC Plant Biology, 23(1), 1–13.

Rontein, D., Basset, G., & Hanson, A. D. (2002). Metabolic engineering of osmoprotectant accumulation in plants. Metabolic Engineering, 4(1), 49–56.

Saini, S., Sharma, I., & Pati, P. K. (2015). Versatile roles of brassinosteroid in plants in the context of its homoeostasis, signaling and crosstalks. Frontiers in Plant Science, 6. 10.3389/fpls.2015.00950.

Saltveit, M. E. (2001). Chilling injury is reduced in cucumber and rice seedlings and in tomato pericarp discs by heat-shocks applied after chilling. Postharvest Biology and Technology, 21(2), 169–177.

Sam, O., Núñez, M., Ruiz-Sánchez, M. C., Dell'Amico, J., Falcón, V., De La Rosa, M. C., & Seoane, J. (2001). Effect of a brassinosteroid analogue and high temperature stress on leaf ultrastructure of *Lycopersicon esculentum*. Biologia Plantarum, 44, 213–218.

Sasse, J. M. S. R., Smith, R., & Hudson, I. (1995). Effect of 24-epibrassinolide on germination of seeds of Eucalyptus camaldulensis in saline conditions. In Proceedings-plant growth regulator society of america-annual meeting, vol. 22, pp. 136–141, Plant Growth Regulator Society of America.

Schilling, G., Schiller, C., & Otto, S. (1991). Influence of brassinosteroids on organ relations and enzyme activities of sugar-beet plants. In: H. G. Cutler, T. Yokota, & G. Adam (eds.), Brassinosteroids (pp. 208–219), ACS Publication.

Shu, S., Tang, Y., Yuan, Y., Sun, J., Zhong, M., & Guo, S. (2016). The role of 24-epibrassinolide in the regulation of photosynthetic characteristics and nitrogen metabolism of tomato seedlings under a combined low temperature and weak light stress. Plant Physiology and Biochemistry, 107, 344–353.

Singh, I., & Shono, M. (2005). Physiological and molecular effects of 24-epibrassinolide, a brassinosteroid on thermotolerance of tomato. Plant Growth Regulation, 47, 111–119.

Singh, J. (1993). Effect of epi-brassinolideon gram (Cicer arietinum) plants grownunder water stress in juvenile stage. Indian Journal of Agricultural Sciences, 63, 395–397.

Su, Q., Zheng, X., Tian, Y., & Wang, C. (2020). Exogenous brassinolide alleviates salt stress in *Malus hupehensis* Rehd. by regulating the transcription of NHX-Type Na+ (K+)/H+ antiporters. Frontiers in Plant Science, 11, 38.

Talaat, N. B., & Shawky, B. T. (2013). 24-Epibrassinolide alleviates salt-induced inhibition of productivity by increasing nutrients and compatible solutes accumulation and enhancing antioxidant system in wheat (*Triticum aestivum* L.). Acta Physiologiae Plantarum, 35, 729–740.

Umar, O. B., Ranti, L. A., Abdulbaki, A. S., Bola, A. L., Abdulhamid, A. K., Biola, M. R., & Victor, K. O. (2021). Stresses in plants: Biotic and abiotic. Current Trends in Wheat Research, 139–145.

Upreti, K. K., & Murti, G. S. R. (2004). Effects of brassmosteroids on growth, nodulation, phytohormone content and nitrogenase activity in French bean under water stress. Biologia Plantarum, 48, 407–411.

Vardhini, B. (2019). Does Application of Brassinosteroids mitigate the Temperature Stress in Plants. International Journal of Earth Science and Geology, 1(2), 59–65.

Vardhini, B. V. (2016). Brassinosteroids are potential ameliorators of heavy metal stresses in plants. In: P. Ahmad (eds), Plant Metal Interaction (pp. 209–237), UK: Elsevier.

Vardhini, B. V., & Anjum, N. A. (2015). Brassinosteroids make plant life easier under abiotic stresses mainly by modulating major components of antioxidant defense system. Frontiers in Environmental Science, 2, 67.

Vardhini, B. V., & Rao, S. S. R. (1998). Effect of brassinosteroids on growth, metabolite content and yield of *Arachis hypogaea*. Phytochemistry, 48(6), 927–930.

Vardhini, B. V., Anuradha, S., & Rao, S. S. R. (2006). Brassinosteroids-New class of plant hormone with potential to improve crop productivity. Indian Journal of Plant Physiology, 11(1), 1.

Vardhini, B. V., Sujatha, E., & Rao, S. S. R. (2011). Brassinosteroids: alleviation of water stress in certain enzymes of sorghum seedlings. Journal of Phytology, 3(10), 38–43.

Wang, Z. Y., Bai, M. Y., Oh, E., & Zhu, J. Y. (2012). Brassinosteroid signaling network and regulation of photomorphogenesis. Annual Review of Genetics, 46, 701–724.

Wei, L. J., Deng, X. G., Zhu, T., Zheng, T., Li, P. X., Wu, J. Q., et al. (2015). Ethylene is involved in brassinosteroids induced alternative respiratory pathway in cucumber (*Cucumis sativus* L.) seedlings response to abiotic stress. Frontiers in Plant Science, 6, 982.

Wu, X., Yao, X., Chen, J., Zhu, Z., Zhang, H., & Zha, D. (2014). Brassinosteroids protect photosynthesis and antioxidant system of eggplant seedlings from high-temperature stress. Acta Physiologiae Plantarum, 36(2), 251–261.

Wu, W., Zhang, Q., Ervin, E. H., Yang, Z., & Zhang, X. (2017). Physiological mechanism of enhancing salt stress tolerance of perennial ryegrass by 24-epibrassinolide. Frontiers in Plant Science, 8, 1017.

Xu, Z. H. (2006). Recent progress in Arabidopsis research in China: A preface. Journal of Integrative Plant Biology, 48(1), 1.

Yang, J., Miao, W., & Chen, J. (2021). Roles of jasmonates and brassinosteroids in rice responses to high temperature stress–A review. The Crop Journal, 9(5), 977–985.

Yokota, T., Arima, M., & Takahashi, N. (1982). Castasterone, a new phytosterol with plant-hormone potency, from chestnut insect gall. Tetrahedron Letters, 23(12), 1275–1278.

Zhang, J., Huang, W., Pan, Q., & Liu, Y. (2005). Improvement of chilling tolerance and accumulation of heat shock proteins in grape berries (*Vitis vinifera* cv. Jingxiu) by heat pretreatment. Postharvest Biology and Technology, 38(1), 80–90.

Zhao, G., Xu, H., Zhang, P., Su, X., & Zhao, H. (2017). Effects of 2, 4-epibrassinolide on photosynthesis and Rubisco activase gene expression in Triticum aestivum L. seedlings under a combination of drought and heat stress. Plant Growth Regulation, 81, 377–384.

Zhu, T., Deng, X., Zhou, X., Zhu, L., Zou, L., Li, P., et al. (2016). Ethylene and hydrogen peroxide are involved in brassinosteroid-induced salt tolerance in tomato. Scientific Reports, 6(1), 1–15.

10 Indoleamines in Abiotic Stress

Naveen Goel, Isha Bhat, Amooru Harika, Ambika Rajendran, and Ayyagari Ramlal*

10.1 INTRODUCTION

Abiotic stressors, such as temperature extremes, drought and salt, have a significant impact on plant development and output (Rehaman et al., 2021). Abiotic stressors impacting different phases of plant growth account for over 70% of yield losses among key agricultural plants such as rice, wheat, maize and so on. Furthermore, abiotic pressures are progressing in the current period of global climate change (Arnao & Hernández-Ruiz, 2019a,b) and pose a significant danger to agricultural sustainability globally. It impacts agricultural plant growth and development by reducing germination, reducing photosynthetic activity and carbon absorption to a great extent, decreasing blooming and pollen sterility, and hence functions as a key limiting factor for crop output (Zeng et al., 2022). As a result, it is critical to comprehend the physiological characteristics of particular biomolecules that regulate plant growth and development via crosstalk. Indoleamines are tryptophan derivatives that are known to stimulate the synthesis of a variety of secondary growth metabolites, including phytoalexins, indole glycosylates, melatonin, alkaloids and serotonin. Melatonin (N-acetyl-5-methoxytryptamine) and serotonin (5-hydroxytryptamine) are indoleamines identified in mammals but are now present in practically every form of life, including plants (Arora & Bhatla, 2017). In plants, indoleamines are generated from tryptophan via the primary biosynthetic route, where tryptophan is first converted to tryptamine by tryptophan decarboxylase (TDC) (Erland et al., 2017). Though TDC is known as tryptophan decarboxylase, it is an aromatic amino acid decarboxylase in plants, capable of decarboxylating other aromatic amino acids such as tyramine, making this a complicated and highly controlled procedure (Erland et al., 2017). After converting tryptophan to tryptamine in the indoleamine pathway, tryptamine-5-hydroxylase (T-5-H) transforms tryptamine into serotonin (Akula et al., 2018). Serotonin is subsequently acetylated by serotonin-N-acetyltransferase (SNAT) to produce N-acetyl serotonin (NAS), which is ultimately transformed into melatonin by acetyl serotonin-O-methyltransferase (ASMT). Melatonin and serotonin perform critical functions in the plant life cycle, including germination, seedling development, vegetative growth, floral patterning and reproductive growth, as well as alleviating a wide range of biotic and abiotic challenges (Akula et al., 2018) (Figure 10.1). They are well-recognized as powerful antioxidants that may give favourable physiological responses to environmental challenges since they operate as direct antioxidants as well as upregulate other antioxidant pathways. Melatonin, as a pleiotropic signalling molecule, has been demonstrated to engage in a variety of abiotic stress responses. Furthermore, it is an effective scavenger of both reactive oxygen and reactive nitrogen species (Arnao & Hernández-Ruiz, 2018, 2019a,b). Melatonin may be found in a variety of species in different organs such as leaves, roots, stems, petals, flower buds, fruits and seeds. It is generally synthesized in chloroplasts and mitochondria (Tan & Reiter, 2020). It also serves as an indirect regulator by controlling gene expression through stress-responsive transcription factors. Melatonin is also an auxin-like

* Equally contributed as the first author.

 DOI: 10.1201/9781003335788-14

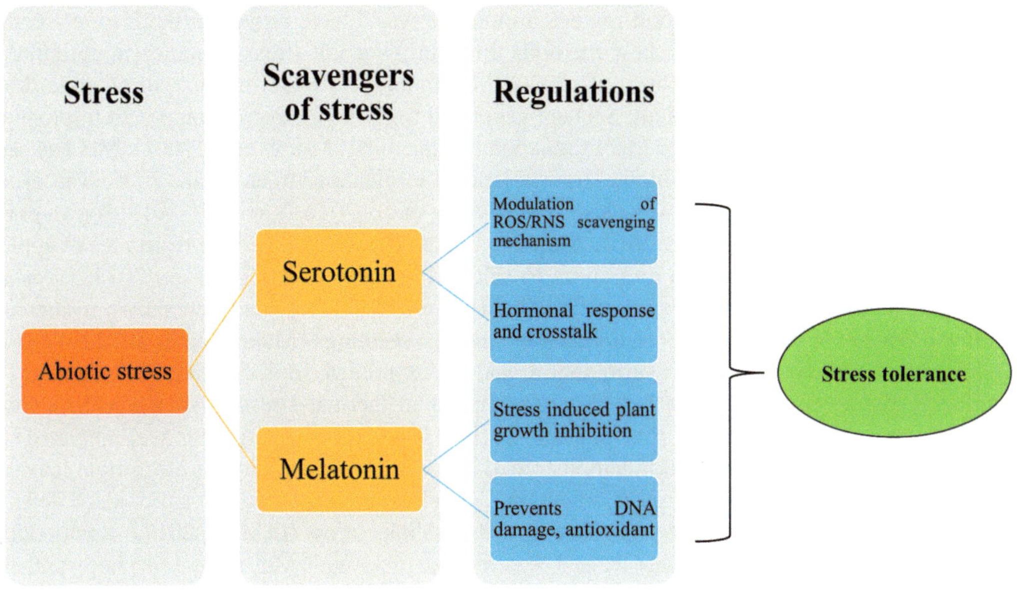

FIGURE 10.1 Basic events during serotonin and melatonin-modulated abiotic stress.

regulator, as it shares a precursor molecule with auxin. Melatonin has received increased scientific interest in recent years due to the rising adverse consequences of climate change, soil salinization and industrial pollution on agriculture, crop output and food chain security. Significant progress has been made in the knowledge of melatonin in plants during the last decade.

As a result, an updated evaluation of current results in indoleamine-mediated abiotic stress tolerance is required. Here, this chapter outlines the functions and effects of indoleamines (melatonin and serotonin) against numerous abiotic stimuli, as well as their applications for enhancing plant stress tolerance.

10.2 PLANT STRESS AND ROLES OF INDOLEAMINES

As a first line of defence against stress, plants have specialized mechanical structures. However, for interaction, detection and adoption, plants produce signalling molecules that enable them to recognize stress and adjust their cellular metabolic machinery accordingly. These signaling molecules work as ligands that bind to particular receptors and are diverse in their structure and function. Despite having a structural resemblance to tryptophan and its precursors, melatonin and serotonin have been detected in higher concentrations in plants subjected to severe conditions. Plants are more tolerant of stressful situations when there is an endogenous increase or exogenous application (Rhodes & Nadolska-Orczyk, 2001).

10.2.1 Chemical Stress

10.2.1.1 Salt Stress

Soil salinity is one of the key environmental stressors that reduce agricultural plant yields (Godfray et al., 2010; Zörb et al., 2019). Worldwide, salt-affected irrigated agricultural land is anticipated to increase from 33% to 50% in the next decade (Jamil et al., 2011; Hasanuzzaman et al., 2013). Increasing soil salinity can substantially reduce food production, as agricultural land produces nearly one-third of the world's food (Godfray et al., 2010; Tilman et al., 2001; Foley et al., 2005). Despite several attempts at water management methods to reduce soil salinity, high salt concentrations in arid and semi-arid regions around the world have proven difficult to overcome, necessitating the development of new methods for improving salt stress tolerance in agricultural plants (Hernández, 2019; van Zelm et al., 2020). Melatonin (N-acetyl-5-methoxytryptamine; Mel) and serotonin (5-hydroxytryptophan; Ser) are generated from the aromatic amino acid tryptophan, with Ser serving as a precursor for Mel production (Reiter, 1991; Murch et al., 2001). Mel has been linked to improved abiotic and biotic stress tolerance in plants (Afreen et al., 2006; Tan et al., 2007a,b; Li et al., 2012; Tiryaki & Kelesh, 2012; Wang et al., 2012; Yin et al., 2013; Bajwa et al., 2014; Li et al., 2017a; Zhao et al., 2018). Mel, in particular, is important in the resistance of apples, soybean, cucumber, watermelon and citrus to salt and drought stress (Li et al., 2012; Li et al., 2017a; Kostopoulou et al., 2015; Lee & Back, 2019). Salinity stress has been shown to increase Mel and Ser levels in sunflower (*Helianthus annuus* L.) seedlings (Mukherjee et al., 2014). Both indoleamines greatly increased germination and were related to differential localization in sunflower tissues, a feature that has also been linked to thermal stress tolerance in St. John's wort (*Hypericum perforatum* L.) (Lee & Back, 2019;). Diverse mechanisms for Mel-mediated salt stress tolerance in plants have been demonstrated, including the prevention of electrolyte leakage, lipid peroxidation, and chlorophyll degradation, as well as the induction of antioxidant defence systems to prevent tissue damage in plants exposed to salinity stress (Li et al., 2017a; Kostopoulou et al., 2015). Pre-treatment of salt-stressed watermelon roots with Mel increased photosynthetic rate by preventing stomatal closure and enhancing light energy absorption and electron transport in photosystem II. Mel also reduced the damage caused by salt-induced oxidative stress by improving redox equilibrium and activating antioxidant enzymes (Li et al., 2017a). Similarly, in salt-stressed rapeseed (*Brassica napus* L.) seedling roots, the application of Mel in conjunction with nitric oxide-releasing compounds reversed seedling growth inhibition and re-established ion homo-eostasis, which was associated with reduced ROS production, a decrease in Na^{+}/K^{+} ratio and modulation of the antioxidant defence genes (Zhao et al., 2018). Multiple signaling pathways are involved in salinity responses, which are integrated into a complex regulatory network that includes the ABA metabolic pathway (Tuteja, 2007; Sah et al., 2016; Zhang et al., 2020). Under high salinity and drought conditions, ABA biosynthesis increases dramatically and mediates the expression of transcription factor genes involved in biotic and abiotic stress responses, including those responding to water dehydration stress in an ABA-dependent and independent manner (Ding et al., 2011; Rock, 2000; Yamaguchi & Blumwald, 2005). Based on Mel's active function in salt stress alleviation, it was suggested that Mel and Ser may effectively reduce salt stress in *Arabidopsis* and that pre-treatment with these compounds will result in changes in the transcription of various drought, salinity and ABA-related genes.

Mel response appears to be mediated by ABA-responsive and antioxidant defence genes, whereas Ser response appears to be mediated by antioxidant defence genes in an ABA-independent way. These protective actions of indoleamines showed that these chemicals operate as inductive signals in preconditioning the plant to resist abiotic stress, a process comparable to the involvement of Mel and Ser in plant morphogenesis and heat stress (Erland et al., 2018).

10.2.1.2 Metal Stress

Heavy metals like copper and zinc are required for regular plant development, but excessive amounts of heavy metals are harmful. Heavy metals also promote oxidative degradation of biomolecules by beginning free radical-mediated chain reactions that culminate in lipid peroxidation, protein oxidation and nucleic acid oxidation. Mel and its precursors may bind to a variety of hazardous metals, including aluminium with Mel, tryptophan and Ser; cadmium with Mel and tryptophan; copper with Mel and Ser; Fe^{3+} with Mel and Ser; Fe^{2+} with tryptophan only; lead with Mel, tryptophan and Ser; and zinc with Mel and tryptophan (Limson et al., 1998). According to electrochemical experiments, Mel may bind to both Cu^{2+} and Cul^{+}, causing free radical damage (Parmar et al., 2002). Mel has been shown to affect biological systems not just by directly quenching free radicals, but also by chelating harmful metals (Flora et al., 2013), the possible links between Mel supplementation and plant environmental tolerance have been documented. The tolerance of pea plants to copper contamination is boosted to a large amount by the addition of Mel to the soil in pea plants treated with high levels of copper in the soil, increasing their chances of survival (Tan et al., 2007a,b). *Brassica oleracea* L. seeds are protected against copper, a particularly poisonous metal when incubated with exogenous Mel solutions (Posmyk et al., 2008). Chemical stress stimulates Mel production in barley roots, greatly increasing their Mel concentration. Cai et al. (2017) show that the transcription factor heat-shock factor Ala (HsfAla) confers cadmium tolerance to tomato plants, in part by inducing Mel biosynthesis during Cd stress. Ser has also been demonstrated to be beneficial in indirectly decreasing the effects of heavy metal exposure in plants. It has a relatively high potential for binding cadmium to form stable complexes.

10.2.2 Environmental Stress

10.2.2.1 Drought Stress

Plant growth is hampered by drought owing to photosynthetic decline. Excess sodium ions interfere with important metabolic processes, resulting in the negative consequences of excessive salt (Zhang & Blumwald, 2001). Mel up-controls the expression of C-repeat-binding factors. Mel buildup is caused by drought-priming increased antioxidant capability in both chloroplasts and mitochondria. Melatonin's drought stress mitigation potential and specific mechanisms of Mel-induced glutathione and ascorbic acid accumulation, along with Mel treatment, significantly increased drought tolerance of wheat seedlings, as evidenced by increased antioxidant capacity and decreased endogenous ROS level (Cui et al., 2017) Furthermore, Mel treatment of maize and cucumber seeds improved seedling growth and the yield of plants developed from them, particularly those subjected to water stress (Zhang et al., 2013). When compared to wild kinds, the Mel-enriched transgenic *Arabidopsis* was more drought tolerant (Zuo et al., 2014). Exogenous Mel also increased tomato plant resistance to alkaline stress (Liu et al., 2015). Applying Mel to tomato plants improves root vigour, lowers stress-related damage to PSII response centres, mitigates the detrimental impact of dryness by regulating the antioxidant system and reduces the plant's cellular content of harmful chemicals (Sun et al., 2021; Altaf et al., 2022). Mel application to apple (*Malus domestica* Borkh. cv. Golden Delicious and cv. Hanfu) leaves reduces drought-induced photosynthetic inhibition (Wang et al., 2012; Wang et al., 2013). It has been seen before that melatonin plays a role in protecting against chlorophyll deterioration, photosynthetic capability and stomata configurations under other stresses like drought and heat (Wang et al., 2012; Xu et al., 2016).

10.2.2.2 Heat Stress

Extreme temperatures degrade or even kill plants by interfering with membrane fluidity and enzyme function. Heat stress treatment significantly increased endogenous Mel levels in *Arabidopsis* leaves, and exogenous Mel treatment enhanced thermo tolerance in *Arabidopsis*, showing the involvement of HSFAIs-activated heat-responsive genes in Mel-mediated thermo tolerance in *Arabidopsis* (Shi et al., 2015). Mel promotes class Al heat-shock factors (HSFAIs), which may have a role in thermal tolerance in *Arabidopsis*. Mel also improves cellular protein preservation in tomato plants by inducing heat-shock proteins (HSPS) and autophagy to refold or destroy denatured proteins under heat stress. Furthermore, HsfA up-regulates Mel biosynthesis in tomato plants to provide cadmium resistance (Cai et al., 2017).

10.2.2.3 Radiation Stress

Radiation stress damages macromolecules such as DNA and proteins, as well as produces ROS and interferes with biological activities. Reversing the inhibitory effects of light and high temperature can significantly reduce the germination of photosensitive and thermosensitive species. Mel identifies and reacts to UV light in Benth seeds (Tiryaki & Keles, 2012). UV produces stress and needs Mel defence measures like increased pigment production and antioxidant activity at high levels. At low levels,

UV is a useful indication for detecting the direction of sunlight (Afreen et al., 2006; Byeon & Back, 2014). Mel may also have a function in light sensing, possibly through COP (constitutive photomorphogenesis protein) receptors, as this pathway has been shown to interact with the UV receptor UVRS (Sanchez-Barcelo et al., 2016). Mel resistance to photodamage and UV-B has also been observed in *Glycyrrhiza* (Afreen et al., 2006) and *Eichhornia* (Tan et al., 2007c).

10.2.2.4 Cold Stress

Cold stress is a major abiotic stressor that limits crop growth, productivity and plant geographical dispersal, particularly in temperate and high-elevation environments (Andaya & Mackill, 2003). It also affects many physiological, biochemical, molecular and metabolic processes, such as membrane fluidity, enzyme activity and metabolism balance (Bajwa et al., 2014). According to a study, exogenous Mel confers cold tolerance in plants. Cold-induced oxidative stress was alleviated by increased Mel accumulation and antioxidant enzyme activity activation in distant untreated tissues. The effects of foliar and rhizospheric Mel pre-treatment on cold stress resistance in untreated leaves and roots were studied (Li et al., 2017b). Mel given to *Vigna radiata* seedlings has been shown to boost plant resistance mechanisms during cold acclimation by enhancing the activity of phenylalanine ammonia lyase (PAL) following re-warming. Cucumber seedlings treated with Mel germinated more quickly after chilling stress (Posmyk et al., 2009). It has an intriguing function as a defender of cold-induced apoptosis in carrot suspension cells (Lei et al., 2004). Pre-treatment with melatonin reduced apoptosis induced by cold temperature in cultured carrot suspension cells (Lei et al., 2004). Mel therapy nearly reduced cold stress-induced shrinking and disruption of carrot cell plasma membranes. According to previous research, melatonin treatment can improve cold tolerance in *Arabidopsis* (Bajwa et al., 2014), wheat (Turk et al., 2014), Bermuda grass (Hu et al., 2016) and *Elymus nutans* (Fu et al., 2017). Recently, the potential involvement of Mel in rice development, photosynthesis and cold stress response has been studied (Han et al., 2017). Also, pre-treatment reduced the negative effects of cold stress and hastened recovery, mostly by increasing the photosynthetic and antioxidant capacity in melon leaves (Zhang et al., 2017).

10.2.3 Role of Indoleamines in Other Stresses

Waterlogging is a situation brought on by too much water or flooding that results in anaerobic conditions in the surrounding area and the roots. Plants grow adventitious roots and aerenchyma tissues in wet conditions to build up the air in the tissues and counteract the effects of anoxia (Kar, 2011; Nishiuchi et al., 2012; Osakabe et al., 2014). Serotonin and melatonin have been discovered to protect plants from several stressors, and the mechanism for this is ROS hunting. Increased ROS production has been observed in plants during drought and wet conditions, and it is thought that this is crucial for plant survival (Cruz de Carvalho, 2008). There are many distinct types of ROS, but the most prevalent ones include nitric oxide (NO), superoxide anion (O•22), hydrogen peroxide (H_2O_2), hydroxyl radical (•OH), singlet oxygen ($^1O2$) and peroxynitrite anion (ONOO2). And melatonin, an amphiphilic substance, may be able to eliminate all of these from cells (Tan et al., 2013). According to a study by Zheng et al., (2017), melatonin irrigation at the roots or spraying on leaves was successful in alleviating the waterlogging stress in apple (*M. baccata* (Linn.) Borkh.) seedlings. They noticed a marked reduction in leaf chlorosis brought on by waterlogging stress. Their follow-up research revealed a substantial increase in endogenous melatonin production under the stress of waterlogging. The stress of waterlogging inhibits the antioxidant enzymes. Exogenous melatonin supplementation, however, is effective in reviving their activity (Zheng et al., 2017). Additionally, the principal cause of losses in plant growth during waterlogging conditions caused by anoxia is a nutrient shortage (Steffens et al., 2005). In response to this, Li et al. (2016b) discovered that melatonin effectively buffers the nutrient shortage (e.g., K and Na).

Tables 10.1 and 10.2 summarize the roles of Mel and Ser under different abiotic stresses.

TABLE 10.1
Role of melatonin in plants under abiotic stresses

Abiotic Stress	Effect of Melatonin	References
Drought	Regulation of the mRNA expression of various stress-responsive genes, such as COR15A, RD22 and KIN1, sucrose accumulation. Reduced reactive oxygen species (ROS), higher antioxidant metabolism, high concentration of amino acids, organic acids, sugars and sugar alcohols.	Ma et al., 2018
Salt	Increased plant height, leaf length/width and stem diameter. Increased gas exchange parameters and relative water content. Reduced abscisic acid and increased salicylic acid content.	Shukla et al., 2021
Heavy metals	Prevents DNA damage, efficacious antioxidant and decreases superoxide radical accumulation.	Galano et al., 2018; Mansoor et al., 2023
Cold stress 4 °C	Increased accumulation of MeJA and H_2O_2, Increased tolerance	Talaat et al., 2021
Paraquat	Increased photosynthetic pigments, improved functioning of the photosynthetic apparatus. Increased water content	Szafrańska et al., 2017
pH stress	Enhanced antioxidants Pas Content and Increased MDA	Gong et al., 2017
High temperature	Down-regulation of ABA biosynthesis genes Decreased oxidative Stress parameters	Zhang et al., 2017
Drought + cold	Better water status, modulating Antioxidant systems	Li et al., 2016a
Fluoride	Decreased oxidative stress, Increased Proline accumulation	Banerjee & Roychoudhury, 2019

TABLE 10.2
Role of serotonin in plants under abiotic stresses

Abiotic Stress	Effects of Serotonin	References
Salinity	Alleviates the growth inhibition of seedlings and significantly promotes the accumulation of the fresh and dry weights of roots and shoots. Effectively activates antioxidant enzyme system through improving the catalase (CAT), superoxide dismutase (SOD) and peroxidase (POD) activities, and the expression of *POD7*, *CAT3* and *Cu-SOD* genes was also up-regulated. Possess the ability to scavenge reactive oxygen species, osmotic pressure regulation and promote growth.	Liu et al., 2021
Heavy metals	Indirectly decreases the effects of heavy metal exposure in plants. It has a relatively high potential for binding cadmium to form stable complexes.	Cai et al., 2017
Temperature	Induces the growth and biomass under abiotic stress by regulating gene expression and metabolism associated with auxin-responsive pathways. Increases secondary metabolites without adverse effects on cell division and biomass production	Kumar et al., 2021

10.3 CONCLUSION AND FUTURE PERSPECTIVE

It is well known now about the functions of indoleamines in coping with different abiotic stresses. Two types of indoleamines, i.e., melatonin and serotonin, play an important role in every aspect, covering the crosstalk with other hormones, and their scavenging mechanisms of ROS and RNS. Melatonin is an endogenous free radical scavenger. It may directly scavenge H_2O_2 and help in the maintenance of intracellular H_2O_2 concentrations at steady-state levels. This might be due to the inhibition of H_2O_2 accumulation and enhanced catalase and POD activities by endogenous melatonin supplementation. However, it is still required to focus on its application procedure to get some biotechnological work patented for better results. it is very important to work on signaling aspects of this, as involved in the crosstalk with other hormones and biomolecules. It is still necessary to characterize the molecular network, and transcriptome analysis of the genes governing the regulation of indoleamines in the form of two different compounds i.e., melatonin and serotonin. Various signaling cascades associated with abiotic stress-induced regulation of serotonin and melatonin biosynthesis and their interactions with other biomolecules (hormones) shall provide new areas for research.

REFERENCES

Afreen, F., Zobayed, S. M., & Kozai, T. (2006). Melatonin in *Glycyrrhiza uralensis*: Response of plant roots to spectral quality of light and UV-B radiation. Journal of Pineal Research, 41(2), 108–115. 10.1111/j.1600-079X.2006.00337.x

Akula, R., Gill, S. S., & Ravishankar, G. A. (2018). Protective role of indoleamines (serotonin and melatonin) during abiotic stress in plants. In: A. Ramakrishna, & S. S. Gill (eds.), Metabolic Adaptations in Plants During Abiotic Stress (pp. 221–228). CRC Press.

Altaf, M. A., et al. (2022). Melatonin improves drought stress tolerance of tomato by modulating plant growth, root architecture, photosynthesis, and antioxidant defense system. Antioxidants, 11, 309. 10.3390/antiox11020309.

Andaya, V. C., & Mackill, D. J. (2003). QTLs conferring cold tolerance at the booting stage of rice using recombinant inbred lines from a japonica x indica cross. TAG. Theoretical and Applied Genetics, 106(6), 1084–1090. 10.1007/s00122-002-1126-7

Arnao, M. B., & Hernández-Ruiz, J. (2018). Melatonin and its relationship to plant hormones. Annals of Botany, 121(2), 195–207. 10.1093/aob/mcx114

Arnao, M. B., & Hernández-Ruiz, J. (2019a). Melatonin: A new plant hormone and/or a plant master regulator?. Trends in Plant Science, 24(1), 38–48. 10.1016/j.tplants.2018.10.010

Arnao, M. B., & Hernández-Ruiz, J. (2019b). Role of melatonin to enhance phytoremediation capacity. Applied Sciences, 9(24), 5293. 10.3390/app9245293

Arora, D., & Bhatla, S. C. (2017). Melatonin and nitric oxide regulate sunflower seedling growth under salt stress accompanying differential expression of Cu/Zn SOD and Mn SOD. Free Radical Biology & Medicine, 106, 315–328. 10.1016/j.freeradbiomed.2017.02.042

Bajwa, V. S., Shukla, M. R., Sherif, S. M., Murch, S. J., & Saxena, P. K. (2014). Role of melatonin in alleviating cold stress in *Arabidopsis thaliana*. Journal of Pineal Research, 56(3), 238–245. 10.1111/jpi.12115

Banerjee, A., & Roychoudhury, A. (2019). Melatonin application reduces fluoride uptake and toxicity in rice seedlings by altering abscisic acid, gibberellin, auxin and antioxidant homeostasis. Plant Physiology and Biochemistry, 145, 164–173. 10.1016/j.plaphy.2019.10.033

Byeon, Y., & Back, K. (2014). Melatonin synthesis in rice seedlings in vivo is enhanced at high temperatures and under dark conditions due to increased serotonin N-acetyltransferase and N-acetylserotonin methyltransferase activities. Journal of Pineal Research, 56(2), 189–195. 10.1111/jpi.12111

Cai, S. Y., Zhang, Y., Xu, Y. P., Qi, Z. Y., Li, M. Q., Ahammed, G. J., Xia, X. J., Shi, K., Zhou, Y. H., Reiter, R. J., Yu, J. Q., & Zhou, J. (2017). HsfA1a upregulates melatonin biosynthesis to confer cadmium tolerance in tomato plants. Journal of Pineal Research, 62(2), 10.1111/jpi.12387. 10.1111/jpi.12387

Cruz de Carvalho, M. H. (2008). Drought stress and reactive oxygen species: Production, scavenging and signaling. Plant Signaling & Behavior, 3(3), 156–165. 10.4161/psb.3.3.5536

Cui, G., Zhao, X., Liu, S., Sun, F., Zhang, C., & Xi, Y. (2017). Beneficial effects of melatonin in overcoming drought stress in wheat seedlings. Plant Physiology and Biochemistry, 118, 138–149. 10.1016/j.plaphy.2017.06.014

Ding, Y., Avramova, Z., & Fromm, M. (2011). The *Arabidopsis* trithorax-like factor ATX1 functions in dehydration stress responses via ABA-dependent and ABA-independent pathways. The Plant Journal, 66(5), 735–744. 10.1111/j.1365-313X.2011.04534.x

Erland, L. A. E., Saxena, P. K., & Murch, S. J. (2017). Melatonin in plant signalling and behaviour. Functional Plant Biology, 45(2), 58–69. 10.1071/FP16384

Erland, L. A. E., Shukla, M. R., Singh, A. S., Murch, S. J., & Saxena, P. K. (2018). Melatonin and serotonin: Mediators in the symphony of plant morphogenesis. Journal of Pineal Research, 64(2), 10.1111/jpi.12452. 10.1111/jpi.12452

Flora, S. J., Shrivastava, R., & Mittal, M. (2013). Chemistry and pharmacological properties of some natural and synthetic antioxidants for heavy metal toxicity. Current Medicinal Chemistry, 20(36), 4540–4574. 10.2174/0929867311320990146

Foley, J. A., DeFries, R., Asner, G. P., Barford, C., Bonan, G., Carpenter, S. R., … & Snyder, P. K. (2005). Global consequences of land use. Science, 309(5734), 570–574.

Fu, J., Wu, Y., Miao, Y., Xu, Y., Zhao, E., Wang, J., Sun, H., Liu, Q., Xue, Y., Xu, Y., & Hu, T. (2017). Improved cold tolerance in *Elymus nutans* by exogenous application of melatonin may involve ABA-dependent and ABA-independent pathways. Scientific Reports, 7, 39865. 10.1038/srep39865

Galano, A., Tan, D. X., & Reiter, R. (2018). Melatonin: a versatile protector against oxidative DNA damage. Molecules, 23, 530. 10.3390/molecules23030530.

Godfray, H. C., Beddington, J. R., Crute, I. R., Haddad, L., Lawrence, D., Muir, J. F., Pretty, J., Robinson, S., Thomas, S. M., & Toulmin, C. (2010). Food security: the challenge of feeding 9 billion people. Science, 327(5967), 812–818. 10.1126/science.1185383

Gong, X., Shi, S., Dou, F., Song, Y., & Ma, F. (2017). Exogenous melatonin alleviates alkaline stress in *Malus hupehensis* Rehd. by regulating the biosynthesis of polyamines. Molecules, 22(9), 1542. 10.3390/molecules22091542

Han, Q. H., Huang, B., Ding, C. B., Zhang, Z. W., Chen, Y. E., Hu, C., Zhou, L. J., Huang, Y., Liao, J. Q., Yuan, S., & Yuan, M. (2017). Effects of melatonin on anti-oxidative systems and photosystem II in cold-stressed rice seedlings. Frontiers in Plant Science, 8, 785. 10.3389/fpls.2017.00785

Hasanuzzaman, M., Nahar, K., & Fujita, M. (2013). Plant response to salt stress and role of exogenous protectants to mitigate salt-induced damages. In: P. Ahmad, M. Azooz, & M. Prasad (eds.), Ecophysiology and Responses of Plants under Salt Stress (pp. 25–87), Springer, New York, NY. 10.1007/978-1-4614-4747-4_2

Hernández J. A. (2019). Salinity Tolerance in Plants: Trends and Perspectives. International Journal of Molecular Sciences, 20(10), 2408. 10.3390/ijms20102408

Hu, Z., Fan, J., Xie, Y., Amombo, E., Liu, A., Gitau, M. M., Khaldun, A. B. M., Chen, L., & Fu, J. (2016). Comparative photosynthetic and metabolic analyses reveal mechanism of improved cold stress tolerance in bermudagrass by exogenous melatonin. Plant Physiology and Biochemistry, 100, 94–104. 10.1016/j.plaphy.2016.01.008

Jamil, A., Riaz, S., Ashraf, M., & Foolad, M. R. (2011). Gene expression profiling of plants under salt stress. Critical Reviews in Plant Sciences, 30(5), 435–458.

Kar, R. K. (2011). Plant responses to water stress: role of reactive oxygen species. Plant Signaling & Behavior, 6(11), 1741–1745. 10.4161/psb.6.11.17729

Kostopoulou, Z., Therios, I., Roumeliotis, E., Kanellis, A. K., & Molassiotis, A. (2015). Melatonin combined with ascorbic acid provides salt adaptation in *Citrus aurantium* L. seedlings. Plant Physiology and Biochemistry, 86, 155–165. 10.1016/j.plaphy.2014.11.021

Kumar, G., Saad, K. R., Arya, M., Puthusseri, B., Mahadevappa, P., Shetty, N. P., & Giridhar, P. (2021). The synergistic role of serotonin and melatonin during temperature stress in promoting cell division, ethylene and isoflavones biosynthesis in glycine max. Current Plant Biology, 26, 100206. 10.1016/j.cpb.2021.100206.

Lee, H. J., & Back, K. (2019). 2-Hydroxymelatonin confers tolerance against combined cold and drought stress in tobacco, tomato, and cucumber as a potent anti-stress compound in the evolution of land plants. Melatonin Research, 2(2), 35–46.

Lei, X. Y., Zhu, R. Y., Zhang, G. Y., & Dai, Y. R. (2004). Attenuation of cold-induced apoptosis by exogenous melatonin in carrot suspension cells: the possible involvement of polyamines. Journal of Pineal Research, 36(2), 126–131. 10.1046/j.1600-079x.2003.00106.x

Li, C., Wang, P., Wei, Z., Liang, D., Liu, C., Yin, L., Jia, D., Fu, M., & Ma, F. (2012). The mitigation effects of exogenous melatonin on salinity-induced stress in *Malus hupehensis*. Journal of Pineal Research, 53(3), 298–306. 10.1111/j.1600-079X.2012.00999.x

Li, X., Tan, D. X., Jiang, D., & Liu, F. (2016a). Melatonin enhances cold tolerance in drought-primed wild-type and abscisic acid-deficient mutant barley. Journal of Pineal Research, 61(3), 328–339. 10.1111/jpi.12350

Li, C., Liang, B., Chang, C., Wei, Z., Zhou, S., & Ma, F. (2016b). Exogenous melatonin improved potassium content in *Malus* under different stress conditions. Journal of Pineal Research, 61(2), 218–229. 10.1111/jpi.12342

Li, H., Chang, J., Chen, H., Wang, Z., Gu, X., Wei, C., Zhang, Y., Ma, J., Yang, J., & Zhang, X. (2017a). Exogenous melatonin confers salt stress tolerance to watermelon by improving photosynthesis and redox homeostasis. Frontiers in Plant Science, 8, 295. 10.3389/fpls.2017.00295

Li, H., et al. (2017b). Local melatonin application induces cold tolerance in distant organs of Citrullus lanatus L. via long distance transport. Scientific Reports, 7. 10.1038/srep40858.

Limson, J., Nyokong, T., & Daya, S. (1998). The interaction of melatonin and its precursors with aluminium, cadmium, copper, iron, lead, and zinc: an adsorptive voltammetric study. Journal of Pineal Research, 24(1), 15–21. 10.1111/j.1600-079x.1998.tb00361.x

Liu, J., Wang, W., Wang, L., & Sun, Y. (2015). Exogenous melatonin improves seedling health index and drought tolerance in tomato. Plant Growth Regulation, 77, 317–326.

Liu, Y., Ding, X., Lv, Y., Cheng, Y., Li, C., Yan, L., Tian, S., & Zou, X. (2021). Exogenous serotonin improves salt tolerance in rapeseed (*Brassica napus* L.) seedlings. Agronomy, 11(2), 400. 10.3390/agronomy11020400

Ma, X., Zhang, J., Burgess, P., Rossi, S., & Huang, B. (2018). Interactive effects of melatonin and cytokinin on alleviating drought-induced leaf senescence in creeping bentgrass (*Agrostis stolonifera*). Environmental and Experimental Botany, 145, 1–11.

Mansoor, S., et al. (2023). Heavy metal induced oxidative stress mitigation and ROS scavenging in plants. Plants, 12, 3003. 10.3390/plants12163003.

Mukherjee, S., David, A., Yadav, S., Baluška, F., & Bhatla, S. C. (2014). Salt stress-induced seedling growth inhibition coincides with differential distribution of serotonin and melatonin in sunflower seedling roots and cotyledons. Physiologia Plantarum, 152(4), 714–728. 10.1111/ppl.12218

Murch, S. J., Campbell, S. S., & Saxena, P. K. (2001). The role of serotonin and melatonin in plant morphogenesis: regulation of auxin-induced root organogenesis in in vitro-cultured explants of St. John's wort (*Hypericum perforatum* L.). In Vitro Cellular & Developmental Biology-Plant, 37, 786–793.

Nishiuchi, S., Yamauchi, T., Takahashi, H., Kotula, L., & Nakazono, M. (2012). Mechanisms for coping with submergence and waterlogging in rice. Rice (New York, N.Y.), 5(1), 2. 10.1186/1939-8433-5-2

Osakabe, Y., Osakabe, K., Shinozaki, K., & Tran, L. S. (2014). Response of plants to water stress. Frontiers in Plant Science, 5, 86. 10.3389/fpls.2014.00086

Parmar, P., Limson, J., Nyokong, T., & Daya, S. (2002). Melatonin protects against copper-mediated free radical damage. Journal of Pineal Research, 32(4), 237–242. 10.1034/j.1600-079x.2002.01859.x

Posmyk, M. M., Kuran, H., Marciniak, K., & Janas, K. M. (2008). Presowing seed treatment with melatonin protects red cabbage seedlings against toxic copper ion concentrations. Journal of Pineal Research, 45(1), 24–31. 10.1111/j.1600-079X.2007.00552.x

Posmyk, M. M., Bałabusta, M., Wieczorek, M., Sliwinska, E., & Janas, K. M. (2009). Melatonin applied to cucumber (*Cucumis sativus* L.) seeds improves germination during chilling stress. Journal of Pineal Research, 46(2), 214–223. 10.1111/j.1600-079X.2008.00652.x

Rehaman, A., Mishra, A. K., Ferdose, A., Per, T. S., Hanief, M., Jan, A. T., & Asgher, M. (2021). Melatonin in plant defense against abiotic stress. Forests, 12, 1404.

Reiter R. J. (1991). Pineal melatonin: cell biology of its synthesis and of its physiological interactions. Endocrine Reviews, 12(2), 151–180. 10.1210/edrv-12-2-151

Rhodes, D., & Nadolska-Orczyk, A. (2001). Plant stress physiology. Encyclopedia of Life Science.

Rock C. D. (2000). Tansley Review No. 120: Pathways to abscisic acid-regulated gene expression. The New Phytologist, 148(3), 357–396. 10.1046/j.1469-8137.2000.00769.x

Sah, S. K., Reddy, K. R., & Li, J. (2016). Abscisic Acid and Abiotic Stress Tolerance in Crop Plants. Frontiers in Plant Science, 7, 571. 10.3389/fpls.2016.00571

Sanchez-Barcelo, E. J., Mediavilla, M. D., Vriend, J., & Reiter, R. J. (2016). Constitutive photomorphogenesis protein 1 (COP1) and COP9 signalosome, evolutionarily conserved photomorphogenic proteins as possible targets of melatonin. Journal of Pineal Research, 61(1), 41–51. 10.1111/jpi.12340

Shi, H., Tan, D. X., Reiter, R. J., Ye, T., Yang, F., & Chan, Z. (2015). Melatonin induces class A1 heat-shock factors (HSFA1s) and their possible involvement of thermotolerance in *Arabidopsis*. Journal of Pineal Research, 58(3), 335–342. 10.1111/jpi.12219

Shukla, M. R., Bajwa, V. S., Freixas-Coutin, J. A., & Saxena, P. K. (2021). Salt stress in *Arabidopsis thaliana* seedlings: Role of indoleamines in stress alleviation. Melatonin Research, 4(1), 70–83.
Steffens, D., Hutsch, B. W., Eschholz, T., Losak, T., & Schubert, S. (2005). Water logging may inhibit plant growth primarily by nutrient deficiency rather than nutrient toxicity. Plant Soil and Environment, 51(12), 545.
Sun, C., Liu, L., Wang, L., Li, B., Jin, C., & Lin, X. (2020). Melatonin: A master regulator of plant development and stress responses. Journal of Integrative Plant Biology, 63, 126–145. 10.1111/jipb.12993.
Szafrańska, K., Reiter, R. J., & Posmyk, M. M. (2017). Melatonin Improves the Photosynthetic Apparatus in Pea Leaves Stressed by Paraquat via Chlorophyll Breakdown Regulation and Its Accelerated de novo Synthesis. Frontiers in Plant Science, 8. 10.3389/fpls.2017.00878.
Talaat, N. B. (2021). Polyamine and nitrogen metabolism regulation by melatonin and salicylic acid combined treatment as a repressor for salt toxicity in wheat (*Triticum aestivum* L.) plants. Plant Growth Regulation, 95(3), 315–329.
Tan, D. X., Manchester, L. C., Terron, M. P., Flores, L. J., & Reiter, R. J. (2007a). One molecule, many derivatives: a never-ending interaction of melatonin with reactive oxygen and nitrogen species?. Journal of Pineal Research, 42(1), 28–42. 10.1111/j.1600-079X.2006.00407.x
Tan, D. X., Manchester, L. C., Helton, P., & Reiter, R. J. (2007b). Phytoremediative capacity of plants enriched with melatonin. Plant Signaling & Behavior, 2(6), 514–516. 10.4161/psb.2.6.4639
Tan, D. X., Manchester, L. C., Di Mascio, P., Martinez, G. R., Prado, F. M., & Reiter, R. J. (2007c). Novel rhythms of N^1-acetyl-N^2-formyl-5-methoxykynuramine and its precursor melatonin in water hyacinth: Importance for phytoremediation. The FASEB Journal, 21(8), 1724–1729.
Tan, D. X., Manchester, L. C., Liu, X., Rosales-Corral, S. A., Acuna-Castroviejo, D., & Reiter, R. J. (2013). Mitochondria and chloroplasts as the original sites of melatonin synthesis: a hypothesis related to melatonin's primary function and evolution in eukaryotes. Journal of Pineal Research, 54(2), 127–138. 10.1111/jpi.12026
Tan, D. X., & Reiter, R. J. (2020). An evolutionary view of melatonin synthesis and metabolism related to its biological functions in plants. Journal of Experimental Botany, 71(16), 4677–4689. 10.1093/jxb/eraa235
Tilman, D., Fargione, J., Wolff, B., D'Antonio, C., Dobson, A., Howarth, R., Schindler, D., Schlesinger, W. H., Simberloff, D., & Swackhamer, D. (2001). Forecasting agriculturally driven global environmental change. Science (New York, N.Y.), 292(5515), 281–284. 10.1126/science.1057544
Tiryaki, I., & Keles, H. (2012). Reversal of the inhibitory effect of light and high temperature on germination of *Phacelia tanacetifolia* seeds by melatonin. Journal of Pineal Research, 52(3), 332–339. 10.1111/j.1600-079X.2011.00947.x
Turk, H., Erdal, S., Genisel, M., Atici, O., Demir, Y., & Yanmis, D. (2014). The regulatory effect of melatonin on physiological, biochemical and molecular parameters in cold-stressed wheat seedlings. Plant Growth Regulation, 74, 139–152. 10.1007/s10725-014-9905-0.
Tuteja N. (2007). Abscisic Acid and abiotic stress signaling. Plant Signaling & Behavior, 2(3), 135–138. 10.4161/psb.2.3.4156
van Zelm, E., Zhang, Y., & Testerink, C. (2020). Salt tolerance mechanisms of plants. Annual Review of Plant Biology, 71, 403–433. 10.1146/annurev-arplant-050718-100005
Wang, P., Yin, L., Liang, D., Li, C., Ma, F., & Yue, Z. (2012). Delayed senescence of apple leaves by exogenous melatonin treatment: toward regulating the ascorbate-glutathione cycle. Journal of Pineal Research, 53(1), 11–20. 10.1111/j.1600-079X.2011.00966.x
Wang, P., Sun, X., Li, C., Wei, Z., Liang, D., & Ma, F. (2013). Long-term exogenous application of melatonin delays drought-induced leaf senescence in apple. Journal of Pineal Research, 54(3), 292–302.
Xu, W., Cai, S. Y., Zhang, Y., Wang, Y., Ahammed, G. J., Xia, X. J., Shi, K., Zhou, Y. H., Yu, J. Q., Reiter, R. J., & Zhou, J. (2016). Melatonin enhances thermotolerance by promoting cellular protein protection in tomato plants. Journal of Pineal Research, 61(4), 457–469. 10.1111/jpi.12359
Yamaguchi, T., & Blumwald, E. (2005). Developing salt-tolerant crop plants: challenges and opportunities. Trends in plant science, 10(12), 615–620. 10.1016/j.tplants.2005.10.002
Yin, L., Wang, P., Li, M., Ke, X., Li, C., Liang, D., Wu, S., Ma, X., Li, C., Zou, Y., & Ma, F. (2013). Exogenous melatonin improves *Malus* resistance to *Marssonina* apple blotch. Journal of Pineal Research, 54(4), 426–434. 10.1111/jpi.12038
Zeng, W., Mostafa, S., Lu, Z., & Jin, B. (2022). Melatonin-mediated abiotic stress tolerance in plants. Frontiers in Plant Science, 13, 847175. 10.3389/fpls.2022.847175

Zhang, H. X., & Blumwald, E. (2001). Transgenic salt-tolerant tomato plants accumulate salt in foliage but not in fruit. Nature Biotechnology, 19(8), 765–768. 10.1038/90824

Zhang, N., Zhao, B., Zhang, H. J., Weeda, S., Yang, C., Yang, Z. C., Ren, S., & Guo, Y. D. (2013). Melatonin promotes water-stress tolerance, lateral root formation, and seed germination in cucumber (*Cucumis sativus* L.). Journal of Pineal Research, 54(1), 15–23. 10.1111/j.1600-079X.2012.01015.x

Zhang, Y. P., Yang, S. J., & Chen, Y. Y. (2017). Effects of melatonin on photosynthetic performance and antioxidants in melon during cold and recovery. Biologia Plantarum, 61(3), 571–578.

Zhang, L., Xie, J., Wang, L., Si, L., Zheng, S., Yang, Y., Yang, H., & Tian, S. (2020). Wheat TabZIP8, 9, 13 participate in ABA biosynthesis in NaCl-stressed roots regulated by TaCDPK9-1. Plant Physiology and Biochemistry, 151, 650–658. 10.1016/j.plaphy.2020.03.039

Zhao, G., Zhao, Y., Yu, X., Kiprotich, F., Han, H., Guan, R., Wang, R., & Shen, W. (2018). Nitric Oxide Is required for melatonin-enhanced tolerance against salinity stress in rapeseed (brassica napus L.) seedlings. International Journal of Molecular Sciences, 19(7), 1912. 10.3390/ijms19071912

Zheng, X., Zhou, J., Tan, D. X., Wang, N., Wang, L., Shan, D., & Kong, J. (2017). Melatonin improves waterlogging tolerance of *Malus baccata* (Linn.) Borkh. seedlings by maintaining aerobic respiration, photosynthesis and ROS migration. Frontiers in Plant Science, 8, 483. 10.3389/fpls.2017.00483

Zörb, C., Geilfus, C. M., & Dietz, K. J. (2019). Salinity and crop yield. Plant Biology (Stuttgart, Germany), 21 Suppl 1, 31–38. 10.1111/plb.12884

Zuo, B., Zheng, X., He, P., Wang, L., Lei, Q., Feng, C., Zhou, J., Li, Q., Han, Z., & Kong, J. (2014). Overexpression of *MzASMT* improves melatonin production and enhances drought tolerance in transgenic *Arabidopsis thaliana* plants. Journal of Pineal Research, 57, 408–417. 10.1111/jpi.12180.

11 Jasmonic Acid

A Critical Player in Abiotic Stress

Ayyagari Ramlal, Richa Jain, Naina Miglani, Ananya Anurag Anand, Apoorva Verma, Nisha Sogan, Sourav Singh Deo, and Aparna Nautiyal

11.1 INTRODUCTION

To feed a rapidly growing population, resources, mainly food, are required to feed and sustain life and as per studies, the output must be increased by 70% for a supplementary 2.3 billion people by 2050 (Wani & Sah, 2014; Medendorp et al., 2022). The biotic (parasites, pathogens, herbivores etc.) and abiotic stresses (salt, drought, chill etc.) affect plants' growth and metabolism to a considerable extent. Modern agriculture faces numerous abiotic stresses such as salinity, drought, chilling, and heat and biotic stresses as major constraints affecting crop yield (Fahad et al., 2015). Plants cannot simply relocate to a different habitat to escape abiotic pressures. Therefore, they have advanced mechanisms to subsidize the offensive stressful environment by modifying their developmental and physiological mechanisms (Raza et al., 2021a). The developments include changes in morphological and growth patterns (growth flexibility) as well as physiological and biochemical processes against several stresses (Maggio et al., 2018; Karlova et al., 2021). Phytohormones, generally regarded as plant growth regulators, are compounds that are derived from precursors through various pathways and can act either regionally (at the site of their synthesis) or transported to other locations within the plant body to intervene in growth and development responses of the plant under the ambient and stressful conditions (Bhattacharya, 2021).

Jasmonic acid (JA) is lipid-derived cyclopentenone and with its derivatives referred to as jasmonates (methyl jasmonates, MeJA)) (Koo, 2018; Eng et al., 2021). It is an important molecule in the regulation of many physiological processes in plant growth and development, especially the mediation of plant responses to biotic and abiotic stress (Lalotra et al., 2020; Jha et al., 2022). JA has a biosynthesis that must be understood to fully comprehend how it works. This stress-related hormone is a cyclopentane fatty acid and is synthesized from alpha-linolenic acid via a pathway called an octadecanoid pathway, three organelles are involved in this biosynthetic pathway, in plastid α-linolenic acid is first converted to 12-oxophytodienoic acid (OPDA) which undergoes β-oxidation to form JA in peroxisome. This JA then exits the peroxisome and gets metabolized to JA-isoleucine and methyl jasmonates. This octadecanoid pathway is generally catalyzed by enzymes such as phospholipase, lipoxygenase, allene oxide synthase, allene oxide cyclase and lastly oxophytodienoic acid reductase. JA typically participates in the physiological and molecular responses to abiotic stresses such as activation of the antioxidant system such as superoxide anion radical, peroxidase, NADPH oxidase etc. (Patel et al., 2022), accumulation of amino acids (isoleucine and methionine) (Ku et al., 2018), soluble sugars and regulation of the stomatal opening and closing. The expression of JA-associated genes (*JAZ, AOS1, AOC, LOX2* and *COI1*) (Hu et al., 2017; Hossain et al., 2022) interactions with other plant hormones (ABA, ET, SA, GA, IAA and BR) (Yang et al., 2019) and interactions with TFs (MYC2 and bHLH148) are involved in the molecular responses (Asif et al., 2021).

DOI: 10.1201/9781003335788-15

The exogenous application shows inhibition of primary root growth, leaf expansion and hypocotyl elongation and thus a significant reduction in seedling growth. While plants such as rice (Kang et al., 2005), barley (Walia et al., 2007) and wheat (Qiu et al., 2014) showed relief from salt injury upon its exogenous application. JA also responded in case of heavy metal stress in plants such as *Arabidopsis thaliana* (L.) Heynh. (thereon *Arabidopsis*) (Maksymiec et al., 2005; Poonam et al., 2013) *Wolffia arrhiza* (L.) Horkel ex Wimm. (Piotrowska et al., 2009) and *Cajanus cajan* (L.) Huth (Poonam et al., 2013). It lightens the effect of heavy metal stress by preventing biosorption, plant growth restoration and primary metabolite formation (Piotrowska et al., 2009). JA collaborates with other phytohormone signalling pathways in complex network crosstalk but does not act autonomously (Raza et al., 2021a). A lot of literature has been published throughout the years regarding its association with other phytohormones through the integration of regulatory transcription factors (TFs) and associated genes. The JAZ-MYC module contributes significantly to the JA signalling pathway, together with other plant hormones like salicylic acid, ethylene and abscisic acid (ABA). Treatment of ABA on *Arabidopsis* showed increased expression of JA biosynthesis (Kim et al., 2021). However, JA and ethylene antagonized and coordinately regulated plant stress responses (Zhu, 2014; Zhu & Lee, 2015). JA works both synergistically and antagonistically to help plants endure environmental stress (Jang et al., 2020).

The present chapter focuses on JA functions in pathways that mediate responses to abiotic stress. In addition, crosstalk between diverse phytohormones which participate in complex signalling networks with JAs in response to abiotic stress is also briefed.

11.2 ROLE OF JA IN DIFFERENT TYPES OF STRESS

11.2.1 Salinity

Basic metabolism in plants may be hampered by salt stress, which also causes oxidative stress, starvation, membrane abnormalities and genotoxicity (Wang et al., 2020; Kumar et al., 2021). By raising the levels of antioxidant molecules and the activity of antioxidant enzymes, JA may promote tolerance to salt stress. JAs can also increase salt tolerance by boosting endogenous hormones. JA levels rise during the initial stage of salt stress and may play an indirect role in the indirect suppression of leaf development in salt-sensitive genotypes (Aslam, et al., 2021). Qiu et al. (2014) showed that 3 days of exposure to exogenous JA (2 mM) dramatically reduced the levels of malondialdehyde (MDA) and hydrogen peroxide in wheat seedlings, increasing their ability to withstand salt stress (150 mM). Also, it was observed that the transcription and the activities of superoxide dismutase (SOD), peroxidase (POD), catalase (CAT) and ascorbate peroxidase also dramatically elevated (Qiu et al., 2014). These findings suggest that JA might improve salt stress tolerance by raising antioxidant enzymatic activities.

11.2.2 Drought

In many significant agricultural regions throughout the world, climate change is contributing to global warming and more frequent and/or extreme drought episodes (Isah, 2019). One of the main causes of crop yield decline and even crop failure is the effect of drought stress on crops, which can reduce yields of numerous crops by more than 50%. In general, drought stress retards growth, photosynthetic rates and leaf withering. In addition, oxidative processes, membrane lipid buildup and the production of antioxidant enzymes can all be triggered by drought stress.

By controlling stomatal opening and closing in *Arabidopsis*, JA can reduce water loss. Following a drought stress, endogenous JA concentrations rise quickly and if the stress is sustained, return to baseline levels. In addition, several genes and TFs linked to drought stress are expressed after drought stress (Balbi & Devoto, 2008; Ali & Baek, 2020). The JA signalling pathway has regulators, usually repressors, called jasmonate ZIM-domain proteins (JAZ). Fu et al.

(2017) have shown that *OsJAZ1*, specifically the ABA and JA signalling pathways, plays a negative regulatory role in rice drought stress tolerance (Fu et al., 2017).

A basic helix-loop-helix protein (*OsbHLH148*) was discovered by Seo et al. (2011) and operates as a transcriptional regulator and upregulates *OsDREB1* and *OsJAZ*, which are associated with the JA signalling pathway and responses to drought stress, respectively. Similarly, Yun-Xia et al. (2010) demonstrated that temporary JA buildup in a drought-tolerant *Prunus armeniaca* L. might encourage leaf senescence, stop excessive water loss and increase plant survival in soil-dry conditions. On the other hand, the exogenous administration of JAs could reduce the harm caused by drought stress in *P. armeniaca*. MeJA applied topically to soybean leaves can boost their ability to withstand water stress and reveal higher concentrations of sugars, phenolic compounds and flavonoids. These findings suggest that a plant's ability to tolerate drought stress is influenced by both endogenous and exogenous JAs.

11.2.3 Freezing

There are two different kinds of cold stress depending on the temperature: chilling stress, which is applied at low positive temperatures around +5 and freezing stress, which is applied at low negative temperatures below 0 (Kolaksazov et al., 2013). Plants have developed a variety of strategies for tolerating cold stress. A multitude of stress phytohormones and plant growth regulators, including ABA, JA, SA and ethylene can mediate these pathways. It is widely known that stress phytohormones function by initiating pathways involving phospho-protein cascades, which in turn cause the expression of genes essential for developing a resistance to cold stress.

Cell dehydration and extracellular ice crystal production are the results of low temperature or cold stress. The response of plants to cold stress involves JA signalling in a significant way (Yang et al., 2019). After bananas were kept in the cold, the MYC TFs and many genes (*MaCBF1*, *MaCBF2*, *MaKIN2*, *MaCOR1*, *MaRD2*, *MaRD5* etc.) that respond to cold were activated (*Musa acuminata*). By boosting the production of antioxidants and activating some defence compounds, MeJA could reduce the effects of cold stress on tomato, loquat, pomegranate, mango, guava, cowpea and peach (e.g., phenolic compounds and other heat shock proteins).

Allene oxide synthase1 (*AOS1*), *DAD1*, allene oxide cyclase (*AOC*), *LOX2* and *AOS1* are among the JA synthesis-related genes whose expressions are induced under low-temperature conditions (Hu et al., 2017). This bioactive JA-Ile is then produced, activating the JA receptor COI1 to bind to JAZ1, leading to JAZ1's degradation. To increase plant cold tolerance, the ICE-CBF transcriptional regulation cascade signalling pathway is then triggered and the expression of cold-regulated genes is stimulated. These findings imply that JAs can reduce cold harm by supporting the antioxidant system and active defence components.

11.2.4 Micronutrient Toxicity

According to many reports, JAs can shield plants from the negative consequences of micronutrient poisoning (Ali & Baek, 2020). As observed in tomato, wheat, barley and apple rootstock, a high boron concentration is detrimental to plant growth and development. Through the activation of the antioxidant defence enzymes (CAT, POD and SOD) and the inhibition of lipid peroxidation, exogenous MeJA treatment may be able to reduce the toxicity of boron in plants. Additionally, JAs are essential for plants to respond defensively to lead (Pb) stress. When seeds were primed with JA, Pb absorption was reduced and tomato plant development was boosted.

11.2.5 Heavy Metals

The modernizing agri-industry has caused an increased level of heavy metals in soil which has caused toxicity not just in plants but has accumulated and affected humans as well. Plants have ways to deal with these and amongst all the methods, the role of JA is very well studied (Raza et al., 2021b). Under toxicity, JA and MeJA activate antioxidant defence systems in plants, JA prevents absorption of heavy metals like cadmium (Cd), and exogenous application of MeJA causes the plant to accumulate more polyamines in their cell that help to rescue growth inhibition by Cd. JA also helps in synthesizing chlorophyll, enzymes of which are inhibited by Cd. Nickel (Ni) toxicity is a persistent problem that causes the shortening of root and shoot length; however, the exogenous application of JA helps in overcoming Ni-toxicity in soybean seedlings (Sirhindi et al., 2016).

11.2.6 Ozone Stress

Ozone causes severe damage like skin damage, disruption of stomatal opening and closing and burning down inner tissues. However, JA can regulate stomata opening and closing, thus preventing ozone damage to plants. In another case study, JA is found to prevent cell death and hypersensitive response to ozone exposure to plants (Kangasjärvi, 2005).

11.2.7 Light Stress

Not much work has been done on how JA helps plants in dealing with light stress. However, some studies have shown that the JA level in plant cells increases with UVB treatment, thus enhancing the plant defence mechanism, but how high JA helps plants in maintaining the photosynthesis rate in varying light intensities is not well understood (Meena et al., 2019).

11.2.8 Heat Stress

Thermal stress is known for its role in contributing to a plant's reduced agricultural yield as it severely affects the physiology of a crop species, declines photosynthetic rate, influences inhibition of enzymes and elevated respiration rate and also causes membrane damage (Janni, 2020; Li, 2020). The role of JA in thermotolerance has been found significant in several major plant crops. For instance, in rice, it was reported that the application of JA influences the antioxidant system, which not only helps in increasing the rate of spikelet opening but also contributes to improved spikelet flowering rate (Yang, 2020). Additionally, the exogenous treatment of JA on tomato plants also proved useful as it has the potential to rescue tomato plants from undergoing stigma exertion. Moreover, the significance of JA was also found prevalent in perennial ryegrass, where JA works by inducing antioxidant defence machinery and performing osmotic adjustments (Su et al., 2021).

11.2.9 Waterlogging Stress

Another abiotic stress affecting plant growth is waterlogging stress, whose primary effect includes a reduction in the oxygen levels of the soil, which ultimately can lead to the accumulation of toxic compounds, diffusion of gas and altered rates of light intensity (Fukao et al., 2019). A study on the treatment of JA on pepper plants (under the threat of waterlogging stress) highlighted the importance of JA in combating this stress as the application of JA contributed to increasing chlorophyll content, reducing the levels of relative conductivity and improving the activities of enzymes and other osmolytes (Ouli-Jun et al., 2017). In addition to this, JA-induced waterlogging stress tolerance was also reported in other crops, including citrus and soybean, as documented in the study by Raza et al. (2021a).

11.3 CONCLUSION

Future research may identify the components of signalling pathways and the manufacture of JAs that plants use to interpret varied environmental cues. Future research will further clarify the molecular mechanisms by which JA moves through the transporter, how resources are allocated between activities associated with development and defence and if JA interacts positively or negatively with other hormonal signalling pathways. These studies will deepen our understanding of the molecular processes that underlie JA's responses to biotic and abiotic stressors. It is becoming clearer as research on the manufacture and function of plant hormones advances that a wide variety of metabolites associated with various phytohormones play a role in plant defence and interactions with the environment. Monitoring the dose/response ratio of phytohormones is still a difficult undertaking, since the hormone levels obtained must be reasonable to maintain a balance between the beneficial effects of plant hormones on various abiotic stress tolerance and the detrimental impacts on growth and development. This issue can be avoided by using conditional promoters to drive gene expression at specific developmental stages, in specific tissues or organs or in response to specific environmental cues. This will enable the development of transgenic crops that can withstand various abiotic stresses while suffering only minimal yield losses. More information regarding the effects of phytohormones on plants' responses to salt stress may be revealed by exploring the molecular pathways that have been identified by molecular biology and proteomic analyses involving the perception of plant hormones.

11.4 FUTURE PERSPECTIVES

Future research that takes advantage of crucial understandings about the function and control of JA under a variety of stressors can therefore produce encouraging outcomes. Furthermore, the mechanisms of JA signalling and how they contribute to signalling crosstalk at the level of the organ, tissue or cell are still poorly understood. Future research may identify the components of signalling pathways and the manufacture of JAs that plants use to interpret varied environmental cues. Future research may further clarify the molecular mechanisms by which JA moves through the transporter, how resources are allocated between activities associated with development and defence, and if JA interacts positively or negatively with other hormonal signalling pathways. These studies will deepen our understanding of the molecular processes that underlie JA's responses to biotic and abiotic stressors. It is becoming clearer as research on the manufacture and function of plant hormones advances that a wide variety of metabolites associated with various phytohormones play a role in plant defence and interactions with the environment. Monitoring the dose/response ratio of phytohormones is still a difficult undertaking since the hormone levels obtained must be reasonable to maintain a balance between the beneficial effects of plant hormones on various abiotic stress tolerance and the detrimental impacts on growth and development. This issue can be avoided by using conditional promoters to drive gene expression at specific developmental stages, in specific tissues or organs or in response to specific environmental cues. This will enable the development of transgenic crops that can withstand various abiotic stresses while suffering only minimal yield losses. More information regarding the effects of phytohormones on plants' responses to salt stress may be revealed by exploring the molecular pathways that have been identified by molecular biology and proteomic analyses involving the perception of plant hormones. Future research that takes advantage of crucial understandings about the function and control of JA under a variety of stressors can therefore produce encouraging outcomes. Furthermore, the mechanisms of JA signalling and how they contribute to signalling crosstalk at the level of the organ, tissue or cell are still poorly understood.

REFERENCES

Ali, M. S., & Baek, K. H. (2020). Jasmonic acid signaling pathway in response to abiotic stresses in plants. International Journal of Molecular Sciences, 21(2), 621.

Asif, A., Baig, M. A., & Siddiqui, M. B.. (2021). Role of Jasmonates and Salicylates in Plant Allelopathy. In: T. Aftab, & M. Yusuf (eds.) Jasmonates and Salicylates Signaling in Plants. Signaling and Communication in Plants (pp. 115–127), Springer, Cham. 10.1007/978-3-030-75805-9_6

Aslam, S., Gul, N., Mir, M. A., Asgher, M., Al-Sulami, N., Abulfaraj, A. A., & Qari, S. (2021). Role of jasmonates, calcium, and glutathione in plants to combat abiotic stresses through precise signaling cascade. Frontiers in Plant Science, 12, 668029.

Balbi, V., & Devoto, A. (2008). Jasmonate signalling network in *Arabidopsis thaliana*: crucial regulatory nodes and new physiological scenarios. New Phytologist, 177(2), 301–318. 10.1111/j.1469-8137.2007.02292.x.

Bhattacharya, A. (2021). Role of Plant Growth Hormones During Soil Water Deficit: A Review. In: A. Bhattacharya (ed.), Soil Water Deficit and Physiological Issues in Plants (pp. 489–583), Springer, Singapore. 10.1007/978-981-33-6276-5_6

Eng, F., Marin, J. E., Zienkiewicz, K., Gutiérrez-Rojas, M., Favela-Torres, E., & Feussner, I. (2021). Jasmonic acid biosynthesis by fungi: derivatives, first evidence on biochemical pathways and culture conditions for production. PeerJ, 9, e10873.

Fahad, S., Hussain, S., Matloob, A., Khan, F. A., Khaliq, A., Saud, S., et al. (2015). Phytohormones and plant responses to salinity stress: a review. Plant Growth Regulation, 75, 391–404.

Fahad, S. , et al. (2015). Crop plant hormones and environmental stress. In: E. Lichtfouse (eds), Sustainable Agriculture Reviews, vol. 15, Cham: Springer. https://doi.org/10.1007/978-3-319-09132-7_10

Fukao, T., Barrera-Figueroa, B. E., Juntawong, P., & Peña-Castro, J. M. (2019). Submergence and waterlogging stress in plants: a review high-lighting research opportunities and understudied aspects. Frontiers in Plant Science, 10, 340.

Fu, J., Wu, H., Ma, S., Xiang, D., Liu, R., & Xiong, L. (2017). OsJAZ1 attenuates drought resistance by regulating JA and ABA signaling in rice. Frontiers in Plant Science, 8, 2108.

Hossain, A., Pamanick, B., Venugopalan, V. K., Ibrahimova, U., Rahman, M. A., Siyal, A. L., et al. (2022). Emerging roles of plant growth regulators for plants adaptation to abiotic stress-induced oxidative stress. In: M. Naeem, & T. Aftab (eds.), Emerging plant growth regulators in agriculture (pp. 1–72), Academic Press.

Hu, Y., Jiang, Y., Han, X., Wang, H., Pan, J., & Yu, D. (2017). Jasmonate regulates leaf senescence and tolerance to cold stress: crosstalk with other phytohormones. Journal of Experimental Botany, 68(6), 1361–1369.

Isah, T. (2019). Stress and defense responses in plant secondary metabolites production. Biological research, 52.

Jang, G., Yoon, Y., & Choi, Y. D. (2020). Crosstalk with jasmonic acid integrates multiple responses in plant development. International Journal of Molecular Sciences, 21(1), 305.

Janni, M. (2020). Molecular and genetic bases of heat stress responses in op plants and breeding for increased resilience and productivity. Journal of Experimental Botany, 71(13), 3780–3802. 10.1093/jxb/eraa034

Jha, P., Sharaya, R., Kundu, P., Chhikara, A., Kaushik, S., Sidhu, A., et al. (2022). Understanding the Role of Jasmonic Acid in Growth, Development, and Stress Regulation in Plants. In: R. Akula, & G. Sirhindi (eds.), Jasmonates and Brassinosteroids in Plants: Metabolism, Signaling, and Biotechnological Applications (pp. 127–140). CRC Press.

Kang, D. J., Seo, Y. J., Lee, J. D., Ishii, R., Kim, K. U., Shin, D. H., et al. (2005). Jasmonic acid differentially affects growth, ion uptake and abscisic acid concentration in salt-tolerant and salt-sensitive rice cultivars. Journal of Agronomy and Crop Science, 191(4), 273–282.

Kangasjärvi, J., Jaspers, P., & Kollist, H. (2005). Signalling and cell death in ozone-exposed plants. Plant, Cell & Environment, 28(8), 1021–1036.

Karlova, R., Boer, D., Hayes, S., & Testerink, C. (2021). Root plasticity under abiotic stress. Plant Physiology, 187(3), 1057–1070.

Kim, H., Seomun, S., Yoon, Y., & Jang, G. (2021). Jasmonic acid in plant abiotic stress tolerance and interaction with abscisic acid. Agronomy, 11(9), 1886.

Kolaksazov, M., Laporte, F., Ananieva, K., Dobrev, P., Herzog, M., & Ananiev, E. D. (2013). Effect of chilling and freezing stresses on jasmonate content in *Arabis alpina*. Bulgarian Journal of Agricultural Science, 19(2), 15–17.

Koo, A. J. (2018). Metabolism of the plant hormone jasmonate: a sentinel for tissue damage and master regulator of stress response. Phytochemistry Reviews, 17(1), 51–80.

Ku, Y. S., Sintaha, M., Cheung, M. Y., & Lam, H. M. (2018). Plant hormone signaling crosstalks between biotic and abiotic stress responses. International Journal of Molecular Sciences, 19(10), 3206.

Kumar, G., Singh, S., Singh, R., & Mishra, R. (2021). Role of Physical Agents in Inducing Genotoxicity and Oxidative Stress in Plants. In: Z. Khan, M. Y. K. Ansari, & D. Shahwar (eds.), Induced Genotoxicity and Oxidative Stress in Plants (pp. 65–102). Springer, Singapore. 10.1007/978-981-16-2074-4_3

Lalotra, S., Hemantaranjan, A., Yashu, B. R., Srivastava, R., & Kumar, S. (2020). Jasmonates: An Emerging Approach in Biotic and Abiotic Stress Tolerance. In: A. Gonzalez, M. Rodriguez, & N. G. Sağlam (eds.), Plant Science: Structure, Anatomy and Physiology in Plants Cultured In Vivo and In Vitro (pp. 47–60). InTech Open.

Li, N. (2020). Plant hormone-mediated regulation of heat tolerance in response to global climate change. Frontiers in Plant Science, 11. 10.3389/fpls.2020.627969

Maggio, A., Bressan, R. A., Zhao, Y., Park, J., & Yun, D. J. (2018). It's hard to avoid avoidance: Uncoupling the evolutionary connection between plant growth, productivity and stress "tolerance". International Journal of Molecular Sciences, 19(11), 3671.

Maksymiec, W., Wianowska, D., Dawidowicz, A. L., Radkiewicz, S., Mardarowicz, M., & Krupa, Z. (2005). The level of jasmonic acid in Arabidopsis thaliana and Phaseolus coccineus plants under heavy metal stress. Journal of Plant Physiology, 162(12), 1338–1346.

Medendorp, J., DeYoung, D., Thiagarajan, D. G., Duckworth, R., & Pittendrigh, B. (2022). A Systems Perspective of the Role of Dry Beans and Pulses in the Future of Global Food Security: Opportunities and Challenges. In: M. Siddiq, & M. A. Uebersax (eds.), Dry Beans and Pulses: Production, Processing, and Nutrition (pp. 531–550), John Wiley & Sons Ltd, Hoboken, NJ.

Meena, M., Divyanshu, K., Kumar, S., Swapnil, P., Zehra, A., Shukla, V., et al. S. (2019). Regulation of L-proline biosynthesis, signal transduction, transport, accumulation and its vital role in plants during variable environmental conditions. Heliyon, 5(12).

Ouli-Jun, Chao-Hui, Z., Zhou-Bin, L., Ge, W., Bo-Zhi, Y., & Xue-Xiao, Z. (2017). Mitigation of waterlogging-induced damages to pepper by exogenous MeJA. Pakistan Journal of Botany, 49(3), 1127–1135.

Patel, M., Fatnani, D., & Parida, A. K. (2022). Potassium deficiency stress tolerance in peanut (*Arachis hypogaea*) through ion homeostasis, activation of antioxidant defense, and metabolic dynamics: Alleviatory role of silicon supplementation. Plant Physiology and Biochemistry, 182, 55–75.

Piotrowska, A., Bajguz, A., Godlewska-Żyłkiewicz, B., Czerpak, R., & Kamińska, M. (2009). Jasmonic acid as modulator of lead toxicity in aquatic plant Wolffia arrhiza (Lemnaceae). Environmental and Experimental Botany, 66(3), 507–513.

Poonam, S., Kaur, H., & Geetika, S. (2013). Effect of jasmonic acid on photosynthetic pigments and stress markers in Cajanus cajan (L.) Millsp. seedlings under copper stress. American Journal of Plant Sciences, 4, 817–823. 10.4236/ajps.2013.44100.

Qiu, Z., Guo, J., Zhu, A., Zhang, L., & Zhang, M. (2014). Exogenous jasmonic acid can enhance tolerance of wheat seedlings to salt stress. Ecotoxicology and Environmental Safety, 104, 202–208.

Raza, A., Charagh, S., Zahid, Z., Mubarik, M. S., Javed, R., Siddiqui, M. H., & Hasanuzzaman, M. (2021a). Jasmonic acid: a key frontier in conferring abiotic stress tolerance in plants. Plant Cell Reports, 40(8), 1513–1541.

Raza, A., Charagh, S., Najafi-Kakavand, S., Siddiqui, M. H. (2021b). The Crucial Role of Jasmonates in Enhancing Heavy Metals Tolerance in Plants. In: T. Aftab, & M. Yusuf (eds.), Jasmonates and Salicylates Signaling in Plants. Signaling and Communication in Plants (pp. 159–183), Springer, Cham. 10.1007/978-3-030-75805-9_8

Seo, J. S., Joo, J., Kim, M. J., Kim, Y. K., Nahm, B. H., Song, S. I., et al. (2011). *OsbHLH148*, a basic helix-loop-helix protein, interacts with OsJAZ proteins in a jasmonate signaling pathway leading to drought tolerance in rice. The Plant Journal, 65(6), 907–921.

Sirhindi, G., Mir, M. A., Abd-Allah, E. F., Ahmad, P., & Gucel, S. (2016). Jasmonic acid modulates the physio-biochemical attributes, antioxidant enzyme activity, and gene expression in *Glycine max* under nickel toxicity. Frontiers in Plant Science, 7, 591.

Su, Y., Huang, Y., Dong, X., Wang, R., Tang, M., Cai, J., et al. (2021). Exogenous methyl jasmonate improves heat tolerance of perennial ryegrass through alteration of osmotic adjustment, antioxidant defense, and expression of jasmonic acid-responsive genes. Frontiers in Plant Science, 12, 664519.

Walia, H., Wilson, C., Condamine, P., Liu, X., Ismail, A. M., & Close, T. J. (2007). Large-scale expression profiling and physiological characterization of jasmonic acid-mediated adaptation of barley to salinity stress. Plant, Cell & Environment, 30(4), 410–421.

Wang, J., Song, L., Gong, X., Xu, J., & Li, M. (2020). Functions of jasmonic acid in plant regulation and response to abiotic stress. International Journal of Molecular Sciences, 21(4), 1446.

Wani, S. H., & Sah, S. K. (2014). Biotechnology and abiotic stress tolerance in rice. Journal of Rice Research, 2(2), e105.

Yang, J., Duan, G., Li, C., Liu, L., Han, G., Zhang, Y., & Wang, C. (2019). The crosstalks between jasmonic acid and other plant hormone signaling highlight the involvement of jasmonic acid as a core component in plant response to biotic and abiotic stresses. Frontiers in Plant Science, 10, 1349.

Yang, J. (2020). Jasmonates alleviate spikelet-opening impairment caused by high temperature stress during anthesis of photo-thermo-sensitive genic male sterile rice lines. Food and Energy Security, 9(4). 10.1002/fes3.233

Yun-Xia, G., Li-Jun, Z., Feng-Hai, L., Zhi-Bin, C., Che, W., Yun-Cong, Y., et al. (2010). Relationship between jasmonic acid accumulation and senescence in drought-stress. African Journal of Agricultural Research, 5(15), 1978–1983.

Zhu, Z. (2014). Molecular basis for jasmonate and ethylene signal interactions in Arabidopsis. Journal of Experimental Botany, 65(20), 5743–5748.

Zhu, Z., & Lee, B. (2015). Friends or foes: new insights in jasmonate and ethylene co-actions. Plant and Cell Physiology, 56(3), 414–420.

12 Roles of Karrikins in Abiotic Stress

Madhu Rani, Anshul Sharma, and Kirti Nain

12.1 INTRODUCTION

Agriculture activity faces new challenges in the 21st century due to climate change and global warming (IPCC Geneva Report, 2019). The excessive use of chemical fertilizers and pesticides causes soil toxicity and nutrient deficiency in crop production. Due to global warming, many types of stresses in plants also affect plant growth. Stress in plants is defined as any type of external condition that causes several changes, such as molecular changes, physiological changes and biochemical changes, resulting in reduced plant growth as well as productivity (Rani et al., 2021; Collier et al., 2017). It includes both biotic as well as abiotic stress. Biotic stress is caused by pathogens, insects, viruses, fungi and herbivores and abiotic stress includes flooding, heat, salt, drought, light intensity, pH and pollutants. Among these stresses, abiotic stress is most common and devastating in plants. Abiotic stress increases senescence and plant death (Wollenweber et al., 2003). As a result of this, food insecurity and global temperature increases by 2–4.5°C due to the emission of greenhouse gases by the end of the 21st century (Change et al., 2014; Peng et al., 2019). Drought, salt stress, osmotic stress and extreme temperature are the most deteriorating abiotic stress in different crops and vegetables. In recent decades, climate is also changing due to a rise in global warming and a rise in GHG (greenhouse gas) emissions. Some anthropogenic activities like mining, excessive use of groundwater and dumping of heavy metals pollutants in rivers are also the main factor of abiotic stress. All these factors are a matter of great concern in plants and the urgent need is to solve these. One of the important aspects to combat abiotic stress is the use of plant-based hormones. Many plant hormones like karrikins (KARs), jasmonic acid and abscisic acid also contain some genes that can help combat abiotic stress in crops. KARs are chemically called butenolides based on a four-carbon heterocyclic lactone ring structure with low molecular weight organic compounds when plant material is burned. Karrikins, nowadays, are an important plant hormone to act as a potential bio-activator that promotes plant growth and alters the mechanism of the plant to combat abiotic stress in plants (Khatoon et al., 2020). A KAR is a smoke-derived plant hormone carrying a butanolide moiety that helps plants in seed germination, photosynthetic yield and seed development and is tolerant to oxidative stress (Kulkarni et al., 2011).

12.2 IMPACT OF ABIOTIC STRESS

Abiotic stress in plants can alter the physiological, molecular and biochemical levels under different external conditions like heat, salt, drought etc. (Amin et al., 2019). Active concentration of solutes comes up in the vacuole at a physiological level and shows the different types of abiotic stress, like drought, freezing and salinity (Harfouche et al., 2014). At a biochemical level, several metabolites, like carbohydrates, polyamines and amino acids, and different secondary metabolites like polyphenols, flavanols and also several antioxidants signaling molecule show abiotic stress tolerance in plants. The metabolism of carbohydrates and their derivatives lead to an increased concentration of β-amylase activity to starch and fructose that is stored in different organs of plants like roots, amyloplasts and shoots and then converted to glucose and fructose, acting as an

DOI: 10.1201/9781003335788-16

important source of energy. Amino acids like proline were found in plants under abiotic stress. These amino acids act as osmolytes, contributing to the osmotic balance of cells and forming part of antioxidant defense. Also, the secondary cell wall of plants constitutes phenylpropanoids that act as reactive oxygen species (ROS) and scavengers, along with UV radiation to protect ions and develop plant tolerance to abiotic stress (Cheynier et al., 2013). Under abiotic stress, carotenoids are produced, which helps in protection in the thylakoid membrane in plants (Espinoza et al., 2013). ROS that are produced in abiotic stress lead to cellular impairment through enzymes like catalase (CAT), superoxide dismutase (SOD), glutathione peroxidase (GPX), dehydro ascorbate reductase (DHAR), ascorbate peroxidase activity (APX) vitamin E and C glutathione S-transferase (GST), glutathione, carotenoids and flavonoids that help to reduce this damage (Gill & Tuteja et al., 2010). The adaptive response of crops to abiotic stress includes an increase in ATP production, repairing chloroplast structure and promotion of cell softening and expansion; all these factors lead to the stability of cell membrane. Also, combined stress compared to each type of stress was found to contain a greater amount of ROS. The changes at the level of O_2 and H_2O_2 by-products of lipid oxidation; expression of enzymes such as SOD, APX, GST, AOX, CAT, GPX, glutathione reductase and peroxidase; and many osmoprotectants, such as betaine, sucrose, glycine, proline and trehalose. Many antioxidants, such as ascorbate, alkaloids, carotenoids, flavonols, tocopherol and phenolic compounds, are found in large amounts during combined abiotic stresses (Li et al., 2014; Jin et al., 2015; Martinez et al., 2016; Carvalho et al., 2016; Pandey et al., 2015; Rasmussen et al., 2013). During heat and drought stress, plants open their stomata to cool off the leaves.

12.3 FUNCTIONS OF KARRIKINS IN ABIOTIC STRESS

Karrikins are known to be involved in abiotic stress tolerance (Figures 12.1, 12.2 and 12.3, Table 12.1). Karrikins positively control photomorphogenesis and seed germination, while they negatively control hypocotyl elongation in many plants. It breaks dormancy in the presence of favorable circumstances, such as those brought on by postfire rainfall. Recently, it was shown that *Arabidopsis* was protected from abiotic stress via the karrikins-KAI2 signaling system, which also provided stress tolerance and inhibited germination under adverse conditions. By promoting cuticle formation, stomatal closure, cell membrane integrity and anthocyanin production, the KAI2 protein confers drought resistance on the leaves of *Arabidopsis* plants. KAR1 has previously been utilized to improve tomato seed germination and seedling vigor when exposed to salt (Jain et al., 2008). Karrikin-containing smoke was shown to promote seed germination and reduce salt stress tolerance in maize, indicating that the chemicals in smoke mixed with water may be able to influence ion homeostasis and regulate various physiological processes. These findings suggested that, depending on the species, karrikins may play a variety of roles in the control of seed germination in response to abiotic stimuli. The abiotic stressors may generate a large rise in ROS and excessive buildup may result in serious cellular oxidative damage. It has been suggested that karrikin signaling genes like *KAI2* and *MAX2* are involved in the reduction of abiotic stress (Yang et al., 2019).

1. KAI2 encourages the formation of anthocyanin; KAI2 mutants are unable to accumulate this antioxidant, which inhibits ROS linked to a number of abiotic stresses.
2. KAI2 encourages the development of the cuticle; KAI2 overexpression has a thicker cuticle compared to KAI2 mutants.

Along with this, karrikins break the dormancy of seeds of species (Californian poppy, *Eschscholzia californica* and *Cistus* sp.) that follow forest fires, which can be found in a variety of plant families (*Anigozanthos flavidus* DC. (Haemodoraceae), *Cistus* sp L., *Helianthemum* sp. Mill. (Cistaceae) (Waters et al., 2013). Despite repeated cycles of wetting and drying, as well as periodic cycles of changes in light and temperature, seeds can stay dormant in the soil for decades. Karrikins break

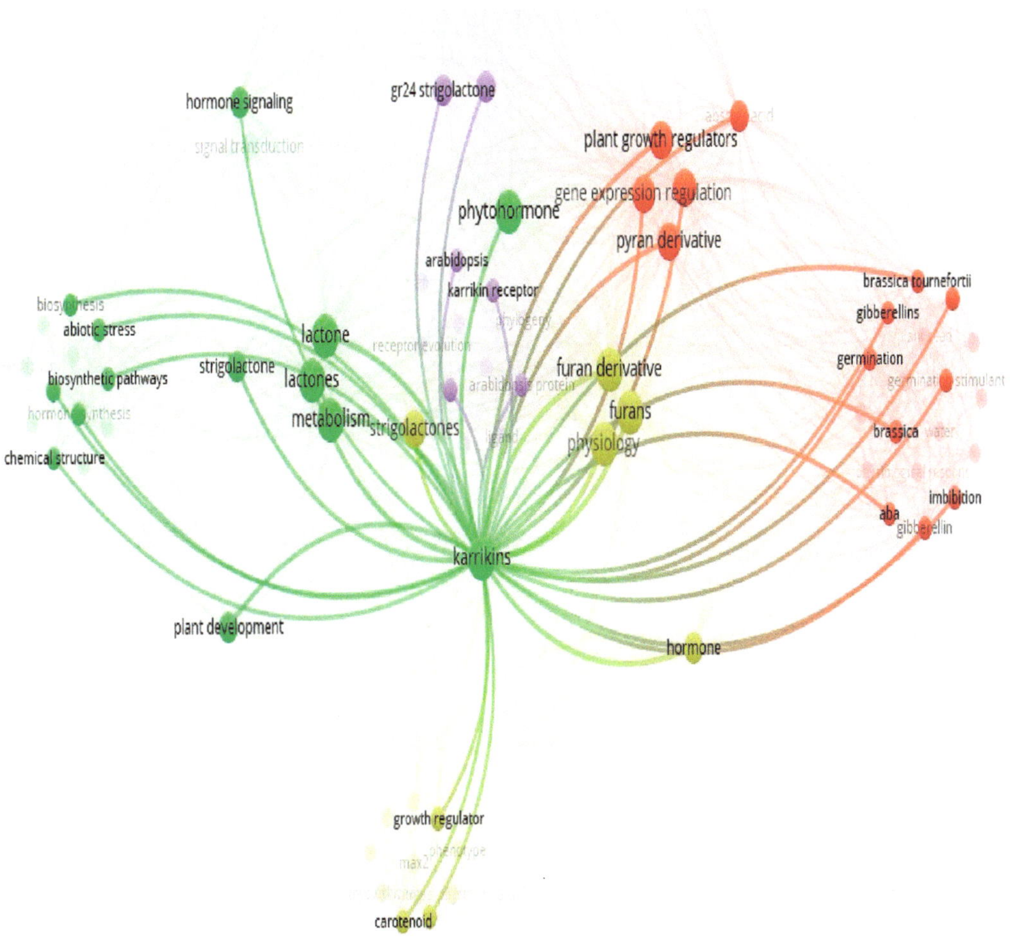

FIGURE 12.1 A karrikin hormone is related to abiotic stress, plant development, plant growth regulators and physiology, as they are interrelated with one another. The interaction between any two bubbles shows the correlation with each other. Different colors are assigned to each publication's citations. The size of each bubble represents the number of citations. The larger the bubble size, the greater the number of citations.

dormancy in the presence of favorable circumstances, such as those brought by postfire rainfall. Along with this Zhao et al. (2020) reported the combined effect of KARs in chilling-stress-exposed tea plants.

12.3.1 Drought Stress

Genetic screening has found the KAR receptor and several associated signaling elements, but only a small number of publications have discussed the upstream and downstream KAR signaling elements and their functions in drought tolerance. It is reported that the reverse genetics method is used to characterize the roles of KAR UPREGULATED F-BOX 1 (KUF1) in drought tolerance in *Arabidopsis* (Tian et al., 2022). In comparison to wild-type (WT) plants, kuf1 mutant plants showed greater tolerance to the effects of drought stress. Transcriptomic-level investigation on the *KUF1* gene shows that it is negatively regulated during drought tolerance. By lowering the stomatal aperture and cuticular permeability, it was found that kuf1 plants prevented excessive leaf water loss. Additionally, stomatal closure, seed germination, primary root growth and other plant processes on kuf1 plants displayed greater sensitivity. Also, transcriptome analyses of kuf1 and WT rosette leave before and, following dehydration, show that

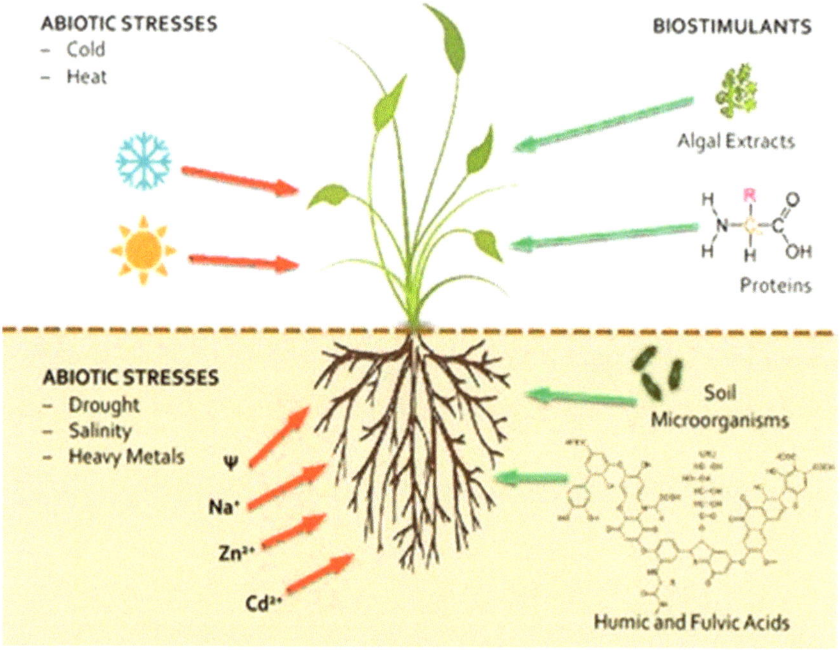

FIGURE 12.2 The effect of different types of abiotic stresses like drought, salinity, cold, heat and heavy metals are detected in different parts of plants like roots and leaves. Also, the role of biostimulants like humic and fulvic acid plays an important role with soil microorganisms in the prevention of abiotic stress.

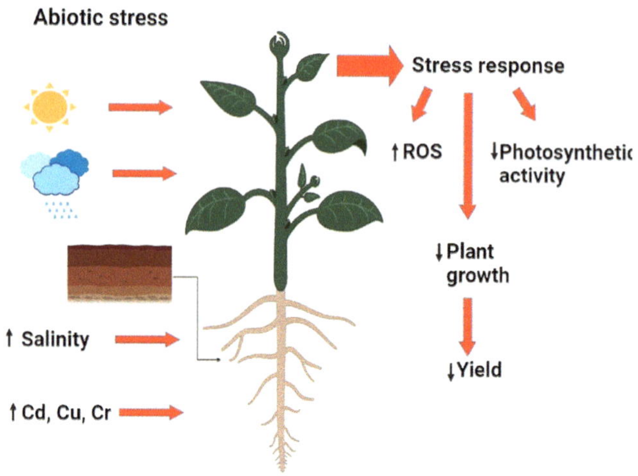

FIGURE 12.3 Different types of abiotic stresses like salinity stress, heavy metals like cadmium (Cd), copper (Cu), chromium (Cr) and heat stress and their response includes an increase in reactive oxygen species (ROS), decrease in physiological activity and yield in plants.

different variations in gene expression of genes linked to mainly lipid and fat metabolism (WAX ESTER SYNTHASE), stomatal closure (OPEN STOMATA1) and ABA response (such as ABA-RESPONSIVE ELEMENT 3) were also associated with disparities in various drought tolerance-related attributes. The root hair densities and root/shoot ratios of the kuf1 plants with mutations were higher than those of WT plants, indicating that they were able to absorb more water. Collectively,

TABLE 12.1
Effect of karrikins on plant growth and metabolism under stress conditions

Plant	Effect of Stress on Plants	Type of Stress	Reference
Triticum aestivum	Effect on salinity toxicity genes like TaSOS4, TaBADH	Salinity stress	Hayat et al., 2021
Oryzae sativum	Karrikin signaling complex like D14l, D3, Os SMAX1 induce mesocotyl elongation	Shade stress	Zheng et al., 2020
Arabidopsis thaliana	KAI2 gene regulates germination potential and regulation of different abiotic stress	Multiple Abiotic stress	Wang et al., 2016
Arabidopsis thaliana	KAI2 gene help to improve drought stress by stomatal aperture and reducing membrane damage	Drought stress	Li et al., 2017
Triticum aestivum	KAR genes help in the reduction of the Na+ gene and increase in the K+ gene under salt stress	Salinity stress	Shah et al., 2021a, b
Brassica oleracea	KAR genes help in membrane stability, stomatal aperture, photosynthetic activity and osmotic potential under Cd stress	Cd stress	Shah et al., 2021a, b
Brassica alboglabra	KAR genes help in the reduction of heat stress by reducing lipid peroxidation and also help in leaf water content, membrane stability and antioxidant potential were increased	Cd, heat stress	Ahmad et al., 2021

these findings show how KUF1 modulates morphological changes, ABA actions and numerous physiological variables to negatively limit tolerance to drought, and they also suggest that genetically modifying KUF1 in crops may help them be resistant to drought (Tian et al., 2022).

12.3.2 Salt and Osmotic Stresses

Under abiotic stress conditions, karrikins are said to promote seed germination, control seedling development and boost seedling vigor in plants. However, the exact mechanism by which karrikins reduce abiotic stress is still unknown. The research studies observed that the key oil plant *Sapium sebiferum* may greatly reduce both salt stress and drought by using karrikins (KAR1). Under salt and osmotic stress conditions, a nanomolar (nM) dose of KAR1 supplementation in growth media was sufficient to restore seed germination. Under abiotic stress conditions, one nanomolar of KAR1 enhanced seedling biomass, lengthened taproots and increased the number of lateral roots, indicating that KAR1 is an effective plant stress reliever. Key antioxidative enzymes, such as superoxide dismutase, were more active in seedlings treated with KAR1 when they are subjected to abiotic stressors. Furthermore, the metabolome study revealed that KAR1 therapy dramatically increased the number of organic acids and amino acids, which is crucial for maintaining redox homeostasis in the presence of stress. This finding suggests that karrikins may reduce abiotic stressors by controlling redox homeostasis. The expression of several ABA signaling genes, including *SNF1-RELATED PROTEIN KINASE2.6*, *ABI3*, *ABI5* and *SNF1-RELATED PROTEIN KINASE2.3* was altered under abiotic stress conditions, indicating potential interactions between karrikins and ABA signaling in the stress responses (Shah et al., 2020).

Salinity poses a significant barrier to sustainable agricultural productivity and crop yields, specifically in arid and semiarid countries, where it costs the industry billions of dollars yearly (Munns & Tester, 2008; Shabala & Cuin, 2008). According to Shahrajabian and Sun (2022), it was observed that salinity had a negative impact on the black cumin plants' seed oil output and photosynthetic characteristics. Additionally, it made electrolyte leakage and ROS production more intense. Karrikin pretreatment of black cumin seeds and calcium foliar spray during seedling growth protected *Nigella sativa* against saline stress by reducing oxidative stress and enhancing oil output and seed composition. According to Ahmad et al. (2023), black cumin plants treated with

NaCl, first treated with KARs and then followed by foliar calcium application, further enhanced proline content, decreased production of H_2O_2 buildup and enhanced the activity of several antioxidant enzymes such as SOD, GR, CAT and APX (Sharifi & Bidabadi, 2020).

12.3.3 Shade Stress

Another common type of stress that plants contain is low light-intensity or shade stress as the ratio of red/far red light reaching the lower vegetation is significantly decreased affecting seedling growth, plant development and seed germination (Casal, 2013). Also, compared to heliophytes plants, sciophytes show very low resistance to abiotic stress (De Wit et al., 2013). Different types of research studies on KARs have shown their ability to restrict the shade-evading response by increasing the content of photosynthetic pigments, mainly chlorophyll and hypocotyl elongation (Nelson et al., 2010). Repression induced by KAR genes with the biosynthesis of IAA, as detected in *Arabidopsis,* showed the improvement of shade stress in seedlings done by KARs, which is used to inhibit the biosynthesis of IAA (Meng et al., 2017).

12.3.4 Heat Stress

Heat stress is one of the most destructive abiotic stresses that leads to a huge amount of yield loss. It induces a high amount of reactive oxygen species that inhibit or negatively affect the synthesis of proteins, lipids and nucleic acids. The role of the main phytochrome, which is a phytochrome interacting factor (PIF), and also several heat shock proteins (HSP) in plants is detected in roots and meristematic tissue. High temperature limits water use efficiency, photosynthesis, stomatal conductivity, growth of plants and water contents (Faroq et al., 2019). Karrikin receptor KAI2 has an important role in seedlings of high-temperature stress (Abdelrahman et al., 2022). Several studies have found that the karrikin receptor increases the activity of several antioxidant enzymes like SOD, APX and POD to combat heat stress, which alleviates ROS in plants (Prerostova & Vankova, 2023).

12.4 CONCLUSION

Abiotic stress in plants causes many types of changes, including chloroplast expansion, stomatal opening, root hair and hypocotyl elongation. These changes can alter plant growth, so some stress hormones like karrikins are effective in terms of combating stress. Due to climate change and global warming, plants also follow some adaptations in terms of their morphology to prevent water loss and photosynthesis. By using stress hormones like abscisic acid and strigolactone alone, or in combination with karrikins, plants can be resistant to many types of abiotic stress as well as biotic stress. Along with this, karrikins are also useful in terms of seed germination, seed growth, root development, photosynthetic yield and mitigation of different abiotic stress. This smoke-derived hormone made from plants can increase the tolerance of plants to many abiotic stresses like salinity, drought, low light intensity and heavy metal toxicity etc. So, karrikins can be one of the potential phytohormones to combat stress in plants.

REFERENCES

Abdelrahman, M., Mostofa, M. G., Tran, C. D., El-Sayed, M., Li, W., Sulieman, S., ... & Tran, L. S. P. (2022). The karrikin receptor karrikin insensitive2 positively regulates heat stress tolerance in *Arabidopsis thaliana*. Plant and Cell Physiology, 63(12), 1914–1926.

Ahmad, A., Shahzadi, I., Mubeen, S., Yasin, N. A., Akram, W., Khan, W. U., & Wu, T. (2021). Karrikinolide alleviates BDE-28, heat and Cd stressors in Brassica alboglabra by correlating and modulating biochemical attributes, antioxidative machinery and osmoregulators. Ecotoxicology and Environmental Safety, 213, 112047.

Ahmad, B., Qadir, S. U., Dar, T. A., Alam, P., Yousuf, P. Y., & Ahmad, P. (2023). Karrikins: smoke-derived phytohormones from stress alleviation to signaling. Journal of Plant Growth Regulation, 42(8), 4784–4796.

Amin, A. B., Rathnayake, K. N., Yim, W. C., Garcia, T. M., Wone, B., Cushman, J. C., & Wone, B. W. (2019). Crassulacean acid metabolism abiotic stress-responsive transcription factors: a potential genetic engineering approach for improving crop tolerance to abiotic stress. Frontiers in Plant Science, 10, 129.

Carvalho, L. C., Coito, J. L., Goncalves, E. F., Chaves, M. M., & Amancio, S. (2016). Differential physiological response of the grapevine varieties Touriga Nacional and Trincadeira to combined heat, drought and light stresses. Plant Biol (Stuttg), 18(Suppl 1), 101–111.

Casal, J. J. (2013). Photoreceptor signaling networks in plant responses to shade. Annual Review of Plant Biology, 64, 403–427.

Change, I. C. (2014). Impacts, adaptation and vulnerability. Part A: global and sectoral aspects. Contribution of working group II to the fifth assessment report of the intergovernmental Panel on Climate Change, 1132.

Cheynier, V., Comte, G., Davies, K. M., Lattanzio, V., & Martens, S. (2013). Plant phenolics: recent advances on their biosynthesis, genetics, and ecophysiology. Plant Physiology and Biochemistry, 72, 1–20.

Collier, R. J., Renquist, B. J., & Xiao, Y. (2017). A 100-Year Review: Stress physiology including heat stress. Journal of Dairy Science, 100(12), 10367–10380.

De Wit, M., Spoel, S. H., Sanchez-Perez, G. F., Gommers, C. M., Pieterse, C. M., Voesenek, L. A., & Pierik, R. (2013). Perception of low red: far-red ratio compromises both salicylic acid-and jasmonic acid-dependent pathogen defences in A rabidopsis. The Plant Journal, 75(1), 90–103.

Espinoza, A., San Martín, A., López-Climent, M., Ruiz-Lara, S., Gómez-Cadenas, A., & Casaretto, J. A. (2013). Engineered drought-induced biosynthesis of α-tocopherol alleviates stress-induced leaf damage in tobacco. Journal of Plant Physiology, 170(14), 1285–1294.

Farooq, M., Hussain, M., Ul-Allah, S., & Siddique, K. H. (2019). Physiological and agronomic approaches for improving water-use efficiency in crop plants. Agricultural Water Management, 219, 95–108.

Gill, S. S., & Tuteja, N. (2010). Reactive oxygen species and antioxidant machinery in abiotic stress tolerance in crop plants. Plant Physiology and Biochemistry, 48(12), 909–930.

Harfouche, A., Meilan, R., & Altman, A. (2014). Molecular and physiological responses to abiotic stress in forest trees and their relevance to tree improvement. Tree Physiology, 34(11), 1181–1198.

Hayat, N., Afroz, N., Rehman, S., Bukhari, S. H., Iqbal, K., Khatoon, A., & Nawaz, G. (2021). Plant-derived smoke ameliorates salt stress in wheat by enhancing expressions of stress-responsive genes and antioxidant enzymatic activity. Agronomy, 12(1), 28.

IPCC. Summary for Policymakers. In Climate Change and Land; An IPCC Special Report on Climate Change, Desertification, Land Degradation, Sustainable Land Management, Food Security, and Greenhouse Gas Fluxes in Terrestrial Ecosystems; IPCC: Geneva, Switzerland, 2019; Available online: https://www.ipcc.ch/srccl/ (accessed on 18 November 2019).

Jain, N., Ascough, G. D., & Van Staden, J. (2008). A smoke-derived butenolide alleviates $HgCl_2$ and $ZnCl_2$ inhibition of water uptake during germination and subsequent growth of tomato–Possible involvement of aquaporins. Journal of Plant Physiology, 165(13), 1422–1427.

Jin, R., Wang, Y., Liu, R., Gou, J. and Chan, Z. (2015) Physiological and metabolic changes of Purslane (*Portulaca oleracea* L.) in response to drought, heat, and combined stresses. Frontiers in Plant Science, 6, 1123

Khatoon, A., Rehman, S. U., Aslam, M. M., Jamil, M., & Komatsu, S. (2020). Plant-derived smoke affects biochemical mechanism on plant growth and seed germination. International Journal of Molecular Sciences, 21(20), 7760.

Kulkarni, M. G., Light, M. E., & Van Staden, J. (2011). Plant-derived smoke: old technology with possibilities for economic applications in agriculture and horticulture. South African Journal of Botany, 77(4), 972–979.

Li, W., Nguyen, K. H., Chu, H. D., Ha, C. V., Watanabe, Y., Osakabe, Y., et al. (2017). The karrikin receptor KAI2 promotes drought resistance in *Arabidopsis thaliana*. PLoS Genetics, 13(11), e1007076.

Li, L., Li, M., Yu, L., Zhou, Z., Liang, X., Liu, Z., et al. (2014). The FLS2-associated kinase BIK1 directly phosphorylates the NADPH oxidase RbohD to control plant immunity. Cell Host & Microbe, 15(3), 329–338.

Martinez, V., Mestre, T. C., Rubio, F., Girones-Vilaplana, A., Moreno, D. A., Mittler, R. and Rivero, R. M. (2016) Accumulation of flavonols over hydroxycinnamic acids favors oxidative damage protection under abiotic stress. Frontiers in Plant Science, 7, 838.

Meng, Y., Shuai, H., Luo, X., Chen, F., Zhou, W., Yang, W., & Shu, K. (2017). Karrikins: regulators involved in phytohormone signaling networks during seed germination and seedling development. Frontiers in Plant Science, 7, 2021.

Munns, R., & Tester, M. (2008). Mechanisms of salinity tolerance. Annual Reviews in Plant Biology, 59, 651–681.

Nelson, D. C., Flematti, G. R., Riseborough, J. A., Ghisalberti, E. L., Dixon, K. W., & Smith, S. M. (2010). Karrikins enhance light responses during germination and seedling development in *Arabidopsis thaliana*. Proceedings of the National Academy of Sciences, 107(15), 7095–7100.

Pandey, P., Ramegowda, V. and Senthil-Kumar, M. (2015) Shared and unique responses of plants to multiple individual stresses and stress combinations: physiological and molecular mechanisms. Frontiers in Plant Science, 6, 723.

Peng, W., Berry, E. M., Ferranti, P., & Anderson, J. R. (2019). Encyclopedia of food Security and Sustainability.

Prerostova, S., & Vankova, R. (2023). Phytohormone-Mediated Regulation of Heat Stress Response in Plants. In: G. J. Ahammed, J. Yu (eds.), Plant Hormones and Climate Change (pp. 167–206), Springer Nature Singapore, Singapore. 10.1007/978-981-19-4941-8_8

Rani, S., Kumar, P., & Suneja, P. (2021). Biotechnological interventions for inducing abiotic stress tolerance in crops. Plant Gene, 27, 100315.

Rasmussen, S., Barah, P., Suarez-Rodriguez, M. C., Bressendorff, S., Friis, P., Costantino, P., Bones, A. M., Nielsen, H. B., & Mundy, J. (2013) Transcriptome responses to combinations of stresses in *Arabidopsis*. Plant Physiology, 161, 1783–1794.

Shabala, S., & Cuin, T. A. (2008). Potassium transport and plant salt tolerance. Physiologia Plantarum, 133(4), 651–669.

Shahrajabian, M. H., & Sun, W. (2022). Sustainable approaches to boost yield and chemical constituents of aromatic and medicinal plants by application of biostimulants. Recent Advances in Food Nutrition & Agriculture, 13(2), 72–92.

Shah, F. A., Wei, X., Wang, Q., Liu, W., Wang, D., Yao, Y., et al. (2020). Karrikin improves osmotic and salt stress tolerance via the regulation of the redox homeostasis in the oil plant Sapium sebiferum. Frontiers in Plant Science, 11, 216.

Shah, F. A., Ni, J., Tang, C., Chen, X., Kan, W., & Wu, L. (2021a). Karrikinolide alleviates salt stress in wheat by regulating the redox and K+/Na+ homeostasis. Plant Physiology and Biochemistry, 167, 921–933.

Shah, F. A., Ni, J., Yao, Y., Hu, H., Wei, R., & Wu, L. (2021b). Overexpression of karrikins receptor gene *Sapium sebiferum* KAI2 promotes the cold stress tolerance via regulating the redox homeostasis in *Arabidopsis thaliana*. Frontiers in Plant Science, 12, 657960. 10.3389/fpls.2021.657960

Sharifi, P., & Bidabadi, S. S. (2020). Protection against salinity stress in black cumin involves karrikin and calcium by improving gas exchange attributes, ascorbate–glutathione cycle and fatty acid compositions. SN Applied Sciences, 2, 1–14.

Tian, H., Watanabe, Y., Nguyen, K. H., Tran, C. D., Abdelrahman, M., Liang, X., et al. (2022). KARRIKIN UPREGULATED F-BOX 1 negatively regulates drought tolerance in *Arabidopsis*. Plant Physiology, 190(4), 2671–2687.

Wang, H., Wang, H., Shao, H., & Tang, X. (2016). Recent advances in utilizing transcription factors to improve plant abiotic stress tolerance by transgenic technology. Frontiers in plant science, 7, 67.

Waters, M. T., Scaffidi, A., Flematti, G. R., & Smith, S. M. (2013). The origins and mechanisms of karrikin signalling. Current Opinion in Plant Biology, 16(5), 667–673.

Wollenweber, B., Porter, J. R., & Schellberg, J. (2003). Lack of interaction between extreme high-temperature events at vegetative and reproductive growth stages in wheat. Journal of Agronomy and Crop Science, 189(3), 142–150.

Yang, T., Lian, Y., & Wang, C. (2019). Comparing and contrasting the multiple roles of butenolide plant growth regulators: strigolactones and karrikins in plant development and adaptation to abiotic stresses. International Journal of Molecular Sciences, 20(24), 6270.

Zhao, M., Wang, L., Wang, J., Jin, J., Zhang, N., Lei, L., et al. (2020). Induction of priming by cold stress via inducible volatile cues in neighboring tea plants. Journal of Integrative Plant Biology, 62(10), 1461–1468.

Zheng, J., Hong, K., Zeng, L., Wang, L., Kang, S., Qu, M., … & Xiong, G. (2020). Karrikin signaling acts parallel to and additively with strigolactone signaling to regulate rice mesocotyl elongation in darkness. Plant Cell, 32(9), 2780–2805.

13 Nitric Oxide and Its Roles in Dealing with Abiotic Stress

Ayyagari Ramlal, Apoorva Verma, Ananya Anurag Anand, Naina Miglani, Ashlesha Manta, and Ambika Rajendran

13.1 INTRODUCTION

Plants, being immobile, are frequently exposed to abiotic stresses like drought, light stress, salt and heavy metal stress etc. All of these hinder their growth and productivity. As such, plants have developed various methods to deal with abiotic stresses, like stress hormones, expression of stress-related genes, ROS scavenging enzymes, metabolite production etc. (Verma et al., 2016; Nguyen et al., 2018). One of the key molecules involved in abiotic stress tolerance in plants is nitric oxide (NO). NO is a free radical, gaseous, short-lived, uncharged, redox molecule associated with signalling, crosstalk with stress regulators and plant abiotic-stress responses (Toledo Jr & Augusto, 2012; Zhou et al., 2021). In high concentrations, NO can damage DNA and cell membranes, reduce respiration and photosynthesis, disturb normal metabolism and hinder plant development (Qiao & Fan, 2008). Upon generation, it can modulate various transcription factors and proteins, an example of which is S-nitrosylation. It has also been known to react with hemes, hydrogen peroxide (H_2O_2), thiols and different kinds of proteins to generate downstream signalling molecules that further regulate physiological and biochemical activities in the cell (Forman et al., 2004). This review chapter will cover a brief overview of NO and elucidate its role in plants in abiotic stress tolerance.

13.2 NO SYNTHESIS

In land plants, NO is mainly synthesized by the enzyme nitrate reductase (NR). This enzyme, which reduces nitrate (NO_3) to nitrite (NO_2) using NAD(P)H, also catalyzes an electron transfer from NAD(P)H to nitrite, resulting in the formation of NO inside the cytoplasm (Planchet & Kaiser, 2006). NR is found in most plants and plant organs with their activity being the highest in plant leaves (Campbell, 1988). Inhibitors such as tungstate, sodium azide and potassium cyanide have been reported to suppress NO production in plants (Bright et al., 2006; Sang et al., 2008). NO synthesis is also possible via the oxidation of L-Arginine through an NADPH-dependent pathway via NO synthase (NOS)-like activity (Barroso et al., 1999; Besson-Bard et al., 2009; Foresi et al., 2010). Enzymatic molecules such as cytochrome P450, xanthine oxidase and copper amine oxidase 1 have also been reported to be involved in NO synthesis (Asgher et al., 2017). Polyamines like spermine and spermidine have also been shown to stimulate NO biosynthesis in plants (Fancy et al., 2017). Plants can also produce NO in non-enzymatic ways in which NO_2 is reduced to NO under acidic conditions in the presence of carotenoids and light inside the apoplastic space (Qiao & Fan, 2008).

13.3 S-NITROSYLATION

S-nitrosylation is a post-translational protein modification that involves the reversible covalent attachment of a NO moiety to the thiol group of a cysteine residue to form S-nitrosothiol (SNO) (Hess et al., 2005; Kovacs & Lindermayr, 2013). However, not all cysteine residues in a protein

DOI: 10.1201/9781003335788-17

undergo S-nitrosylation. Only those cysteines that are surrounded by a particular amino-acid motif, the "SNO motif," are nitrosylated (Akhtar et al., 2012). SNOs are also quite labile since the covalent link between the cysteine thiol (SH) and the NO moiety is unstable (Wolhuter et al., 2018). There have been a large number of substrates reported for S-nitrosylation, thus making it a key regulatory factor in several signalling pathways, protein activity, translocation and protein function (Anand & Stamler, 2012; Kovacs & Lindermayr, 2013; Plenchette et al., 2015). For instance, SNO modification of the major subunit of ribulose-1,5-bisphosphate carboxylase/ oxygenase (Rubisco) under conditions of cold stress causes its carboxylase activity to be inhibited (Abat & Deswal, 2009). In peas (*Pisum sativum* L.), ascorbate peroxidase (APX), which is involved in the control of a cellular hydrogen peroxide (H_2O_2) concentration, is S-nitrosylated, which in turn promotes the enzyme's activity (Fancy et al., 2017). S-nitrosylation of the cysteine-rich metal-binding peptide phytochelatin (PC2, PC3 and PC4) in cadmium-stressed *Arabidopsis* cells, however, has had no impact on its functionality (Elviri et al., 2010). In plants, S-nitrosylation has a protective role in stress conditions as it prevents the irreversible carbonylation of proteins caused by harsh saline conditions. It was found that plants that were not pre-treated with NO showed low levels of S-nitrosylation because of a rise in ROS which induced protein carbonylation (Kovacs et al., 2016; Tola et al., 2021). S-nitrosylation is also involved in the regulation of the ascorbate glutathione cycle, which is an essential antioxidant system for the control of H_2O_2 levels under unfavourable conditions in plant cells (Asada, 1992; Noctor & Foyer, 1998; Shigeoka et al., 2002; Begara-Morales et al., 2016).

13.4 TYPES OF STRESS

Abiotic stresses such as extreme temperatures, salinity, drought and heavy metal stress are major factors limiting plant growth and development (He et al., 2018; Waqas et al., 2019). These stressors greatly hamper crop productivity and yield, thus mounting a need to produce crops with improved stress tolerance. Therefore, it is important to understand the different molecular mechanisms that plants have evolved against these stresses. Over the past decade, NO has emerged as a significant modulator in plant abiotic stress tolerance (Shi et al., 2012). This section will shed light on some of the major abiotic stresses experienced by plants as well as the role of NO as a regulator in response to these stresses.

13.4.1 Salt Stress

Salinity is considered to be one of the most common abiotic constraints and has been estimated to affect 800 million hectares of arable land across the world (Butcher et al., 2016). High salt concentrations in the soil can disrupt a variety of plant processes with the most common among them being the reduced ability of plants to take up water and minerals from the soil ultimately affecting their biochemical and physiological activities. Various methods have been implemented to overcome the extremities of salinity stress including alternate irrigation practices and application of different compounds and hormones. Over the last few years, NO has gained a lot of attention in combating salt stress.

Plants tend to produce NO under salt stress, mainly by the reductive pathway, which involves the reduction of nitrate to nitrite by the enzyme called nitrate reductase (NR), following which the nitrite is again reduced to NO by NR (Shang et al., 2022). NO has established its role as a salinity stress buster among different plant species majorly in the leguminous crop, chickpea which is known to be very sensitive to salt stress. The supply of nitric oxide to chickpeas under salt stress showed a significant improvement in shoot length, root length and shoot dry weight under high salinity conditions (Hasanuzzaman et al., 2011; Mostofa et al., 2015; Ahmad et al., 2016). Additionally, the accumulation of photosynthetic pigments was also observed to have increased upon supply of NO in high salinity conditions (Ahmad et al., 2016). According to studies, NO was also reported to have improved the cytoplasmic viscosity and increased the osmotic pressure of plant cells (Ahmad et al., 2016; Nabi et al., 2019). It also works as an antioxidant and a ROS scavenger in sorghum (Jasid et al., 2008; Arora et al., 2016). In wheat, exogenous applications of NO increased tolerance against salinity-induced oxidative damage (Hasanuzzaman et al., 2011). As per another finding, NO was seen associating with sulphur and showed to have enhanced the photosynthetic rate under salt stress conditions (Fatma et al., 2016). Interestingly, the NO application also showed its effect on seed germination as it helped to improve the germination index, vigour index and imbibition percentage of wheat seeds (Patel et al., 2017). NO's association with various other phytohormones under salt stress conditions is still a fresh intake for researchers. For example, in rice and maize, NO was shown to be positively regulating abscisic acid (ABA) levels in salt-stressed soil, while both NO and ethylene accumulation increased in root apices of tomatoes (Poór et al., 2014; Huang et al., 2020; Shang et al., 2022). It has also been reported that NO's association with brassinosteroids increased plant adaptation to salinity stress (Shang et al., 2022).

13.4.2 Drought Stress

Water is one of the most important factors responsible for the proper growth and yield of crops. So, a water deficit soil can pose a big threat to crop productivity. Drought stress has the potential of leading to oxidative stress by disrupting cellular redox homeostasis in plants and thus causing cell injury (Lau et al., 2021). Adapting to drought stress by developing various drought-avoiding/escaping mechanisms such as deeper roots, thick or waxy cuticle, sunken stomata or deposition of suberin have generally proven useful and they may or may not be associated with NO (Seleiman et al., 2021). NO has been shown to aid in coping with drought stress in a variety of plant species, including grains, legumes, fruit trees, medicinal plants and vegetables (Lau et al., 2021). For example, in watermelon, it was found that exogenous applications of NO significantly decreased the malondialdehyde accumulation due to drought stress as well as increased root fresh weight and length (Hamurcu et al., 2020). A detailed study performed in hull-less barley, in which sodium nitroprusside (SNP) acts as a NO donor, was found more effective in alleviating drought stress in comparison to other treatments. The NO application was found to have significantly reduced a drought-induced increase in leaf H_2O_2 content, MDA content and membrane permeability. Furthermore, the NO application under drought stress was also reported to increase proline contents in leaf cells, hinting that the buildup of osmolytes is associated with osmoregulation under drought conditions, allowing increased uptake of water (Gan et al., 2015). ROS accumulation and production are generally a consequence of water-deficit soil (Cruz de Carvalho, 2008). NO acts as a stress buster by not only reducing ROS levels by its antioxidant activity but by also reducing lipid-free radicals and superoxide anions (Halliwell & Gutteridge, 2006). Overall, it can be concluded that both exogenous and endogenous applications of NO have proven to be successful in combating drought and protecting plants from various oxidative injuries (Nabi et al., 2019).

13.4.3 Chilling Stress

Among the several abiotic stresses encountered by plants, cold stress is known to be a key factor affecting the growth and productivity of crops worldwide. It causes an imbalance of metabolites, cellular dysfunction, membrane damage and disruption of enzyme activity (Sánchez-Vicente et al., 2019). Recent studies on plant cold stress tolerance focus on exploring efficient methods to deal with the same; thus bringing NO into the light. NO levels have generally been found to increase in various plants as a response to cold stress and are formed mainly by the activity of nitrate reductase. An exogenous application of NO via SNP to melon (*Cucumis melo* L.) seedlings under cold stress showed an increase in chlorophyll and soluble solutes levels, thus improving seedling growth (Diao et al., 2022). SNP treatment also showed reduced relative conductivity, increased accumulation of proline, soluble sugars and protein and improved the activity of enzymes like catalase and superoxide dismutase in tea leaves (Wang et al., 2021).

In an interesting finding, SNP and osmopriming were proved to be effective in improving seed germination under low-temperature conditions as they increased amylase activity and soluble sugars, which ultimately resulted in better germination (Amooaghaie & Nikzad, 2013). Several other studies have demonstrated crosstalks of NO with other phytohormones in dealing with low-temperature stress (Freschi, 2013; Nawaz et al., 2017). In maize seedlings, NO, salicylic acid and ABA interaction significantly improved chilling injury by inducing antioxidant activity (Esim & Atici, 2016; Zhang et al., 2021). In bananas, NO has been reported to increase the activities of the enzymes diamine oxidase and glutamate decarboxylase, which lead to γ-aminobutyric acid accumulation, thereby aiding against chilling injury (Wang et al., 2016).

13.4.4 Heat Stress

High temperature causes lipid peroxidation, injuries to cell membranes, inactivation of enzymes, degradation of proteins and DNA damage in plants (Suzuki & Mittler, 2006). NO is known to enhance the protection of the cell membrane and reduce chlorophyll damage during heat stress. Plants pre-treated with exogenous NO showed high photosynthesis rates due to an increase in their chlorophyll content (Gautam et al., 2021). Such plants also show an increase in transcription of small heat shock protein 26 (sHsp26) and ROS scavenging enzymes like superoxide dismutase (SOD), peroxidase (POD), catalase (CAT) and ascorbate peroxidase (APX) (Uchida et al., 2002; Song et al., 2006). Thus, the application of NO or NO donors like SNP enhances cell viability, reduces ion leakage and promotes growth and other beneficial activities that were hindered by heat stress (Siddiqui et al., 2011).

13.4.5 Heavy Metal Stress

The NO treatment has been observed to reduce heavy metal toxicity and decrease ROS production in plants. In wheat, pre-treatment with NO was reported to alleviate oxidative stress due to aluminium (Al^{3+}) toxicity by increasing the activity of antioxidant enzymes APX, CAT and SOD and by decreasing H_2O_2 and MDA synthesis (Zhang et al., 2008). NO was also found to be an effective modulator against copper (Cu) toxicity (Yu et al., 2005; Singh et al., 2008). In tomato plants, the NO application via SNP induced several ROS scavenging enzymes and reduced the accumulation of H_2O_2 and growth hindrances induced by copper chloride (Cui et al., 2009). NO was also reported to be instrumental in alleviating cadmium (Cd) toxicity. In rice, the exogenous application of NO increased pectin and hemicellulose contents as well as Cd distribution in the cell walls of roots and decreased Cd accumulation in leaves as a means of Cd tolerance (Xiong et al., 2009).

13.5 CONCLUSION AND FUTURE PERSPECTIVES

NO has a highly significant part in plant defence against different kinds of stresses. Its role as a signalling molecule in abiotic stress tolerance has made it the focus of various stress acclimatization research in the scientific community. NO mainly participates in signal transduction by the protein post-translational modification (PTMs) and S-nitrosylation. Protein S-nitrosylation has been observed to affect a variety of functional parameters such as enzymatic activity, subcellular localization, protein-protein interactions and protein stability. A recent study has also found that several other PTMs like phosphorylation, acetylation, ubiquitylation etc. are also affected by S-nitrosylation through crosstalks. Due to these revelations about the role of S-nitrosylation in cellular functioning, there have been growing efforts to identify *S*-nitrosylated proteins in plants. While it has been recognized that S-nitrosylation is an integral part of plant abiotic stress responses, much of the molecular mechanisms behind it remain to be established. The

S-nitroso glutathione reductase enzyme, which regulates S-nitroso glutathione metabolism, has also been known to control cellular S-nitrosylation levels. Nevertheless, its role during abiotic stress and the development of plant immunity remains to be elucidated. The regulatory role of NO in the management of these NO-dependent PTMs also remains uncertain. Even though NO has been identified in almost all types of abiotic stress (salt, drought, heat, chilling, heavy metal) tolerance mechanisms in plants, the molecular mechanisms underlying the various signalling pathways is yet to be fully understood. As the integral role of NO in plant tolerance to abiotic stress emerges, insights into the sources and factors affecting its levels in plants as well as the fundamental mechanisms associated with its regulatory and signalling functions might aid in crop breeding to develop stress-tolerant varieties, ultimately leading to the improvement of crop growth and yield. This becomes particularly important, owing to the rising global consumption and declining agronomic conditions due to climate change.

REFERENCES

Abat, J. K., & Deswal, R. (2009). Differential modulation of S-nitrosoproteome of *Brassica juncea* by low temperature: change in S-nitrosylation of Rubisco is responsible for the inactivation of its carboxylase activity. Proteomics, 9(18), 4368–4380. 10.1002/pmic.200800985.

Ahmad, P., Abdel Latef, A. A., Hashem, A., Abd_Allah, E. F., Gucel, S., & Tran, L. S. P. (2016). Nitric oxide mitigates salt stress by regulating levels of osmolytes and antioxidant enzymes in chickpea. Frontiers in Plant Science, 7, 347. 10.3389/fpls.2016.00347.

Akhtar, M. W., Sunico, C. R., Nakamura, T., & Lipton, S. A. (2012). Redox regulation of protein function via cysteine S-nitrosylation and its relevance to neurodegenerative diseases. International Journal of Cell Biology, 2012. 10.1155/2012/463756.

Amooaghaie, R., & Nikzad, K. (2013). The role of nitric oxide in priming-induced low-temperature tolerance in two genotypes of tomato. Seed Science Research, 23(2), 123–131. 10.1017/s0960258513000068.

Anand, P., & Stamler, J. S. (2012). Enzymatic mechanisms regulating protein S-nitrosylation: implications in health and disease. Journal of Molecular Medicine, 90, 233–244. 10.1007/s00109-012-0878-z.

Arora, D., Jain, P., Singh, N., Kaur, H., & Bhatla, S. C. (2016). Mechanisms of nitric oxide crosstalk with reactive oxygen species scavenging enzymes during abiotic stress tolerance in plants. Free Radical Research, 50(3), 291–303. 10.3109/10715762.2015.1118473.

Asada, K. (1992). Ascorbate peroxidase–a hydrogen peroxide-scavenging enzyme in plants. Physiologia Plantarum, 85(2), 235–241. 10.1111/j.1399-3054.1992.tb04728.x.

Asgher, M., Per, T. S., Masood, A., Fatma, M., Freschi, L., Corpas, F. J., & Khan, N. A. (2017). Nitric oxide signaling and its crosstalk with other plant growth regulators in plant responses to abiotic stress. Environmental Science and Pollution Research, 24, 2273–2285. 10.1007/s11356-016-7947-8.

Barroso, J. B., Corpas, F. J., Carreras, A., Sandalio, L. M., Valderrama, R., Palma, J., et al. (1999). Localization of nitric-oxide synthase in plant peroxisomes. Journal of Biological Chemistry, 274(51), 36729–36733. 10.1074/jbc.274.51.36729.

Begara-Morales, J. C., Sánchez-Calvo, B., Chaki, M., Valderrama, R., Mata-Pérez, C., Padilla, M. N., et al. (2016). Antioxidant systems are regulated by nitric oxide-mediated post-translational modifications (NO-PTMs). Frontiers in Plant Science, 7, 152. 10.3389/fpls.2016.00152.

Besson-Bard, A., Gravot, A., Richaud, P., Auroy, P., Duc, C., Gaymard, F., et al. (2009). Nitric oxide contributes to cadmium toxicity in *Arabidopsis* by promoting cadmium accumulation in roots and by up-regulating genes related to iron uptake. Plant Physiology, 149(3), 1302–1315. 10.1104/pp.108.133348.

Bright, J., Desikan, R., Hancock, J. T., Weir, I. S., & Neill, S. J. (2006). ABA-induced NO generation and stomatal closure in *Arabidopsis* are dependent on H2O2 synthesis. The Plant Journal, 45(1), 113–122. 10.1111/j.1365-313x.2005.02615.x.

Butcher, K., Wick, A. F., DeSutter, T., Chatterjee, A., & Harmon, J. (2016). Soil salinity: A threat to global food security. Agronomy Journal, 108(6), 2189–2200. 10.2134/agronj2016.06.0368.

Campbell, W. H. (1988). Nitrate reductase and its role in nitrate assimilation in plants. Physiologia Plantarum, 74(1), 214–219. 10.1111/j.1399-3054.1988.tb04965.x.

Cruz de Carvalho, M. H. (2008). Drought stress and reactive oxygen species: production, scavenging and signaling. Plant Signaling & Behavior, 3(3), 156–165. 10.4161/psb.3.3.5536.

Cui, X., Zhang, Y., Chen, X., Jin, H., & Wu, X. (2009). Effects of exogenous nitric oxide protects tomato plants under copper stress. In: 3rd International Conference on Bioinformatics and Biomedical Engineering (pp. 1–7). Institute of Electricals and Electronic Engineers. 10.1109/icbbe.2009.5162740.

Diao, Q. N., Cao, Y. Y., Wang, H., Zhang, Y. P., & Shen, H. B. (2022). Nitric oxide confers chilling stress tolerance by regulating carbohydrate metabolism and the antioxidant defense system in melon (*Cucumis melo* L.) seedlings. Hortscience, 57(10), 1249–1256. 10.21273/hortsci16677-22.

Elviri, L., Speroni, F., Careri, M., Mangia, A., di Toppi, L. S., & Zottini, M. (2010). Identification of in vivo nitrosylated phytochelatins in *Arabidopsis thaliana* cells by liquid chromatography-direct electrospray-linear ion trap-mass spectrometry. Journal of Chromatography A, 1217(25), 4120–4126. 10.1016/j.chroma.2010.02.013.

Esim, N., & Atici, Ö. (2016). Relationships between some endogenous signal compounds and the antioxidantsystem in response to chilling stress in maize (*Zea mays* L.) seedlings. Turkish Journal of Botany, 40(1), 37–44. 10.3906/bot-1408-59.

Fancy, N. N., Bahlmann, A. K., & Loake, G. J. (2017). Nitric oxide function in plant abiotic stress. Plant, Cell & Environment, 40(4), 462–472. 10.1111/pce.12707.

Fatma, M., Masood, A., Per, T. S., & Khan, N. A. (2016). Nitric oxide alleviates salt stress inhibited photosynthetic performance by interacting with sulfur assimilation in mustard. Frontiers in Plant Science, 7, 521. 10.3389/fpls.2016.00521.

Foresi, N., Correa-Aragunde, N., Parisi, G., Calo, G., Salerno, G., & Lamattina, L. (2010). Characterization of a nitric oxide synthase from the plant kingdom: NO generation from the green alga Ostreococcus tauri is light irradiance and growth phase dependent. The Plant Cell, 22(11), 3816–3830. 10.1105/tpc.109.073510.

Forman, H. J., Fukuto, J. M., & Torres, M. (2004). Redox signaling: thiol chemistry defines which reactive oxygen and nitrogen species can act as second messengers. American Journal of Physiology, 287(2), C246–C256. 10.1152/ajpcell.00516.2003.

Freschi, L. (2013). Nitric oxide and phytohormone interactions: current status and perspectives. Frontiers in Plant Science, 4, 398. 10.3389/fpls.2013.00398.

Gan, L., Wu, X., & Zhong, Y. (2015). Exogenously applied nitric oxide enhances the drought tolerance in hulless barley. Plant Production Science, 18(1), 52–56. 10.1626/pps.18.52.

Gautam, H., Sehar, Z., Rehman, M. T., Hussain, A., AlAjmi, M. F., & Khan, N. A. (2021). Nitric oxide enhances photosynthetic nitrogen and sulfur-use efficiency and activity of ascorbate-glutathione cycle to reduce high temperature stress-induced oxidative stress in rice (*Oryza sativa* L.) plants. Biomolecules, 11(2), 305. 10.3390/biom11020305.

Gutteridge, J. M., & Halliwell, B. (2000). Free radicals and antioxidants in the year 2000: a historical look to the future. Annals of the New York Academy of Sciences, 899(1), 136–147. 10.1111/j.1749-6632.2000.tb06182.x.

Halliwell B., & Gutteridge J. M. C. (2006). Free Radicals in Biology and Medicine, 4th Ed. Oxford: Clarendon Press.

Hamurcu, M., Khan, M. K., Pandey, A., Ozdemir, C., Avsaroglu, Z. Z., Elbasan, F., et al. (2020). Nitric oxide regulates watermelon (*Citrullus lanatus*) responses to drought stress. 3 Biotech, 10, 1–14. 10.1007/s13205-020-02479-9.

Hasanuzzaman, M., Hossain, M. A., & Fujita, M. (2011). Nitric oxide modulates antioxidant defense and the methylglyoxal detoxification system and reduces salinity-induced damage of wheat seedlings. Plant Biotechnology Reports, 5, 353–365. 10.1007/s11816-011-0189-9.

He, M., He, C. Q., & Ding, N. Z. (2018). Abiotic stresses: general defenses of land plants and chances for engineering multistress tolerance. Frontiers in Plant Science, 9, 1771. 10.3389/fpls.2018.01771.

Hess, D. T., Matsumoto, A., Kim, S. O., Marshall, H. E., & Stamler, J. S. (2005). Protein S-nitrosylation: purview and parameters. Nature Reviews Molecular Cell Biology, 6(2), 150–166. 10.1038/nrm1569.

Huang, J., Zhu, C., Hussain, S., Huang, J., Liang, Q., Zhu, L., et al. (2020). Effects of nitric oxide on nitrogen metabolism and the salt resistance of rice (*Oryza sativa* L.) seedlings with different salt tolerances. Plant Physiology and Biochemistry, 155, 374–383. 10.1016/j.plaphy.2020.06.013.

Jasid, S., Simontacchi, M., & Puntarulo, S. (2008). Exposure to nitric oxide protects against oxidative damage but increases the labile iron pool in sorghum embryonic axes. Journal of Experimental Botany, 59(14), 3953–3962. 10.1093/jxb/ern235.

Kovacs, I., & Lindermayr, C. (2013). Nitric oxide-based protein modification: formation and site-specificity of protein S-nitrosylation. Frontiers in Plant Science, 4, 137. 10.3389/fpls.2013.00137.

Kovacs, I., Holzmeister, C., Wirtz, M., Geerlof, A., Fröhlich, T., Römling, G., … & Lindermayr, C. (2016). ROS-mediated inhibition of S-nitrosoglutathione reductase contributes to the activation of anti-oxidative mechanisms. Frontiers in Plant Science, 7, 1669. 10.3389/fpls.2016.01669.

Lau, S. E., Hamdan, M. F., Pua, T. L., Saidi, N. B., & Tan, B. C. (2021). Plant nitric oxide signaling under drought stress. Plants, 10(2), 360. 10.3390/plants10020360.

Mostofa, M. G., Fujita, M., & Tran, L. S. P. (2015). Nitric oxide mediates hydrogen peroxide-and salicylic acid-induced salt tolerance in rice (*Oryza sativa* L.) seedlings. Plant Growth Regulation, 77, 265–277. 10.1007/s10725-015-0061-y.

Nabi, R. B. S., Tayade, R., Hussain, A., Kulkarni, K. P., Imran, Q. M., Mun, B. G., & Yun, B. W. (2019). Nitric oxide regulates plant responses to drought, salinity, and heavy metal stress. Environmental and Experimental Botany, 161, 120–133. 10.1016/j.envexpbot.2019.02.003.

Nawaz, F., Shabbir, R. N., Shahbaz, M., Majeed, S., Raheel, M., Hassan, W., & Sohail, M. A. (2017). Cross talk between nitric oxide and phytohormones regulate plant development during abiotic stresses. In: M. El-Esawi (ed.), Phytohormones: Signaling Mechanisms and Crosstalk in Plant Development and Stress Responses (pp. 117–141). IntechOpen. 10.5772/intechopen.69812.

Nguyen, H. C., Lin, K. H., Ho, S. L., Chiang, C. M., & Yang, C. M. (2018). Enhancing the abiotic stress tolerance of plants: from chemical treatment to biotechnological approaches. Physiologia Plantarum, 164(4), 452–466. 10.1111/ppl.12812.

Noctor, G., & Foyer, C. H. (1998). Ascorbate and glutathione: keeping active oxygen under control. Annual Review of Plant Biology, 49(1), 249–279. 10.1146/annurev.arplant.49.1.249.

Patel, P., Kadur Narayanaswamy, G., Kataria, S., & Baghel, L. (2017). Involvement of nitric oxide in enhanced germination and seedling growth of magnetoprimed maize seeds. Plant Signaling & Behavior, 12(12), e1293217. 10.1080/15592324.2017.1293217.

Planchet, E., & Kaiser, W. M. (2006). Nitric oxide production in plants: facts and fictions. Plant Signaling & Behavior, 1(2), 46–51. 10.4161/psb.1.2.2435.

Plenchette, S., Romagny, S., Laurens, V., & Bettaieb, A. (2015). S-Nitrosylation in TNF superfamily signaling pathway: Implication in cancer. Redox Biology, 6, 507–515. 10.1016/j.redox.2015.08.019.

Poór, P., Borbély, P., Kovács, J., Papp, A., Szepesi, Á., Takács, Z., & Tari, I. (2014). Opposite extremes in ethylene/nitric oxide ratio induce cell death in suspension culture and root apices of tomato exposed to salt stress. Acta Biologica Hungarica, 65(4), 428–438. 10.1556/abiol.65.2014.4.7.

Qiao, W., & Fan, L. M. (2008). Nitric oxide signaling in plant responses to abiotic stresses. Journal of Integrative Plant Biology, 50(10), 1238–1246. 10.1111/j.1744-7909.2008.00759.x.

Sánchez-Vicente, I., Fernández-Espinosa, M. G., & Lorenzo, O. (2019). Nitric oxide molecular targets: reprogramming plant development upon stress. Journal of Experimental Botany, 70(17), 4441–4460. 10.1093/jxb/erz339.

Sang, J., Jiang, M., Lin, F., Xu, S., Zhang, A., & Tan, M. (2008). Nitric oxide reduces hydrogen peroxide accumulation involved in water stress-induced subcellular anti-oxidant defense in maize plants. Journal of Integrative Plant Biology, 50(2), 231–243. 10.1111/j.1744-7909.2007.00594.x.

Seleiman, M. F., Al-Suhaibani, N., Ali, N., Akmal, M., Alotaibi, M., Refay, Y., et al. (2021). Drought stress impacts on plants and different approaches to alleviate its adverse effects. Plants, 10(2), 259. 10.3390/plants10020259.

Shang, J. X., Li, X., Li, C., & Zhao, L. (2022). The role of nitric oxide in plant responses to salt stress. International Journal of Molecular Sciences, 23(11), 6167. 10.3390/ijms23116167.

Shi, H. T., Li, R. J., Cai, W., Liu, W., Fu, Z. W., & Lu, Y. T. (2012). In vivo role of nitric oxide in plant response to abiotic and biotic stress. Plant Signaling & Behavior, 7(3), 437–439. 10.4161/psb.19219.

Shigeoka, S., Ishikawa, T., Tamoi, M., Miyagawa, Y., Takeda, T., Yabuta, Y., & Yoshimura, K. (2002). Regulation and function of ascorbate peroxidase isoenzymes. Journal of Experimental Botany, 53(372), 1305–1319. 10.1093/jexbot/53.372.1305.

Siddiqui, M. H., Al-Whaibi, M. H., & Basalah, M. O. (2011). Role of nitric oxide in tolerance of plants to abiotic stress. Protoplasma, 248, 447–455. 10.1007/s00709-010-0206-9.

Singh, H. P., Batish, D. R., Kaur, G., Arora, K., & Kohli, R. K. (2008). Nitric oxide (as sodium nitroprusside) supplementation ameliorates Cd toxicity in hydroponically grown wheat roots. Environmental and Experimental Botany, 63(1-3), 158–167. 10.1016/j.envexpbot.2007.12.005.

Song, L., Ding, W., Zhao, M., Sun, B., & Zhang, L. (2006). Nitric oxide protects against oxidative stress under heat stress in the calluses from two ecotypes of reed. Plant Science, 171(4), 449–458. 10.1016/j.plantsci.2006.05.002.

Suzuki, N., & Mittler, R. (2006). Reactive oxygen species and temperature stresses: a delicate balance between signaling and destruction. Physiologia Plantarum, 126(1), 45–51. 10.1111/j.0031-9317.2005.00582.x.

Tola, A. J., Jaballi, A., & Missihoun, T. D. (2021). Protein carbonylation: Emerging roles in plant redox biology and future prospects. Plants, 10(7), 1451. 10.3390/plants10071451.

Toledo Jr, J. C., & Augusto, O. (2012). Connecting the chemical and biological properties of nitric oxide. Chemical Research in Toxicology, 25(5), 975–989. 10.1021/tx300042g.

Uchida, A., Jagendorf, A. T., Hibino, T., Takabe, T., & Takabe, T. (2002). Effects of hydrogen peroxide and nitric oxide on both salt and heat stress tolerance in rice. Plant Science, 163(3), 515–523. 10.1016/s0168-9452(02)00159-0.

Verma, V., Ravindran, P., & Kumar, P. P. (2016). Plant hormone-mediated regulation of stress responses. BMC Plant Biology, 16, 1-10. 10.1186/s12870-016-0771-y.

Wang, Y., Luo, Z., Mao, L., & Ying, T. (2016). Contribution of polyamines metabolism and GABA shunt to chilling tolerance induced by nitric oxide in cold-stored banana fruit. Food Chemistry, 197, 333–339. 10.1016/j.foodchem.2015.10.118.

Wang, Y., Yu, Q., Li, Y., Li, J., Chen, J., Liu, Z., ... & Eissa, M. A. (2021). Mechanisms of nitric oxide in the regulation of chilling stress tolerance in Camellia sinensis. Horticulturae, 7(10), 410. 10.3390/horticulturae7100410.

Waqas, M. A., Kaya, C., Riaz, A., Farooq, M., Nawaz, I., Wilkes, A., & Li, Y. (2019). Potential mechanisms of abiotic stress tolerance in crop plants induced by thiourea. Frontiers in Plant Science, 10, 1336. 10.3389/fpls.2019.01336.

Wolhuter, K., Whitwell, H. J., Switzer, C. H., Burgoyne, J. R., Timms, J. F., & Eaton, P. (2018). Evidence against stable protein S-nitrosylation as a widespread mechanism of post-translational regulation. Molecular Cell, 69(3), 438–450. 10.1016/j.molcel.2017.12.019.

Xiong, J., An, L., Lu, H., & Zhu, C. (2009). Exogenous nitric oxide enhances cadmium tolerance of rice by increasing pectin and hemicellulose contents in root cell wall. Planta, 230, 755–765. 10.1007/s00425-009-0984-5.

Yu, C. C., Hung, K. T., & Kao, C. H. (2005). Nitric oxide reduces Cu toxicity and Cu-induced NH4+ accumulation in rice leaves. Journal of Plant Physiology, 162(12), 1319–1330. 10.1016/j.jplph.2005.02.003.

Zhang, H., Li, Y. H., Hu, L. Y., Wang, S. H., Zhang, F. Q., & Hu, K. D. (2008). Effects of exogenous nitric oxide donor on antioxidant metabolism in wheat leaves under aluminum stress. Russian Journal of Plant Physiology, 55, 469–474. 10.1134/s1021443708040067.

Zhang, Q., Li, D., Wang, Q., Song, X., Wang, Y., Yang, X., ... & Yang, D. (2021). Exogenous salicylic acid improves chilling tolerance in maize seedlings by improving plant growth and physiological characteristics. Agronomy, 11(7), 1341. 10.3390/agronomy11071341.

Zhou, X., Joshi, S., Khare, T., Patil, S., Shang, J., & Kumar, V. (2021). Nitric oxide, crosstalk with stress regulators and plant abiotic stress tolerance. Plant Cell Reports, 40(8), 1395–1414. 10.1007/s00299-021-02705-5.

14 Responses of Oxylipins to Abiotic Stresses

Amooru Harika, Ayyagari Ramlal, Nguyen Trung Duc, Ambika Rajendran, and Dhandapani Raju

14.1 INTRODUCTION

In the complex field of plant biology, where survival is crucial, plants have developed a wide range of complex defence systems to fend off possible attackers and adjust to constantly shifting environmental conditions (Karban, 2008). Plants have functionally identical defensive mechanisms, unlike animals with their immune systems, which allow them to recognize viruses, stresses and physical harm and launch strong defence responses (Muthamilarasan & Prasad, 2013). An intricate defence orchestra includes oxylipins, a class of chemicals that are among the main players.

Oxylipins are known modulators of plants' biotic/abiotic stress responses and development (Howe & Schilmiller, 2002; Howe et al., 2018; Wasternack & Feussner, 2018; Wasternack & Strnad, 2018; Waadt et al., 2022). Phyto-oxylipins are diverse bioactive lipids derived from the oxidation of polyunsaturated fatty acids, mainly linoleic acid and α-linolenic acid. Jasmonates, divinyl ethers and green leaf volatiles (GLVs) are examples of this class (ul Hassan et al., 2015). Plant oxylipins regulate many aspects of plant development, growth and responses to environmental stimuli.

A great deal of research, aided by precise biochemical dissection and genetic inquiry, has revealed the role of oxylipins in plant biology. They are not passive observers; rather, they actively contribute to strengthening plant defence systems, affecting both biotic and abiotic stress reactions, and are deeply entwined with the growth of plants (Hou et al., 2016). There are intricate connections between oxylipins and several abiotic stressors, each of which poses particular difficulties for plant life and understanding under different stresses such as temperature changes, dehydration and osmoticum (Solhi et al., 2023). These stressors include a wide range of challenges, such as mechanical harm caused by different agents, temperature changes, dehydration, osmotic stress, exposure to ozone and the extreme pressures associated with heavy metal stress. The potential of oxylipins as biocontrol agents has been revealed by recent research, which shows that they can both increase plant resistance and advance environmental sustainability (Christie & Harwood, 2020). The functions of oxylipins in mitigating the effects of high ozone levels and heavy metal stress have been deciphered (Griffiths, 2015). While the literature has extensively scrutinized oxylipins in the context of biotic stress reactions and their contributions to plant defence against pathogens, their equally vital roles in assisting plants in adapting to abiotic stress conditions have not garnered the same level of attention (Siddiqi & Husen, 2019).

This chapter aims to highlight the critical functions that oxylipins play during abiotic stressors. It also presents a story of oxylipins and elucidates their crucial yet often-overlooked functions in plant adaptation to abiotic stressors to uncover new information that can revolutionize the approaches to environmentally responsible farming and sustainable agriculture.

14.2 ABIOTIC STRESS

Abiotic stress refers to adverse environmental conditions that significantly impact plant growth, development, and overall well-being (Oshunsanya et al., 2019). Various abiotic stresses and the

DOI: 10.1201/9781003335788-18

crucial role played by oxylipins in mitigating their effects will be discussed in the subsequent sections.

14.2.1 Drought Stress

A major problem for plants is drought stress, which is caused by either too much transpiration or too little soil water and is frequently exacerbated by high temperatures (Ahluwalia et al., 2021). Plants use a variety of lipid oxidation processes to convert unsaturated fatty acids oxidatively, producing oxylipins, which are essential signaling molecules (Mosblech et al., 2009). Their importance in drought stress responses is highlighted by their diversity and complexity. These compounds have a variety of functions in stress tolerance and plant hydration (Llanes et al., 2018). The regulation of stomatal closure is one of the prominent roles of oxylipins (Simeoni et al., 2022). Oxylipins play a crucial role in lowering transpiration rates during dry spells by controlling the closure of stomata, which are tiny holes in plant leaves (Vahdati & Lotfi, 2013). In plant cells, oxylipins also support osmotic adjustment mechanisms that preserve turgor pressure and general cellular hydration (Colmenero-Flores et al., 2020). Furthermore, reactive oxygen species (ROS) produced under drought stress act as protectors against oxidative damage caused by oxylipins. They serve as ROS scavengers, preventing oxidative damage to plant cells (Czarnocka & Karpiński, 2018). Within the context of molecular signaling, oxylipins play the role of secondary messengers, coordinating the activation of stress-responsive genes that are essential for drought resilience (Liang et al., 2023). They are much more important for plants to adapt to water constraints because of their complex interactions with other signaling pathways.

14.2.2 Wounding

Plants can get wounds from a variety of sources, including biotic factors such as feeding insects and abiotic elements like wind or hail (Cabane et al., 2012). A thorough examination of the methods and causes underlying mechanical injury to plant tissues is necessary to comprehend wounding. Numerous variables, including agricultural techniques, climatic conditions like wind, or external forces like herbivore feeding, can cause mechanical damage (Rouet-Leduc et al., 2021). Plant tissues may sustain damage that compromises their structural and physiological integrity. In the plant's defence mechanism against injury, oxylipins are an essential component (León et al., 2001). Oxylipins serve as messengers, quickly informing other plant sections about the damage (Knieper et al., 2023). Activating defence mechanisms, such as producing chemicals that protect, fortifying cell walls and even starting communication cascades that discourage herbivores, are some examples of these reactions (Wolf, 2022).

Oxylipin signaling includes an intriguing feature called positive feedback loops (Wasternack & Feussner, 2018). They function as enhancers, boosting the oxylipin-mediated reactions to injury. This explains the nuances of these feedback processes and shows how the plant may create a strong defence against additional damage as a result of them (Schilmiller & Howe, 2005). Determining how resilient plants are in the face of painful experiences requires an understanding of these positive feedback loops. A significant part of the healing process following injury is played by phytoprostanes, a class of bioactive lipids produced by oxidative reactions in plants (Christie & Harwood, 2020; Imbusch & Mueller, 2000). The phytoprostanes emphasize their importance in the way plants react to wounds. From signaling pathways to the manufacture of defensive chemicals, these substances are involved in many aspects of wound response (Mansoor et al., 2023). Our knowledge of how plants defend themselves against damage is improved by unraveling the role of phytoprostanes, which may have implications for farming and other fields.

14.2.3 Salinity Stress

Salinity stress is a potent abiotic stressor that raises the concentration of salt in the soil, which interferes with a plant's regular physiological processes (Chauhan et al., 2022). Oxylipins sometimes interplay with other hormones, in particular, jasmonates play in plants' adaptations to this demanding type of abiotic stress (Santino et al., 2013). When soil salinity levels rise to a level where they make it more difficult for plants to absorb water and vital nutrients, it is known as salinity stress (Shrivastava & Kumar, 2015). The difficulties caused by salinity stress are decreased water absorption, ionic imbalances and cellular toxicity due to excess salt (Arif et al., 2020). These difficulties may impede development and growth and in extreme situations, result in plant mortality. A subclass of oxylipins known as jasmonates become important in aiding plants in their defence against salt stress (Wasternack & Hause, 2002). This part explores the precise role that jasmonates play in plant reactions to high soil salt. As signaling molecules, jasmonates set off a series of reactions that work to lessen the negative consequences of salt stress (Taheri et al., 2020). To combat the oxidative stress brought by excess salt, plants control ion transport, adjust osmotic potential and trigger antioxidant mechanisms.

14.2.4 Heat Stress

Heat stress, which is defined as temperatures that are higher than what is ideal for a plant, is a serious hazard to the life and productivity of plants (Wahid et al., 2007). Numerous physiological functions, like as photosynthesis, transpiration and nutrient intake, can be hampered by high temperatures (dos Santos et al., 2022). Heat stress can therefore result in decreased growth, yield losses and even irreparable tissue damage in plants. Oxylipins/jasmonates mitigate the effects of heat stress on plants (Sharma et al., 2023). As signaling molecules, jasmonates coordinate a variety of defensive reactions. They assist plants to withstand the damaging effects of heat by regulating osmoprotectants, heat shock proteins and antioxidant enzymes (Sharma et al., 2020). According to Knieper et al., (2023), RES oxylipins serve as a primitive form of stress resistance against thermotolerance that develops independently of JA signaling. Major chaperones implicated in heat stress acclimation, 12-OPDA, phytoprostanes and MDA stimulate the synthesis of heat shock proteins to combat heat stress (Monte et al., 2020).

14.2.5 Poor Light and Temperature

One of the most prevalent environmental stresses that plants have to deal with is fluctuations in temperature and light (Vonshak & Torzillo, 2003). An increase or decrease in light levels might cause oxylipin buildup Ghanem et al., 2012). The mechanism by which changes in light intensity and quality boost plant oxylipin synthesis. In addition to their possible use in boosting plant resilience, it addresses the particular pathways and enzymes involved in light-related oxylipin accumulation. To mediate stress reactions, oxylipins and jasmonates frequently work together.

14.2.6 Dehydration and Osmotic Stress

Plants face difficult challenges from dehydration and osmotic stress since these conditions impair cellular turgor and general health (Wood, 2015). Oxylipins, along with jasmonates, are involved during osmotic stress. This examines the critical function that jasmonates play in coordinating the plant's defence mechanisms against water deprivation (Souri et al., 2020). It explores the complex gene expression networks and signaling pathways that jasmonates modulate, providing insight into their roles in osmotic stress resilience. Plants frequently collect osmoprotectants, which support cellular integrity and enable them to tolerate osmotic stress (Raza et al., 2021). It elucidates the molecular processes via which jasmonates control the synthesis of osmoprotectants, improving the plant's capacity to flourish in conditions devoid of water.

14.2.7 Ozone Exposure

Even though ozone is essential to life in the Earth's upper atmosphere, large concentrations of it close to the surface can be hazardous (McKenzie et al., 2011). The effects of high ozone levels on plants are discussed in detail in this section, along with the complex roles that oxylipins and jasmonates play in regulating this abiotic stress. High ozone levels are a silent danger to plants, causing obvious leaf damage as well as reduced photosynthesis (Gheorghe & Ion, 2011). Jasmonates become important adversaries against ozone stress within the intricate interplay of plant defence mechanisms (Kazan & Manners, 2011). However, studies are required to understand the roles of oxylipins in mitigating stress caused by ozone in plants.

14.2.8 Heavy Metal Stress

When exposed to contaminated soils, heavy metals represent a serious threat to plants due to their poisonous nature (Clemens, 2006). Plants are at risk due to heavy metal contamination in soils. It explains how heavy metals cause nutritional imbalances, oxidative stress and stunted growth in plants by interfering with basic physiological processes (Nagajyoti et al., 2010).

When exposed to heavy metal stress, plants use oxylipins and defensins as two different lines of defence. Defensins, small cysteine-rich proteins found in plant and animal cells, play a crucial role in many cases of jasmonate-dependent resistance to heavy metals. Most plant defensins have bactericidal and fungicidal activities (Thomma et al., 2002). The oxylipins act as signaling molecules in coordinating the body's defences against the harmful effects of heavy metals. In addition, defensins play a vital role in the plant's defensive repertoire due to their antibacterial characteristics. Intriguingly, the resistance to selenium that results from the antagonistic interaction between JA and ethylene observed during ozone stress is suppressed by the combined action of these two hormones (Tamaoki et al., 2008).

14.2.9 Nutrient Toxicity

Excessive soil nutrient levels can induce nutrient toxicity in plants, resulting in adverse effects such as leaf browning, reduced growth and impaired nutrient absorption (McCauley et al., 2009). Oxylipins play a pivotal role in regulating plant responses to nutrient toxicity by initiating various protective mechanisms, including detoxification and sequestration of surplus nutrients (Kumar et al., 2017). Harnessing the understanding of oxylipin signaling holds significant potential for enhancing nutrient management in agriculture, bolstering crop resistance to soil fluctuations and optimizing nutrient uptake efficiency. This area needs to be explored to understand the mechanism of oxylipins under nutrient-deprived conditions.

14.3 CONCLUSION AND FUTURE PROSPECTS

The roles of oxylipins in aiding plants to adapt to various abiotic stressors, including heat, salinity, drought, dehydration, osmotic stress, ozone exposure and heavy metal toxicity were explored in this chapter. Oxylipins have been shown to enhance plant resilience through their ability to facilitate rapid responses, amplify signaling through feedback mechanisms and interact with other molecules like phytoprostanes and jasmonates. These findings have significant implications for both environmental sustainability and agriculture. A deeper understanding of oxylipins' involvement in abiotic stress responses can lead to innovative farming practices that enhance crop resilience, optimize nutrient management and reduce the environmental impact of agriculture. The use of oxylipins as biocontrol agents may also reduce the need for chemical crop protection methods, promoting sustainability. However, the progress is still in its infancy and there is a need for understanding and exploring the underlying mechanisms and molecular basis of oxylipins during different stresses.

REFERENCES

Ahluwalia, O., Singh, P. C., & Bhatia, R. (2021). A review on drought stress in plants: Implications, mitigation and the role of plant growth promoting rhizobacteria. Resources, Environment and Sustainability, 5, 100032.

Arif, Y., Singh, P., Siddiqui, H., Bajguz, A., & Hayat, S. (2020). Salinity induced physiological and biochemical changes in plants: An omic approach towards salt stress tolerance. Plant Physiology and Biochemistry, 156, 64–77.

Cabane, M., Afif, D., & Hawkins, S. (2012). Lignins and abiotic stresses. In Advances in botanical research, vol. 61, pp. 219–262, Academic Press.

Chauhan, P. K., Upadhyay, S. K., Tripathi, M., Singh, R., Krishna, D., Singh, S. K., & Dwivedi, P. (2022). Understanding the salinity stress on plant and developing sustainable management strategies mediated salt-tolerant plant growth-promoting rhizobacteria and CRISPR/Cas9. Biotechnology and Genetic Engineering Reviews, 1–37.

Christie, W. W., & Harwood, J. L. (2020). Oxidation of polyunsaturated fatty acids to produce lipid mediators. Essays in Biochemistry, 64(3), 401–421.

Clemens, S. (2006). Toxic metal accumulation, responses to exposure and mechanisms of tolerance in plants. Biochimie, 88(11), 1707–1719.

Colmenero-Flores, J. M., Arbona, V., Morillon, R., & Gómez-Cadenas, A. (2020). Salinity and water deficit. In The genus citrus, pp. 291–309, Woodhead Publishing.

Czarnocka, W., & Karpiński, S. (2018). Friend or foe? Reactive oxygen species production, scavenging and signaling in plant response to environmental stresses. Free Radical Biology and Medicine, 122, 4–20.

dos Santos, T. B., Ribas, A. F., de Souza, S. G. H., Budzinski, I. G. F., & Domingues, D. S. (2022). Physiological responses to drought, salinity, and heat stress in plants: a review. Stresses, 2(1), 113–135.

Ghanem, M. E., Ghars, M. A., Frettinger, P., Pérez-Alfocea, F., Lutts, S., Wathelet, J. P., et al. (2012). Organ-dependent oxylipin signature in leaves and roots of salinized tomato plants (Solanum lycopersicum). Journal of Plant Physiology, 169(11), 1090–1101.

Gheorghe, I. F., & Ion, B. (2011). The effects of air pollutants on vegetation and the role of vegetation in reducing atmospheric pollution. In: M. K. Khallaf (ed.), The impact of air pollution on health, economy, environment and agricultural sources (vol. 29, pp. 241–280). InTechOpen.

Griffiths, G. (2015). Biosynthesis and analysis of plant oxylipins. Free Radical Research, 49(5), 565–582.

Hou, Q., Ufer, G., & Bartels, D. (2016). Lipid signalling in plant responses to abiotic stress. Plant, Cell & Environment, 39(5), 1029–1048.

Howe, G. A., & Schilmiller, A. L. (2002). Oxylipin metabolism in response to stress. Current Opinion in Plant Biology, 5(3), 230–236.

Howe, G. A., Major, I. T., & Koo, A. J. (2018). Modularity in jasmonate signaling for multistress resilience. Annual Reviews in Plant Biology, 69(1), 387–415.

Imbusch, R., & Mueller, M. J. (2000). Analysis of oxidative stress and wound-inducible dinor isoprostanes F1 (phytoprostanes F1) in plants. Plant Physiology, 124(3), 1293–1304.

Karban, R. (2008). Plant behaviour and communication. Ecology Letters, 11(7), 727–739.

Kazan, K., & Manners, J. M. (2011). The interplay between light and jasmonate signalling during defence and development. Journal of Experimental Botany, 62(12), 4087–4100.

Knieper, M., Viehhauser, A., & Dietz, K. J. (2023). Oxylipins and reactive carbonyls as regulators of the plant redox and reactive oxygen species network under stress. Antioxidants, 12(4), 814.

Kumar, S. S., Kadier, A., Malyan, S. K., Ahmad, A., & Bishnoi, N. R. (2017). Phytoremediation and rhizoremediation: uptake, mobilization and sequestration of heavy metals by plants. In: D. Singh, H. Singh, & R. Prabha (eds.), Plant-Microbe Interactions in Agro-Ecological Perspectives (pp. 367–394). Springer, Singapore. 10.1007/978-981-10-6593-4_15.

León, J., Rojo, E., & Sánchez-Serrano, J. J. (2001). Wound signalling in plants. Journal of Experimental Botany, 52(354), 1–9.

Liang, Y., Huang, Y., Liu, C., Chen, K., & Li, M. (2023). Functions and interaction of plant lipid signalling under abiotic stresses. Plant Biology, 25(3), 361–378.

Llanes, A., Andrade, A., Alemano, S., & Luna, V. (2018). Metabolomic approach to understand plant adaptations to water and salt stress. In: P. Ahmad, M. A. Ahanger, V. P. Singh, D. K. Tripathi, P. Alam, & M. N. Alyemeni (eds.), Plant metabolites and regulation under environmental stress (pp. 133–144). Academic Press.

Mansoor, S., Manhas, S., Rasool, A., Kaur, N., & Mir, M. A. (2023). Plant Oxylipins: Types and classifications. In: S. M. Shafi, C. Egbuna, & C. O. Adetunji (eds.), Phyto-Oxylipins: Metabolism, Physiological Roles, and Profiling Techniques (pp. 1–9). CRC Press, Taylor & Francis, Boca Raton.

McCauley, A., Jones, C., & Jacobsen, J. (2009). Plant nutrient functions and deficiency and toxicity symptoms. Nutrient management module, Montana State University, 9, 1–16.

McKenzie, R. L., Aucamp, P. J., Bais, A. F., Björn, L. O., Ilyas, M., & Madronich, S. (2011). Ozone depletion and climate change: impacts on UV radiation. Photochemical & Photobiological Sciences, 10(2), 182–198.

Monte, I., Kneeshaw, S., Franco-Zorrilla, J. M., Chini, A., Zamarreño, A. M., García-Mina, J. M., & Solano, R. (2020). An ancient COI1-independent function for reactive electrophilic oxylipins in thermotolerance. Current Biology, 30(6), 962–971.

Mosblech, A., Feussner, I., & Heilmann, I. (2009). Oxylipins: structurally diverse metabolites from fatty acid oxidation. Plant Physiology and Biochemistry, 47(6), 511–517.

Muthamilarasan, M., & Prasad, M. (2013). Plant innate immunity: an updated insight into defense mechanism. Journal of Biosciences, 38, 433–449.

Nagajyoti, P. C., Lee, K. D., & Sreekanth, T. V. M. (2010). Heavy metals, occurrence and toxicity for plants: a review. Environmental Chemistry Letters, 8, 199–216.

Oshunsanya, S. O., Nwosu, N. J., & Li, Y. (2019). Abiotic stress in agricultural crops under climatic conditions. In: M. Jhariya, A. Banerjee, R. Meena, & D. Yadav (eds.), Sustainable Agriculture, Forest and Environmental Management (pp. 71–100). Springer, Singapore. 10.1007/978-981-13-6830-1_3.

Raza, A., Charagh, S., Zahid, Z., Mubarik, M. S., Javed, R., Siddiqui, M. H., & Hasanuzzaman, M. (2021). Jasmonic acid: a key frontier in conferring abiotic stress tolerance in plants. Plant Cell Reports, 40(8), 1513–1541.

Rouet-Leduc, J., Pe'er, G., Moreira, F., Bonn, A., Helmer, W., Shahsavan Zadeh, S. A., … & van der Plas, F. (2021). Effects of large herbivores on fire regimes and wildfire mitigation. Journal of Applied Ecology, 58(12), 2690–2702.

Santino, A., Taurino, M., De Domenico, S., Bonsegna, S., Poltronieri, P., Pastor, V., & Flors, V. (2013). Jasmonate signaling in plant development and defense response to multiple (a) biotic stresses. Plant Cell Reports, 32, 1085–1098.

Schilmiller, A. L., & Howe, G. A. (2005). Systemic signaling in the wound response. Current Opinion in Plant Biology, 8(4), 369–377.

Sharma, L., Priya, M., Kaushal, N., Bhandhari, K., Chaudhary, S., Dhankher, O. P., … & Nayyar, H. (2020). Plant growth-regulating molecules as thermoprotectants: functional relevance and prospects for improving heat tolerance in food crops. Journal of Experimental Botany, 71(2), 569–594.

Sharma, P., Lakra, N., Goyal, A., Ahlawat, Y. K., Zaid, A., & Siddique, K. H. (2023). Drought and heat stress mediated activation of lipid signaling in plants: a critical review. Frontiers in Plant Science, 14, 1216835.

Shrivastava, P., & Kumar, R. (2015). Soil salinity: A serious environmental issue and plant growth promoting bacteria as one of the tools for its alleviation. Saudi Journal of Biological Sciences, 22(2), 123–131.

Siddiqi, K. S., & Husen, A. (2019). Plant response to jasmonates: current developments and their role in changing environment. Bulletin of the National Research Centre, 43(1), 1–11.

Simeoni, F., Skirycz, A., Simoni, L., Castorina, G., de Souza, L. P., Fernie, A. R., et al. (2022). The AtMYB60 transcription factor regulates stomatal opening by modulating oxylipin synthesis in guard cells. Scientific Reports, 12(1), 533.

Solhi, L., Guccini, V., Heise, K., Solala, I., Niinivaara, E., Xu, W., … & Kontturi, E. (2023). Understanding Nanocellulose–Water Interactions: Turning a Detriment into an Asset. Chemical Reviews, 123(5), 1925–2015.

Souri, Z., Karimi, N., Farooq, M. A., & Akhtar, J. (2020). Phytohormonal signaling under abiotic stress. In: D. K. Tripathi, V. P. Singh, D. K. Chauhan, S. Sharma, N. K. Dubey, N. Ramawat (eds.), Plant life under changing environment: Responses and management (pp. 397–466). Academic Press.

Taheri, Z., Vatankhah, E., & Jafarian, V. (2020). Methyl jasmonate improves physiological and biochemical responses of Anchusa italica under salinity stress. South African Journal of Botany, 130, 375–382.

Tamaoki, M., Freeman, J. L., Marquès, L., & Pilon-Smits, E. A. H. (2008). New insights into the roles of ethylene and jasmonic acid in the acquisition of selenium resistance in plants. Plant Signaling & Behavior, 3(10), 865–867.

Thomma, B. P., Cammue, B. P., & Thevissen, K. (2002). Plant defensins. Planta, 216, 193–202.

ul Hassan, M. N., Zainal, Z., & Ismail, I. (2015). Green leaf volatiles: biosynthesis, biological functions and their applications in biotechnology. Plant Biotechnology Journal, 13(6), 727–739.

Vahdati, K., & Lotfi, N. (2013). Abiotic stress tolerance in plants with emphasizing on drought and salinity stresses in walnut. In: K. Vahdati, & C. Leslie (eds.), Abiotic stress: Plant responses and applications in agriculture (vol. 10, pp. 307–365). InTechOpen.

Vonshak, A., & Torzillo, G. (2003). Environmental stress physiology. Handbook of microalgal culture: Biotechnology and Applied Phycology, 57–82.
Waadt, R., Seller, C. A., Hsu, P. K., Takahashi, Y., Munemasa, S., & Schroeder, J. I. (2022). Plant hormone regulation of abiotic stress responses. Nature Reviews Molecular Cell Biology, 23(10), 680–694.
Wahid, A., Gelani, S., Ashraf, M., & Foolad, M. R. (2007). Heat tolerance in plants: an overview. Environmental and Experimental Botany, 61(3), 199–223.
Wasternack, C., & Hause, B. (2002). Jasmonates and octadecanoids: signals in plant stress responses and development. In: Progress in Nucleic Acid Research and Molecular Biology, vol 72, pp. 165–221, Elsevier.
Wasternack, C., & Feussner, I. (2018). The oxylipin pathways: biochemistry and function. Annual Review of Plant Biology, 69, 363–386.
Wasternack, C., & Strnad, M. (2018). Jasmonates: News on occurrence, biosynthesis, metabolism and action of an ancient group of signaling compounds. International Journal of Molecular Sciences, 19(9), 2539.
Wolf, S. (2022). Cell wall signaling in plant development and defense. Annual Review of Plant Biology, 73, 323–353.
Wood, J. M. (2015). Perspectives on: the response to osmotic challenges: bacterial responses to osmotic challenges. The Journal of General Physiology, 145(5), 381.

15 Role of Polyamines in Abiotic Stresses

Roshni Rajamohan, Sumit Sagar, Vidisha Saxena, and Anjali Rajhans

15.1 INTRODUCTION

Plants form a huge majority of the biodiversity and show variations and adaptations to several environmental conditions. Being immobile, plants face a variety of biotic as well as abiotic stresses. The statistical analysis of the last decade reveals the negative impact of abiotic stresses on the productivity of plants. The abiotic stresses include drought, excessive salinity, extremely low temperature, harmful UV radiation and several others. Plants tend to overcome such stresses by making some adaptations in their structure or physiology (Velázquez et al., 2013). Phytohormones are becoming revolutionizing assets for plant stress biology wherein the study of stress hormones can help us improve plant growth and increase crop productivity (El Sabagh et al., 2022; Tyagi et al., 2023).

Plant hormones or phytohormones are regulatory substances synthesized inside the plant body as a result of the biosynthetic pathways. These substances are required in very low amounts and have an impact on certain physiological features (Mukherjee et al., 2022). Polyamines (PAs) are present as positively charged low molecular weight organic compounds found in all living organisms. In plants, polyamines exist primarily as putrescine (Put), spermidine (Spd) and spermine (Spm) (Chen et al., 2019). Some plants have an additional form of polyamine, thermospermine (tSpm), which either exists in place of spermine or along with it. PAs in plant cells are found in free, soluble forms when conjugated with macromolecules (such as phenolic compounds) and insoluble forms when bonded with proteins. The polycationic nature of PAs has proven useful in biological activities (Gill & Tuteja, 2010; Spormann et al., 2021).

Maintenance of crop production under environmental stress conditions is one of the major challenges faced in agriculture. Plants have developed a unique feature of chemical defense mechanisms that involves synthesis of specific plant hormones under various stress conditions of the environment. Polyamines can be regarded as stress messengers in the responses of plants toward stress stimuli. PA plays a crucial role in the defense response of plants toward metal toxicity, chilling temperature, oxidative stress, salinity, drought and waterlogging (Fujita et al., 2015). PA also interacts with other plant hormones to regulate plant development and mitigate abiotic stress. It has also been postulated that PAs have a role in plant tolerance and drought acclimatization by influencing ABA production (Napieraj et al., 2023). By altering the genes that code for the ABA biosynthesis enzymes, PAs can control the ABA content during stress. PAs can raise the cytoplasmic calcium concentration by modifying Ca^{2+} channel activity and inactivating the K^{+} inward rectifier in the membrane, which results in stomatal closure (Figure 15.1).

The interaction of PAs with other plant hormones to mitigate plant abiotic stress is discussed further. The chapter primarily focuses on giving insights about polyamines and also creating awareness about the challenges and future prospects of polyamines as a stress hormone.

 DOI: 10.1201/9781003335788-19

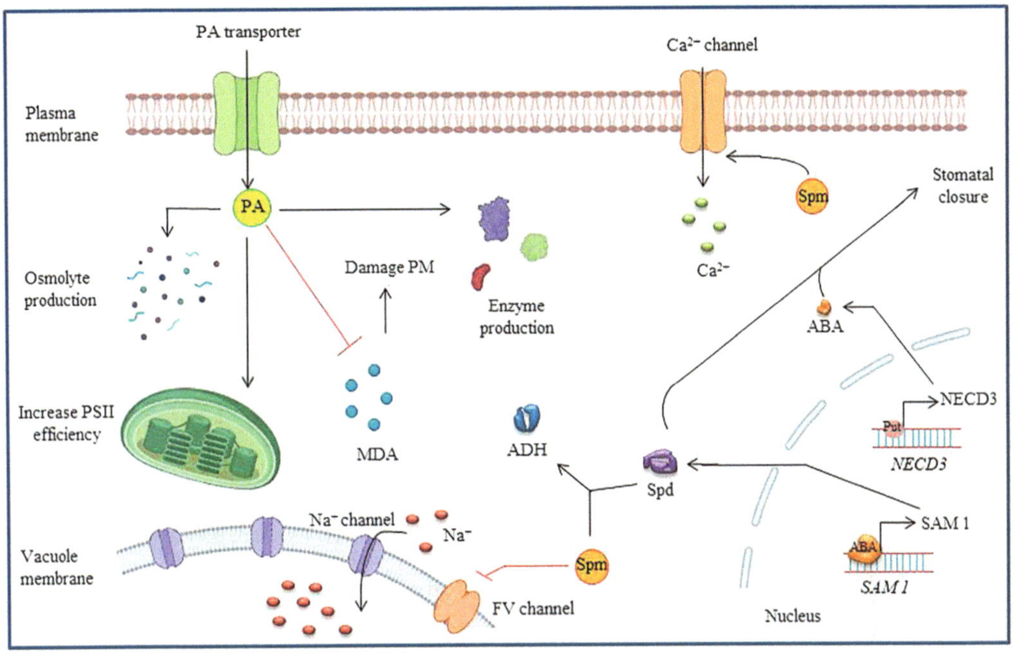

FIGURE 15.1 A model illustrating the PA mediated signaling system that regulates abiotic stress tolerance in plants. PA is involved in the formation of osmolytes and increases PSII efficiency in chloroplasts, resulting in increased photosynthesis and plant survival during drought stress. PAs (Put) supress the salt-induced elevation in MDA, implying that Put may protect the plasma membrane from stress by maintaining membrane integrity. PAs are a crucial component of the plant antioxidant system because they are involved in the production of several enzymes (POD, SOD, CAT) that metabolize ROS produced during abiotic stress. Calcium, an important secondary messenger involved in both abiotic and biotic stress signals, is modulated by Spm. Increased cytoplasmic Ca^{2+} prevents Na^+/K^+ entry into the cytoplasm, promotes Na^+/K^+ influx to the vacuole or suppresses Na^+/K^+ release from the vacuole, all of which increase salt tolerance. Waterlogging causes an increase in PA level that enhances the level of ADH enzyme and prevents the accumulation of excessive acetaldehyde in cells and increase tolerance to waterlogging. During abiotic stress both PAs and ABA stimulate one another. Put induces NECD3 gene that play an important role in ABA biosynthesis, whereas ABA induces SAM1 gene which is involved in Spd biosynthesis. Both PAs and ABA are involved in stress induced stomatal closing. ABA: abscisic acid; CAT: catalase; Ca^{2+}: calcium, FV: fast activating vacuolar channel; MDA: malondialdehyde; Na^+: sodium; NECD3: 9-cis-epoxycarotenoid dioxygenase 3; PA: polyamine; PM: plasma membrane; POD: peroxidase; SAM1: S-adenosyll-methionine synthetase 1; SOD: superoxide dismutase; Spm, spermine; Spd, spermidine.

15.2 DISCOVERY OF POLYAMINES

The discovery of polyamines is dated around 200 years before that of nucleic acids (Bachrach, 2010). In 1678, Leeuwenhoek first described the presence of crystalline substances in the human semen, which were absent in the fresh sample and only developed after several days of standing. The crystals were phosphate derivatives of an unknown new compound (Vauquelin, 1791), which was later identified as an organic base by Schreiner in 1878 (Schreiner, 1878). This compound was eventually given the name "spermine" by Ladenburg and Abel in 1888 (Ladenburg & Abel, 1888). It was the work of Rosenheim in 1924 that ultimately gave us its correct structure and led the first step toward polyamine studies. Rosenheim synthesized putrescine (Put) $[NH_2(CH_2)_4NH_2]$, spermine (Spm) $[NH_2(CH_2)_3NH(CH_2)_4NH(CH_2)_3NH_2]$ and the related base spermidine (Spd) $[NH_2(CH_2)_3NH(CH_2)_4NH_2]$ (Rosenheim, 1924; Bachrach, 2010).

Ciamician and Ravenna (1911) demonstrated that *Datura stramonium* L. contains putrescine, which led to the discovery of polyamines in higher plants for the first time (Ciamician et al., 1911). Putrescine was also found in 1948 in orange juice (Herbst et al., 1948). Smith (1971) concluded that Put, Spd, Spm and agmatine (Agm) are ubiquitous organic cations in higher plants. In 1952, Richards and Coleman found that Put was produced and accumulated under potassium-deficient conditions in barley leaves, which aided in the development of deficiency symptoms (Richards & Coleman, 1952). These studies were extended by Smith, who reported that in barley, Put is derived from arginine via agmatine and N-carbamyl putrescine (Smith, 1971).

In higher plants, the most prevalent PAs are diamine putrescine (Put), triamine spermidine (Spd), tetramine spermine (Spm), thermospermine (Tspm) and cadaverine (Cad). Put is the most prevalent PA in nature and the central product of the PA biosynthetic pathway, being primarily generated by two pathways derived from ornithine (Orn) or arginine (Arg) as a result of the action of ornithine decarboxylase.

15.3 POLYAMINE DISTRIBUTION AND TRANSPORT

Polyamines are found in both eukaryotes and prokaryotes, as well as plant RNA viruses and plant tumours. There are different forms of PAs and mostly found in free form in higher plants the most prevalent ones are Put, Spd, Spm, Tspm and Cad. Other PAs are exclusively present in specific plants or under specific circumstances. Polyamines have tissue- and organ-specific distribution patterns in plants. Put, for example, was discovered to be the most abundant in leaves, with levels three times greater than Spd and Spm, whereas Spd was discovered to be the most abundant PA in other parts like stem, roots and floral parts. Additionally, many PA subtypes have various cellular localization patterns. Put was observed to accumulate in the cytoplasm of carrot cells and Spm in the cell wall. PA distribution patterns may be related to their specific roles. Generally speaking, more active plant growth and metabolism are linked to increased PA production and higher PA contents (Chen et al., 2019). PAs are required for plant growth and development. Therefore, PA transport systems have been a subject of interest. PA transport was discovered in apples for the first time in plant studies (Fujita & Shinozaki, 2015). Put was absorbed by apple leaves and flowers and consequently influenced fruit set, fruit development and yield (Fujita & Shinozaki, 2015). PAs are found largely in plant cell walls and vacuoles and to a smaller extent in mitochondria and chloroplasts (Fujita & Shinozaki, 2015). The transport of paraquat (PQ), one of the most commonly used herbicides worldwide, has been examined to learn more about PA transport mechanisms (Fujita & Shinozaki, 2014). PQ transport in maize roots was competitively hindered by diamines such as Put or cadaverine, but not by Spd or Spm (Anwar et al., 2015).

Many studies found that exogenous application of PAs provided substantial levels of protection against PQ toxicity in a wide range of plant species. It was proposed that the primary cause of a PA protective mechanism was their function as antioxidants (Fujita & Shinozaki, 2014). The gene responsible for this PQ tolerance was found using genome-wide association analysis and F2 mapping. Methyl viologen is another name for PQ; the gene was found resistant to methyl viologen 1 (RMV1). The RMV1 gene encodes AtLAT1, a putative amino acid permease that belongs to the L-type amino acid transporter (LAT) family.

Five genes encoding the LAT family proteins are found in *Arabidopsis*. PA transport activity was observed in at least three AtLAT proteins (RMV1/AtLAT1/AtPUT3, PAR1/AtLAT4/AtPUT2 and AtLAT3/AtPUT1). These proteins had considerable sequence similarity (68–76%); yet, their subcellular distribution was diverse, showing that they were involved in various cellular processes (Fujita & Shinozaki, 2014). RMV1/AtLAT1/AtPUT3 is found in the plasma membrane and involved in PA transport. *Arabidopsis* protoplasts used in transient expression studies show that AtPUT1/AtLAT3 localizes to the endoplasmic reticulum while PAR1/AtLAT4/AtPUT2 localizes to the Golgi apparatus (GA). Rice OsPAR1/OsPUT2 also localizes to the GA. These findings imply that PUT proteins are involved in the intracellular translocation of PAs across the plasma membrane (Fujita & Shinozaki, 2015).

15.4 PHYSIOLOGICAL ROLES

15.4.1 Drought

Given the effects of climate change on plant growth and productivity, drought is one of the biggest threats to the sustainability of food production. Drought responses in plants include morphological, physiological, molecular, hormonal and transcriptional changes that are intricately coordinated and rely on the duration and severity of the water scarcity. Stomata gradually closed after a drought, which was accompanied by a decline in net photosynthesis rate and water-use efficiency. The expression of aquaporins and stress proteins, ion balance, accumulation of osmolytes and osmoprotectants, ROS scavenging by enzymatic and non-enzymatic antioxidant systems, cell membrane stability and ion balance are key factors influencing drought tolerance. Exogenous application of Put by foliar sprayings, improved hydration status, chlorophyll, proline, amino acids, and soluble sugars concentrations in water-stressed wheat plants, which resulted in increased plant height, leaf area and grain yield. Similarly, exogenous Put treatment (foliar sprayings) decreased stomatal density, retained chloroplast structure, and prevented cell plasmolysis, which improved lettuce's ability to use water efficiently and withstand dry conditions. Following physiological and proteomic investigations, PAs may activate many pathways, osmolyte accumulation and antioxidant enzyme systems that improve tolerance to salt and drought conditions (Alcázar et al., 2020).

Drought reduces germination rate and causes seedling establishment to be delayed. Seed priming is a pre-sowing procedure that exposes seeds to a fluid that provides partial hydration but does not allow for radicle emergence. When the primed seeds are sown, the enlargement of the embryos inside the seeds promotes germination by enabling water absorption. It enhances pre-germination metabolic processes, enabling seedlings to emerge more quickly, grow more vigorously and perform better in challenging environments, shielding the seeds from abiotic and biotic challenges during the crucial stage of seedling establishment. Studies have examined the effects of drought stress on white clover seeds showing that priming with Spd increased germination rates and the seed germination process and seedlings showed increased vigour, as evidenced by longer roots and heavier weights. Seed priming with Spd was thought to improve starch metabolism, owing to increased α and β amylase activity. In wheat, seed priming with Spd and Spm boosted levels of different hormones, accelerated starch breakdown and increased the concentration of soluble sugars during seed germination, which may encourage the germination of seeds under drought stress (Li et al., 2014).

The identification of genes encoding enzymes of PA biosynthesis from multiple biological sources, followed by plant transformation using genetic engineering, resulted in the formation of several transgenic plants with better drought tolerance. Transgenic *Arabidopsis* plants overexpressing the *Arabidopsis ADC2* gene had higher Put and were more drought-tolerant than non-transgenic plants. Drought tolerance was shown to be correlated to the overall amount of Put accumulated in the various transgenic lines. Put accumulation under drought is primarily an ABA-dependent metabolic response, as shown by the reduced Put accumulation in ABA-insensitive and ABA-deficient *Arabidopsis* mutant plants (Alcázar et al., 2010).

15.4.2 Salinity

As the external osmotic potential rises, water absorption during seed imbibition decreases. The harmful effects of Na^+ and Cl^- ions on embryo viability can also impact seed germination. Damage to cell organelles and the plasma membrane, impairment of respiration, photosynthesis and protein synthesis are a few of these harmful effects. Enzymes and other macromolecule structures are also disrupted. Salinity causes soil particles to retain water, limiting plant uptake (osmotic impact) and affecting ionic equilibrium, resulting in nutritional deficiencies of K^+ and other ions due to higher Cl^- and Na^+ concentrations (ionic effect) (Acosta-Motos et al., 2017).

In oats under osmotic stress, exposure to numerous osmolytes, including sorbitol, mannitol, proline, betaine and sucrose, led to Put and Spd accumulation in the cell wall and protoplast. Under osmotic stress, ADC transcript levels increase, supporting the ADC activation mechanism (Perez-Amador et al., 2002). Osmotic stress increased the concentrations of Put and diaminopropane (DAP) while decreasing the concentration of Spm in rapeseed, indicating the function of Spm in ADC post-translational regulation (Tiburcio et al., 2003). Many investigations found that exogenous application of various PAs helped to overcome the stress impact to some extent. Exogenous application of PAs resulted in the translocation and storage of Put and other PAs in specific organs, conferring salt stress tolerance in rice (Ndayiragije & Lutts, 2006). Cucumber plants treated with Put had better photosynthetic capacity due to increased PSII photochemical efficiency, which reduced the negative impacts of NaCl (Zhang et al., 2009). Put treatment in lemon suppressed the salt-induced elevation in malondialdehyde (MDA), implying that Put may protect the plasma membrane from stress by maintaining membrane integrity. By regulating ROS scavenging activity, Put was also demonstrated to have a positive impact on the photosynthetic machinery of tea plants cultivated on 50–100 mM NaCl (Xiong et al., 2018). Exogenous Spd application combined with salinity-alkalinity stress reduced the superoxide anion (O_2^-) generation rate and MDA content, while increasing ascorbate-glutathione cycle components, resulting in a reduction of stress-induced damage (Zhang et al., 2016).

Bouchereau et al. (1999) proposed that polyamine responses to salt stress are also ABA-dependent because ABA induces both *ADC2* and spermine synthase (*SPMS*) genes. Interestingly, Alcázar et al. (2006) found stress-responsive, drought-responsive (DRE), low temperature-responsive (LTR) and ABA-responsive elements (ABRE and/or ABRE-related motifs) in the promoters of polyamine biosynthesis genes. This also supports the idea that ABA regulates the expression of some of the genes involved in polyamine production in response to drought and salt treatments (Ahmad et al., 2012).

15.4.3 Nutrient Deficiency

Potassium is one of the most important macronutrients due to its function in the metabolism of plants, growth and stress adaptation (Bano et al., 2020). Deficiency symptoms include wilted or drooped appearance of the plant, short internodes and small leaf blades. Symptoms of deficiency are more visible on the leaves or the seedling. Putrescine accretion has been observed in the leaves of K-deficient barley plants (Richards & Coleman, 1952). It was later determined that the cation-anion equilibrium in plants is maintained due to some undiscovered function of Put (Bouchereau et al., 1999). ADC in *Arabidopsis* controls the Put content during K deficiency (Watson & Malmberg, 1996). The use of ADC (DFMA) and ODC enzyme inhibitors (DFMO) under cadmium stress further showed the involvement of ADC in Put accretion in bean seed (Weinstein et al., 1986). The accretion of different PAs and increased activity of ADC has also been observed when the plant is subjected to other heavy metals such as cadmium (Cd), copper (Cu), nickel (Ni) and zinc (Zn) (Groppa & Benavides, 2008). A correlation between the heavy-metal stress-tolerance mechanism and biosynthesis of PA has also been established. Heavy metal stress, such as copper and chromium, elevated the Put and Spd levels in plants providing them with increased resistance against oxidative damage (Choudhary et al., 2010).

15.4.4 Oxidative Stress

PAs play an important role in modulating antioxidant systems. Under abiotic stress conditions, reactive oxygen species (ROS) are produced at an elevated level leading to oxidative stress that ultimately leads to cell death because of the toxicity caused by lipid peroxidation and membrane damage (Biswas & Mano, 2015). PAs modulate this oxidative stress in two ways. Firstly, it inhibits the auto-oxidation of metals, thereby impairing the supply of electrons for ROS generation (Liu et al., 2015). Secondly, PAs cause enhanced tolerance to abiotic stresses such as drought, heat and cold, which further leads to activation of antioxidant enzymes (Liu et al., 2015). For example, an elevation of POD (peroxidase), SOD (superoxide dismutase) and CAT (catalase) activity was observed in *Poncirus trifoliata* after exogenous application of Spm (Shi et al., 2010).

15.4.5 Waterlogging

Due to global climate change, drought and waterlogging have become more frequent and severe, which has a significant negative influence on crop development and production. Waterlogging is a significant impediment to agriculture (Pan et al., 2021). Approximately 10% of irrigated farmlands worldwide experience recurrent waterlogging (Tyagi et al., 2023). Under waterlogging stress, oxygen availability in the plant rhizosphere decreases. Du et al. (2018) showed that wheat seedling roots exposed to waterlogging stress accumulated substantial levels of Spd and Spm and also increased alcohol dehydrogenase (ADH) activity and alcohol content in tolerant cultivars. Plants have evolved adaptive mechanisms, such as aerenchyma production in the root structure and adventitious root formation, and metabolic adaptations, such as the stimulation of fermentation pathway enzymes to survive under such conditions (Jia et al., 2021). One of the fermentation metabolic pathways is that pyruvate gets converted to acetaldehyde by pyruvate decarboxylase (PDC), which is subsequently converted to alcohol by ADH, while another process is that pyruvate is converted to lactate by lactate dehydrogenase (LDH). To reduce the risk of excessive acetaldehyde in cells and increase tolerance to waterlogging, it is crucial to convert acetaldehyde to alcohol. Increased LDH activity and lactate concentration were unfavourable to the ability of wheat seedlings to acclimatize to stress during logging stress (Du et al., 2018).

15.5 CROSSTALK BETWEEN POLYAMINE AND OTHER HORMONES DURING ABIOTIC STRESS

Abscisic acid is an endogenous anti-transpirant that decreases water loss via the leaf's stomatal pores. PAs and ABA seem to have a mutually beneficial relationship. The 9-cis-epoxy carotenoid dioxygenase 3 (*NCED3*) gene, which plays a role in ABA synthesis, is induced by Put, whereas ABA treatment enhances the expression levels of S-adenosyl L-methionine synthetase 1 (*SAM1*), *SAM3* and spermidine synthase (*SPDS3*) (Tyagi et al., 2023). To regulate stomata, the ABA signaling pathway requires a variety of different components, including transcription factors, protein kinases and phosphatases, G-proteins, ABA receptors and secondary messengers like Ca^{2+}, ROS and NO. Put, Spd and Spm have been shown to modulate stomatal responses by decreasing their aperture and inducing closure. Evidence suggests that polyamines involved in ROS formation and NO signaling interact with each other in ABA-mediated stress responses. ROS production is intimately tied to polyamine catabolic processes because amino oxidases produce H_2O_2, a ROS associated with plant defence and abiotic stress responses. Furthermore, polyamines have been shown to increase NO production in *Arabidopsis*. Both H_2O_2 and NO are involved in the regulation of ABA-induced stomatal movements; hence, NO production is dependent on H_2O_2 production (Alcázar et al., 2010).

Additionally, the synthesis of conjugated PAs in plants has been associated with jasmonic acid (JA). This has been demonstrated by the silencing of the JA-responsive transcription factor R2R3-MYB8 using RNAi, which showed that methyl jasmonic acid (MeJA) triggers PA conjugation and involves R2R3-MYB8. Similarly, MeJA promotes the transcription of Put N-methyltransferase (PMT), an enzyme that converts Put to N-methyl-Put. In mangoes and apples, a rise in free Spd and Spm in MeJA-treated fruits correlates with low-temperature stress tolerance, implying that free

Spd and Spm play a role in fruit ripening and low-temperature stress tolerance (González-Aguilar et al., 2000). Higher amounts of Spd and Spm increase the expression of numerous auxin-regulated genes in tomato fruit, whereas Spm downregulates the expression of several auxin carriers in *Arabidopsis*, including the ARF, Aux/IAA and SAUR genes. These findings show that individual PAs interact differently with plant auxin signaling (Kolotilin et al., 2011).

15.6 REGULATION OF ION CHANNELS IN RESPONSE TO ABIOTIC STRESS

Positively charged polyamines can electrostatically interact with negatively charged proteins, including ion channels. Furthermore, polyamines at physiological concentrations block the fast-activating vacuolar (FV) cation channel in a charge-dependent manner (Spm $^{4+}$ > Spd $^{3+}$ » Put $^{2+}$), at both whole-cell and single-channel levels, demonstrating a direct blocking of the channel by polyamines. In response to various abiotic stressors, such as potassium deficiency, Put levels were dramatically increased (reaching millimolar concentrations), although the levels of Spd and Spm are not much impacted, and this rise of Put may significantly reduce FV channel activity. Under salinity stress, all PA levels rise in amount, and the elevated Spm concentration likely inhibits FV channel activity. These findings can be explained by the fact that PA in plants may affect ion channel activity by directly interacting with the channel proteins and/or the membrane components that are linked with those proteins. PAs may thus influence protein kinase and/or phosphatase activities to modulate ion channel functioning. PAs also help to maintain Ca^{2+} homeostasis. Spm's ability to protect against high salt and drought stress is due to altered control of Ca^{2+} allocation via modulating Ca^{2+} permeable channels. Increased cytoplasmic Ca^{2+} prevents Na^+/K^+ entry into the cytoplasm, promotes Na^+/K^+ influx to the vacuole, or suppresses Na^+/K^+ release from the vacuole, all of which increase salt tolerance (Ahmad et al., 2012).

15.7 CONCLUSION AND FUTURE PERSPECTIVES

Globally, one of the major causes of crop losses are the abiotic stresses. Firstly, to increase our knowledge about the correlation between PA and stress tolerance, it is essential to investigate the function of PAs during abiotic stress which can be done using gene manipulation. A lot of effort is needed to expose the detailed molecular mechanism behind the defensive role of Spd, Spm and Put in abiotic stress tolerance. Some strategies that can be implemented in the future include studying PA biosynthesis and genes involved in it as well as external application of PA to study the degree of tolerance. Further investigation is required at the molecular level to further define this correlation. Consequently, research on PA and stress endurance is at an exciting stage and has created an opening for intensive study to recognize a range of functions of these relatively simple molecules.

REFERENCES

Acosta-Motos, J. R., Ortuño, M. F., Bernal-Vicente, A., Diaz-Vivancos, P., Sanchez-Blanco, M. J., & Hernandez, J. A. (2017). Plant responses to salt stress: adaptive mechanisms. Agronomy, 7(1), 18.

Ahmad, P., Kumar, A., Gupta, A., Hu, X., Azooz, M. M., & Sharma, S. (2012). Polyamines: Role in Plants Under Abiotic Stress. In: M. Ashraf, M. Öztürk, M. Ahmad, & A. Aksoy (eds.), Crop Production for Agricultural Improvement. Springer, Dordrecht. 10.1007/978-94-007-4116-4_19

Alcázar, R., Altabella, T., Marco, F., Bortolotti, C., Reymond, M., Koncz, C., ... & Tiburcio, A. F. (2010). Polyamines: molecules with regulatory functions in plant abiotic stress tolerance. Planta, 231(6), 1237–1249.

Alcázar, R., Bueno, M., & Tiburcio, A. F. (2020). Polyamines: Small amines with large effects on plant abiotic stress tolerance. Cells, 9(11), 2373.

Alcázar, R., Marco, F., Cuevas, J. C., Patron, M., Ferrando, A., Carrasco, P., ... & Altabella, T. (2006). Involvement of polyamines in plant response to abiotic stress. Biotechnology Letters, 28, 1867–1876.

Anwar, R., Mattoo A., and Handa, A. (2015). Polyamine interactions with plant hormones: Crosstalk at several levels. In: T. Kusano, H. Suzuki (eds). Polyamines a universal molecular nexus for growth, survival and specialized metabolism. Tokyo: Springer, pp. 267–303.

Bachrach, U. (2010). The early history of polyamine research. Plant Physiology and Biochemistry, 48(7), 490–495.

Bano, C., Amist, N., & Singh, N. (2020). Role of polyamines in plants abiotic stress tolerance: Advances and future prospects. In D. K. Tripathi, V. P. Singh, D. K. Chauhan, S. Sharma, S. M. Prasad, N. K. Dubey, & N. Ramawat (eds.), Plant Life under Changing Environment (1st ed., Vol. 1, pp. 481–496). Academic Press United States. 10.1016/B978-0-12-818204-8.00021-7

Biswas, M. S., & Mano, J. I. (2015). Lipid peroxide-derived short-chain carbonyls mediate hydrogen peroxide-induced and salt-induced programmed cell death in plants. Plant Physiology, 168(3), 885–898.

Bouchereau, A., Aziz, A., Larher, F., & Martin-Tanguy, J. (1999). Polyamines and environmental challenges: recent development. Plant Science, 140(2), 103–125.

Chen, D., Shao, Q., Yin, L., Younis, A., & Zheng, B. (2019). Polyamine function in plants: Metabolism, regulation on development, and roles in abiotic stress responses. Frontiers in Plant Science, 9, 1945.

Choudhary, S. P., Bhardwaj, R., Gupta, B. D., Dutt, P., Gupta, R. K., Kanwar, M., & Dutt, P. (2010). Changes induced by $Cu2^{+}$ and $Cr6^{+}$ metal stress in polyamines, auxins, abscisic acid titers and antioxidative enzymes activities of radish seedlings. Brazilian Journal of Plant Physiology, 22, 263–270.

Ciamician, G., & Ravenna, C. (1911) cited by Smith, T. A., Historical perspective on research in plant polyamine biology. In: R. D. Slocum, & H. E. Flores (eds.), Biochemistry and Physiology of Polyamines in Plants. CRC Press, Boca Raton, p. 2.

Du, H. Y., Liu, D. X., Liu, G. T., Liu, H. P., & Kurtenbach, R. (2018). Relationship between polyamines and anaerobic respiration of wheat seedling root under water-logging stress. Russian Journal of Plant Physiology, 65, 874–881.

EL Sabagh, A., Islam, M. S., Hossain, A., Iqbal, M. A., Mubeen, M., Waleed, M., ... & Abdelhamid, M. T. (2022). Phytohormones as growth regulators during abiotic stress tolerance in plants. Frontiers in Agronomy, 4, 765068.

Fraire-Velázquez, S., & Balderas-Hernández, V. E. (2013). Abiotic stress in plants and metabolic responses. In: K. Vahdati, & C. Leslie (eds.), Abiotic stress—plant responses and applications in agriculture (pp. 25–48). IntechOpen.

Fujita, M., & Shinozaki, K. (2014). Identification of polyamine transporters in plants: paraquat transport provides crucial clues. Plant and Cell Physiology, 55(5), 855–861.

Fujita, M., & Shinozaki, K. (2015). Polyamine Transport Systems in Plants. In: T. Kusano, H. Suzuki (eds.), Polyamines (pp. 179–185). Springer, Tokyo. 10.1007/978-4-431-55212-3_15.

Gill, S. S., & Tuteja, N. (2010). Polyamines and abiotic stress tolerance in plants. Plant Signaling & Behavior, 5(1), 26–33.

González-Aguilar, G. A., Gayosso, L., Cruz, R., Fortiz, J., Báez, R., & Wang, C. Y. (2000). Polyamines induced by hot water treatments reduce chilling injury and decay in pepper fruit. Postharvest Biology and Technology, 18(1), 19–26.

Groppa, M. D., & Benavides, M. P. (2008). Polyamines and abiotic stress: recent advances. Amino Acids, 34(1), 35–45.

Herbst, E. J., & Snell, E. E. (1948). Putrescine as a Growth factor for Hemophilus parainfluenzae. Journal of Biological Chemistry, 176(2), 989–990.

Jia, W., Ma, M., Chen, J., & Wu, S. (2021). Plant morphological, physiological and anatomical adaption to flooding stress and the underlying molecular mechanisms. International Journal of Molecular Sciences, 22(3), 1088.

Kolotilin, I., Koltai, H., Bar-Or, C., Chen, L., Nahon, S., Shlomo, H., ... & Reuveni, M. (2011). Expressing yeast SAMdc gene confers broad changes in gene expression and alters fatty acid composition in tomato fruit. Physiologia Plantarum, 142(3), 211–223.

Ladenburg, A., & Abel, J. (1888). Ueber das aethylenimin (Spermin?). Berichte der deutschen chemischen Gesellschaft, 21(1), 758–766.

Li, Z., Peng, Y., Zhang, X. Q., Ma, X., Huang, L. K., & Yan, Y. H. (2014). Exogenous spermidine improves seed germination of white clover under water stress via involvement in starch metabolism, antioxidant defenses and relevant gene expression. Molecules, 19(11), 18003–18024.

Liu, J. H., Wang, W., Wu, H., Gong, X., & Moriguchi, T. (2015). Polyamines function in stress tolerance: from synthesis to regulation. Frontiers in Plant Science, 6, 827.

Mukherjee, A., Gaurav, A. K., Singh, S., Yadav, S., Bhowmick, S., Abeysinghe, S., & Verma, J. P. (2022). The bioactive potential of phytohormones: a review. Biotechnology Reports, 35, e00748.

Napieraj, N., Janicka, M., & Reda, M. (2023). Interactions of Polyamines and Phytohormones in Plant Response to Abiotic Stress. Plants, 12(5), 1159.

Ndayiragije, A., & Lutts, S. (2006). Do exogenous polyamines have an impact on the response of a salt-sensitive rice cultivar to NaCl?. Journal of Plant Physiology, 163(5), 506–516.

Pan, J., Sharif, R., Xu, X., & Chen, X. (2021). Mechanisms of waterlogging tolerance in plants: Research progress and prospects. Frontiers in Plant Science, 11, 627331.

Perez-Amador, M. A., Leon, J., Green, P. J., & Carbonell, J. (2002). Induction of the arginine decarboxylase ADC2 gene provides evidence for the involvement of polyamines in the wound response in Arabidopsis. Plant Physiology, 130(3), 1454–1463.

Richards, F. J., & Coleman, R. G. (1952). Occurrence of putrescine in potassium-deficient barley. Nature, 170(4324), 460-460.

Rosenheim, O. (1924). The isolation of spermine phosphate from semen and testis. Biochemical Journal, 18(6), 1253.

Schreiner, P. (1878). Ueber eine neue organische Basis in thierischen Organismen. Justus Liebigs Annalen der Chemie, 194(1), 68–84.

Shi, J., Fu, X. Z., Peng, T., Huang, X. S., Fan, Q. J., & Liu, J. H. (2010). Spermine pretreatment confers dehydration tolerance of citrus in vitro plants via modulation of antioxidative capacity and stomatal response. Tree Physiology, 30(7), 914–922.

Smith, T. A. (1971). The occurrence, metabolism and functions of amines in plants. Biological Reviews, 46(2), 201–241.

Spormann, S., Soares, C., Teixeira, J., & Fidalgo, F. (2021). Polyamines as key regulatory players in plants under metal stress—A way for an enhanced tolerance. Annals of Applied Biology, 178(2), 209–226.

Tiburcio, A. F., Altabella, T., & Cordeiro, A. (2003). Regulators of growth: Polyamines. In: B. Thomas, B. G. Murray, & D. J. Murphy (eds.), Encyclopedia of Applied Plant Sciences 1st ed, (vol. 1, pp. 1042–1051). Elsevier Oxford.

Tyagi, A., Ali, S., Ramakrishna, G., Singh, A., Park, S., Mahmoudi, H., & Bae, H. (2023). Revisiting the role of polyamines in plant growth and abiotic stress resilience: mechanisms, crosstalk, and future perspectives. Journal of Plant Growth Regulation, 42(8), 5074–5098.

Vauquelin, L. N. (1791). Experiences sur le sperme humain. Ann. Chim, 9, 64–80.

Watson, M. B., & Malmberg, R. L. (1996). Regulation of Arabidopsis thaliana (L.) Heynh arginine decarboxylase by potassium deficiency stress. Plant Physiology, 111(4), 1077–1083.

Weinstein, L. H., Kaur-Sawhney, R., Rajam, M. V., Wettlaufer, S. H., & Galston, A. W. (1986). Cadmium-induced accumulation of putrescine in oat and bean leaves. Plant Physiology, 82(3), 641–645.

Xiong, F., Liao, J., Ma, Y., Wang, Y., Fang, W., & Zhu, X. (2018). The protective effect of exogenous putrescine in the response of tea plants (*Camellia sinensis*) to salt stress. HortScience, 53(11), 1640–1646.

Zhang, R. H., Li, J., Guo, S. R., & Tezuka, T. (2009). Effects of exogenous putrescine on gas-exchange characteristics and chlorophyll fluorescence of NaCl-stressed cucumber seedlings. Photosynthesis Research, 100(3), 155–162.

Zhang, Z., Chang, X. X., Zhang, L., Li, J. M., & Hu, X. H. (2016). Spermidine application enhances tomato seedling tolerance to salinity-alkalinity stress by modifying chloroplast antioxidant systems. Russian Journal of Plant Physiology, 63(4), 461–468.

16 Peptide Hormone with Emphasis on Abiotic Stress Tolerance in Plants

K Rajarajan, Meenakshi Jadhav, Varsha C, Sakshi Sahu, Ambika Rajendran, A K Handa, and A Arunachalam

16.1 INTRODUCTION

The world requires a significant boost in agricultural output due to the fast-growing human population. By 2050, productivity must rise by 70% to feed the world's population of an additional 2.3 billion people (Hemathilake & Gunathilake, 2022). However, both biotic and abiotic stressors are significant barriers to agricultural productivity (Sahoo et al., 2023). The main environmental factors that reduce crop productivity are drought, salinity, heat, chilling, freezing and ozone. Drought, salt and heat are the three most prevalent and significant abiotic stressors (Hussain Wani et al., 2013). The intensity of stress, its progression, stage of the plant and biotic and abiotic factors can affect how the plant responds to stress (Feller & Vaseva 2014). Early damage to some crops may be followed by recovery and eventual survival. Crop species or genotypes may differ significantly in their susceptibility to stress or level of resistance to it. Plant growth, membrane integrity, pigment concentration, osmotic adjustment, water relations and photosynthetic activity are just a few of the many ways that abiotic stress affects plants (Sanghera et al., 2011; Pathak et al., 2014). The most critical ecological problem that severely impairs plant growth and the photosynthetic process is drought (Rizwan et al., 2015; Fahad et al., 2017). In addition, salt stress widely affects plants' hormonal system and antioxidant systems, which leads to oxidative stress (Jebara et al., 2005). Also, plants on prolonged exposure to waterlogging conditions exhibit a series of morphological, physiological and biochemical changes such as leaf senescence, chlorosis, necrosis, wilting, stunted growth, biomass production, oxidative stress and cellular damage (Anjum et al., 2015). The main source of carbohydrates, carbon metabolism, is imbalanced during drought stress, which results in partial stomatal closure at carboxylation sites with lower carbon dioxide availability (Hu et al., 2019). The physiology of an individual as well as biochemical-, cellular- and molecular-based processes are included in the mechanisms of abiotic stress tolerance. In a nutshell, strengthening the root system, leaf structure, osmotic balance, comparative water content and stomatal adjustment are thought to be the most important characteristics for crop plants to most abiotic stresses. Additionally, phytohormones like auxin, gibberellin, ethylene, brassinosteroids and peptide compounds as well as calcium are important strategies for coping with stress. These phytohormones include abscisic acid, salicylic acid, jasmonic acid and auxin. Phytohormones-like signaling peptides control a range of a variety of aspects of plant development and growth are regulated by biochemical, physiological and developmental mechanisms (Oh et al., 2018; Fletcher 2020; Kim et al., 2021). Most signaling peptides are either short (5–20 amino acids) or long (40–100 amino acids) and they are translated without any proteolytic processing directly from very small open reading frames or precursor proteins (Oh et al., 2018; Tabata et al., 2014). These signaling peptides include CLAVATA3 (CLV3), phytosulfokine-α, C-terminally encoded peptide (CEP) are involved in drought stress tolerance, cysteine-rich secretory protein, antigen 5,

DOI: 10.1201/9781003335788-20

pathogenesis-related 1 protein (CAP) and rapid alkalinization factor (RALF) are involved in salt stress response and sterility-regulating kinase member 1 (SKM1) is involved in heat stress tolerance. In response to abiotic stress, a variety of peptides that regulate adaptation and tolerance pathways have been discovered. In this discussion, we highlight the roles of short peptides and their signaling mechanisms in coordinating plant responses to drought stress, and we address future perspectives for peptide research in abiotic stress response.

16.2 SIGNALING PEPTIDE AND ITS RESPONSE TO DROUGHT STRESS

During phenological developmental phases, drought stress has an impact on plants. Simultaneously, plants have also evolved with their whole plant response mechanisms to combat abiotic stresses (Figure 16.1). Many physiological processes in plants that affect yield are vulnerable to drought (Farooq et al., 2009). Photosystems 1 and 2 are also negatively impacted by drought stress, as well as photosynthetic pigments such as chlorophyll a, b and carotenoid components (Fu & Huang, 2001). Additionally, it decreases the production of starch in plants by affecting the Calvin cycle enzymes (ribulose phosphatase). Production of Reactive oxygen species (ROS), such as superoxide anions, hydroxyl radicals, hydrogen peroxide and singlet oxygen, is probably one of the first metabolic responses of plants exposed to environmental stressors like water deprivation. Under drought stress, ROS interact with lipids, proteins and nucleic acids to induce oxidative damage, which ultimately impairs normal cellular functioning or even leads to cell death (Lata et al., 2011; Anjum et al., 2011).

Recently, it was discovered that several released peptides regulate plant cellular development. These peptides include the CLAVATA3 (CLV)/EMBRYO-SURROUNDING REGION (CLV), INFLORESCENCE DEFICIENT IN ABSCISSION (IDA) peptide, phytosulfokine-α (PSK-α) and C-terminally encoded peptide (CEP) (Kim et al., 2021). Among the peptides, the CLV peptide is a well-studied plant peptide that is essential for the development of shoot apical meristems (Nir et al., 2014). Recently, it was discovered that the CLE25 peptide regulates the abscisic acid generation and stomatal regulation of transpiration in *Arabidopsis* and transmits signals of water deprivation across

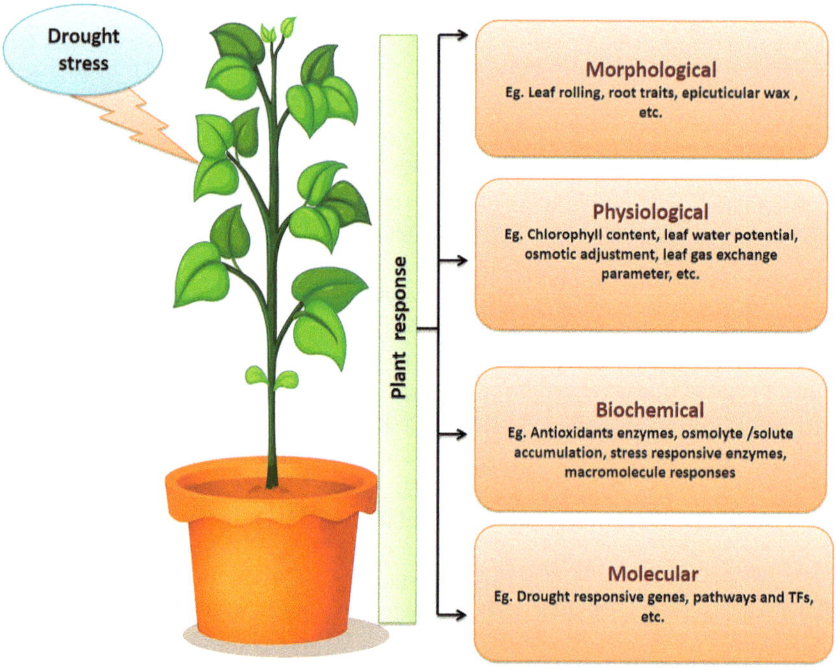

FIGURE 16.1 Whole plant response mechanisms to drought stress.

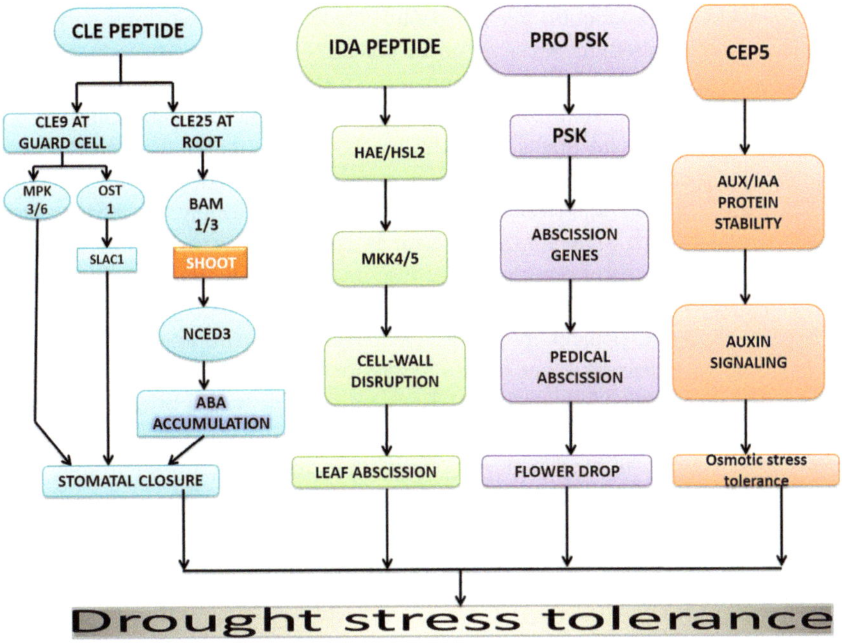

FIGURE 16.2 Drought stress responses by signaling peptides.

vascular tissues. In vascular tissues and roots where these genes are expressed (Takahashi et al., 2018). In comparison to the wild type, the peptide showed increased drought stress tolerance and lower ABA sensitivity. It is located in the nucleus and cytoplasm (Yu et al., 2016). Other peptides include phytosulfokine (PSK), which regulates cell proliferation; rapid alkalinization factor (RALF); LUREs, which direct pollen tube growth; STOMAGEN, which affects stomatal development; and Casparian strip integrity factor (CIF), which is linked to the development of the Casparian strip diffusion barrier. Recently, researchers also identified another peptide, called AtPep3, that is crucial in the stress caused by salinity and drought (Nir et al., 2014; Liang et al., 2016; Yu et al., 2016).

Loss of water increases the expression of CLE25 in the root vasculature and, as a result, peptides move from the roots to the leaves and cause stomatal closure (Takahashi et al., 2018). CLE9 controls stomatal closure, which improves drought tolerance (Zhang et al., 2019). By combining OST1 and SLAC1, CLE9 regulates stomatal closure in response to drought stress, increasing drought stress tolerance. Since guard cells are the only cells where CLE9 is particularly expressed (Zhang et al., 2019), CLE9, unlike CLE25, controls stomatal closure at the local level (Figure 16.2).

According to several studies (Butenko et al., 2003; Stenvik et al., 2006; Cho et al., 2008; Kumpf et al., 2013; Zhu et al., 2019), the INFLORESCENCE DEFICIENT IN ABSCISSION (IDA) is another peptide that utilizes the LRR-RLKs, HAESA (HAE) and HAES to operate in floral organ abscission and lateral root emergence. Drought-induced cauline leaf abscission is caused by IDA (Patharkar & Walker 2016). These organ abscission events are brought on by IDA's induction of cell wall collapse and subsequent cell separation (Stenvik et al., 2006; Zhu et al., 2019).

Phytosulfokine- (PSK-) is a pentapeptide sulphate, and the PSK-PSK receptor (PSKR)/ coreceptor SOMATIC EMBRYOGENESIS RECEPTOR KINASE 3 (SERK3) components aid in primary and lateral root growth (Amano et al., 2007; Ladwig et al., 2015; Wang et al., 2015). In response to osmotic stress by mannitol, four PSK genes—*PSK1*, *PSK3*, *PSK4* and *PSK5*—and three SBT genes—*SBT1.4*, *SBT3.7* and *SBT3.8*—are highly elevated (Stührwohldt et al., 2020). ProPSK1 overexpressed in *Arabidopsis*, and these plants are more tolerant to osmotic stress and show improved root and hypocotyl development. Additionally, *SBT3.8* overexpression increases shoot and root development and enhances osmotic stress tolerance. These findings imply that *SBT3.8* mediates the production of PSK peptides from the precursor protein proPSK1, which helps the body adapt to drought stress (Kim et al., 2021). The drought-induced tomato flower dropping is regulated by phytosulfokine signaling (Reichardt et al., 2020). Premature flower abscission is facilitated by overexpression of phytaspase 2 (phyt2), a subtilisin-like protease that splits proPSK protein in tomato plants. SlPhyt2 knockdown decreases flower drop under drought stress circumstances, suggesting that SlPhyt2 is involved in the drought's ability to abort the development of flowers and fruits (Reichardt et al., 2020). Furthermore, the PSK treatment causes downregulation of genes that maintain the dormant abscission, and overexpression of polygalacturonase genes in tomato pedicel abscission. These results imply that in response to drought stress, cell wall hydrolases in the abscission region trigger abscission when the subtilase SlPhyt2, which is expressed in the pedicel, produces bioactive PSK. Additionally, the peptide with a C-terminal code (CEP) abscission zone causes abscission.

Also, the C-terminally encoded peptides (CEPs) are proteins with a brief C-terminal extension, and N-terminal secretion signal, a variable domain and one or more CEP domains (Ogilvie et al., 2014). CEP5 plays a function in *Arabidopsis*' ability to endure osmotic and drought stress, as well as negatively regulating the development of primary and lateral roots (Roberts et al., 2016; Smith et al., 2020). Further, without changing auxin levels or auxin transport activity, CEP5 stabilizes AUX/IAA proteins, which are negative regulators of AUXIN RESPONSE FACTORS.

16.3 SALINITY STRESS RESPONSES

Salt stress has negative impact of excessive amounts of minerals like Na^+ and/or Cl^- on plants and causes production decline (Munns, 2005). Although it has been established that soil salinity existed before people and agriculture, the issue has only recently become more of a concern (Zhu, 2001). According to Parihar et al. (2014), salt stress harms whole plant levels as it affects germination, growth, photosynthetic pigments and photosynthesis, water relations, nutritional imbalance, oxidative stress and yield. To adapt extreme salinity stress, plants have evolved a complex regulatory system that includes water flux control and cellular osmotic adjustment (Golldack et al., 2014). High salinity induces the production of specific signaling peptides like PEP3 and CAPE1 (Chen et al., 2016). The defence-related peptide PLANT ELICITOR PEPTIDE AtPep3 and its receptor PEP1 RECEPTOR 1 (PEPR1) are also involved in the response of plants to salt, and the PEPR1 loss of function almost completely reverses AtPep3-induced salt resistance in *Arabidopsis*. Additionally, a functional module that links salt stress-induced cell wall changes to salt stress responses is formed by LEUCINE-RICH REPEAT EXTENSINs and RAPID ALKALINIZATION FACTOR (RALF) peptides (Zhao et al., 2018; Xiao & Zhou, 2023).

16.3.1 CAPE Peptides

A member of the cysteine-rich secretory protein, antigen 5 and pathogenesis-related 1 protein (CAP) families, CAP-derived peptide 1 (CAPE1) is generated from the C-terminus of PATHOGENSIS-RELATED PROTEIN1b (PR-1b) (Chen et al., 2014). The fact that CAPE1 significantly increases anti-pathogen responses in tomatoes suggests that PR-1 is involved in immunological signaling (Chen et al., 2014). Salt stress tolerance in *Arabidopsis* is negatively regulated by the 11 amino acid protein AtCAPE1 (Chien et al., 2015). Since the sequence similarities to the tomato CAPE1 precursor and the conserved motif at the C-terminus, nine putative *Arabidopsis* CAPEs were identified to be the precursors of CAPEs and given the moniker PROAtCAPEs (Chien et al., 2015). Among them, salt stress mostly inhibits PROAtCAPE1. However, the sensitive phenotype of the mutant is restored by exogenous application of a synthetic AtCAPE1 peptide or by overexpressing PROAtCAPE1; demonstrating this shows that AtCAPE1 inhibits *Arabidopsis'* ability to tolerate salt stress. AtCAPE1 also inhibits salt-induced genes, including those involved in osmolyte synthesis and detoxification (Chien et al., 2015), as illustrated in Figure 16.3.

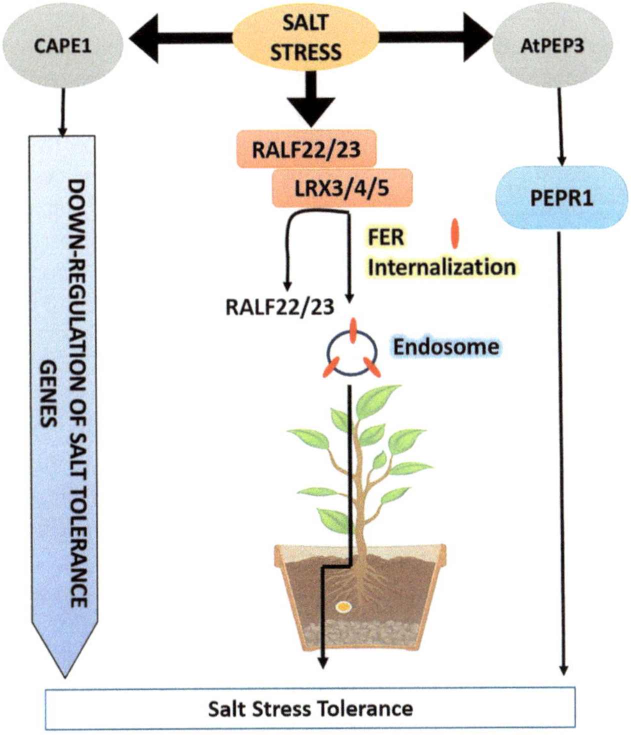

FIGURE 16.3 Salt stress tolerance response mediated by signaling peptides.

16.3.2 RALF Peptide

Peptides known as rapid alkalinization factor (RALF) peptides are 5 kDa cysteine-rich peptides that cause the extracellular compartments of plant cells to rapidly alkalize. As a result, the proton electrochemical potential necessary for solute uptake is decreased, which inhibits cell development (Blackburn et al., 2020). When a plant reproduces, RALF peptides have a role in pollen tube growth and termination, immunological responses, guard cell movement and root expansion (Ge et al., 2017; Mecchia et al., 2017; Blackburn et al., 2020). FER protein and RALF peptides are physically connected (Zhao et al., 2018). When under salt stress, mature RALF22 peptides separate from LRX proteins, which induces FER to internalize via the endosomal pathway (Zhao et al., 2018). Thus, salt-induced cell wall changes are a mechanism by which salt tolerance and plant growth are regulated by RALF22/23-FER and LRX3/4/5. To maintain the integrity of cell walls and prevent root cells from bursting during growth under severe salt stress, FERONIA (FER) is also required (Feng et al., 2018). Under salt stress, pectin cross-link fortification restores the development and cell wall integrity of *fer* mutant seedlings (Feng et al., 2018). Under *in-vitro* conditions the FER extracellular domain binds pectin. It is possible that FER physically interacts with the pectin in the cell wall to boost calcium signaling in response to salt stress to maintain cell wall integrity with pectin in the cell wall. For salinity-induced [Ca^{2+}] transients to maintain cell wall integrity throughout growth recovery, FER is necessary.

16.3.3 AtPEP3

The pattern-triggered immunity inducers that are endogenous and protect against bacteria, fungi and herbivores are called plant elicitor peptides (Peps) (Qin et al., 2011; Bartels & Boller 2015).

AtPep3 is one of eight members of the *Arabidopsis thaliana* precursor Pep (AtPROPEP) family that contributes to the plant's reaction to salinity stress (Nakaminami et al., 2018). AtPROPEP3 has the highest induction in response to the high salinity of the AtPROPEP gene family. Salt stress tolerance is induced by either exogenous application of the synthetic AtPep3 peptide (30 amino acids), which is derived from the C-terminal portion of PROPEP3 (Nakaminami et al., 2018). An exogenous AtPep3 peptide application recovers plant survival only in the pepr2 mutants, despite high-salt treatment significantly reducing plant survival of the pepr1, pepr2 and pepr1/2 mutants (Nakaminami et al., 2018). This suggests that AtPep3 is recognized by the PEPR1 receptor to induce salinity stress tolerance in plants. As a result, the AtPEP3-PEPR1 module may have a dual role in immunological responses as well as salinity stress tolerance.

16.4 HEAT STRESS RESPONSES

Globally, heat stress is the most destructive abiotic stress that inhibits plant growth, metabolism and productivity. Several heat-sensitive biochemical processes are involved in the growth and development of plants (El-Beltagi et al., 2016). Plant responses to heat stress vary depending on the type of plant, its intensity and duration. To cope with heat stress conditions, plants have a variety of adaptation, avoidance or acclimation mechanisms (Khan & Shahwar, 2020). Major tolerance mechanisms are triggered to counteract stress-induced biochemical and physiological changes. These mechanisms make use of proteins, ion transporters, osmoprotectants, antioxidants and other elements involved in signaling cascades and transcriptional control (Hasanuzzaman et al., 2017). It has recently reported that signaling peptides are crucially involved in heat stress tolerance in *Arabidopsis* by normal pollen tube formation mediated by CLE45- STERILITY-REGULATING KINASE MEMBER1 (SKM1)/SMK2 receptor under heat stress condition (Endo et al., 2013).

16.4.1 CLE45 Peptide

Environmental stresses have a detrimental impact on plants (Giorno et al., 2013; De Storme & Geelen, 2014) as reproductive development and, consequently, seed yield. A signaling peptide has now been discovered to play a role in the plant's response to high-temperature stress (Endo et al., 2013), i.e., *Arabidopsis* CLE45-STERILITY-REGULAINGKINASE MEMBER1 (SKM1)/SKM2 receptor module in pollen tube growth (Endo et al., 2013). SKM1 in pollen and heat-inducible CLE45 in pistils seem to function in the same signaling pathway (Figure 16.4). At 30°C but not at room temperature (22°C), RNA interference (RNAi) or the creation of a dominant-negative version of SKM1 that is kinase-dead reduces seed quantity and size. All of these results point to the successful seed formation caused by the CLE45-SKM1/2 pathway, which maintains pollen tube growth under high temperatures.

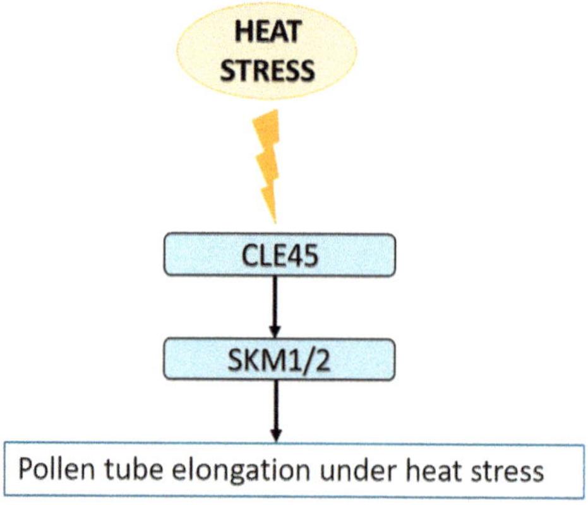

FIGURE 16.4 Pollen tube elongation under heat stress.

16.5 LOW-NUTRIENT STRESS RESPONSES

The most essential nutrients for plants are nitrogen (N) and inorganic phosphate (Pi); their deficiencies have a significant impact on the morphology, physiology, growth and development of plants (Reddy, 2006). It has been determined that peptides like CEPs, CLEs and RGFs play a part in the control of nutrient uptake and have an impact on plant physiology and growth responses. According to Nakaminami et al., (2018), CEP1 functions as a root-to-shoot transportable signal to encourage adjusting N acquisition in the other N-rich regions of the roots. Under N deficit, CEP1 peptides from the roots are transported to the shoots through the xylem, where they are recognized by XIP1/CEPR1 and CEPR2 receptors and produce descending shoot-to-root signals, CEP Downstream 1 (CEPD1) and CEPD2, which are members of the glutaredoxin family. To promote N absorption, CEPDs are delivered to the roots via the phloem and upregulate the nitrate transporter gene NRT2.1. Together with CEPD1/2, CEPD-like 2 from the shoots promotes high-affinity N-uptake in the roots as well as root-to-shoot transfer of nitrate (Ota et al., 2020). CEP3 regulates the nutrient deprivation response that inhibits root growth in the presence of N deficiency, possibly improving seedling survival (Delay et al., 2019).

16.6 WATERLOGGING STRESS

In many areas, waterlogging becomes a major abiotic stress that severely limits crop growth and production (Jackson & Colmer, 2005). About 16% of production areas globally are currently affected by waterlogging, which has become a severe global food production limitation (Ahsan et al., 2007). Simultaneously, plants evolved with their tolerance mechanisms to cope with this combined stress as well (Nakaminami et al., 2018; Nanjo et al., 2011; Xu et al., 2016).

16.7 COLD STRESS/CHILLING INJURY

Chilling or cold stress is one of the most important abiotic stresses of agricultural plants; cold stress comprises both chilling injury (less than 20°C) and freezing injury (less than 0°C), and it affects both plant development and yield (Lang et al., 2005). The survival of crop growth and production being threatened due to cold stress might be significant. It causes poor germination, reduced seedling growth, chlorosis (yellowing of the leaves), restricted leaf growth, wilting and necrosis of tissue. In addition, it affects the development of plants' reproductive systems as well. Cold stress causes serious membrane damage, which is one of its main adverse effects. According to Yadav (2010), the acute dehydration brought on by freezing under cold stress is mostly responsible for this injury. Cold stress has been shown to decrease the fluidity of cellular membranes (Örvar et al., 2000; Sangwan et al., 2002). RLKs and HISTIDINE KINASES (HKs), among other membrane-localized proteins, may be able to detect this change (Zhu, 2016). Chen et al. (2021) reported that many RLKs are essential in controlling plant resistance to cold stressors. Additionally, by inhibiting the activity of MPK3 and MPK6 that is triggered by cold, *Arabidopsis* CALCIUM/CALMODULIN REGULATED RECEPTORLIKE KINASE1 (CRLK1) and CRLK2 positively control cold stress responses (Yang et al., 2020). To regulate freezing tolerance in *Arabidopsis*, the receptor-like cytoplasmic kinase COLD RESPONSIVE PROTEIN KINASE 1 (CRPK1) transmits a cold signal from the plasma membrane to the nucleus via the 14-3-3 and C REPEAT BINDING FACTOR (CBF) proteins (Chen et al., 2021).

16.8 FUTURE PERSPECTIVE AND CONCLUSION

Crop productivity is significantly hampered by drought, which is predicted to get worse soon. Moreover, droughts are intensifying, spreading and becoming more frequent as a result of climate change. So, scientists are working to create drought-tolerant crops and comprehend various mechanisms of drought tolerance. Thus, researchers are striving to develop crops that can withstand droughts and to comprehend the various systems involved in drought tolerance. The core idea of this paper is that plants respond to drought stress by peptide signaling. Many peptides and signaling pathways have been identified in recent years that play a significant role in regulating plant responses to abiotic stresses. However, most of the processes by which peptides elicit drought tolerance in plants are poorly understood and need further research due to their high level of complexity. Additionally, there are a lot of concerns involving peptide signaling during abiotic stress responses at the molecular level. There are still unanswered questions regarding the conversion of peptide precursors into mature peptides and the secretion and delivery of mature peptides to the target tissues via long-distance peptide signaling pathways. An investigation is required into how peptide-receptor interactions under abiotic stress result in certain developmental regulations. To develop stress resistance, it will also be interesting to look into how signaling peptides are connected to traditional phytohormone pathways. To increase crop stress tolerance and ensure the sustainability of agriculture, new strategies will be made available through the identification of plant peptides and signaling pathways.

REFERENCES

Ahsan, N., Lee, D. G., Lee, S. H., Lee, K. W., Bahk, J. D., & Lee, B. H. (2007). A proteomic screen and identification of waterlogging-regulated proteins in tomato roots. Plant and Soil, 295, 37–51.

Amano, Y., Tsubouchi, H., Shinohara, H., Ogawa, M., & Matsubayashi, Y. (2007). Tyrosine-sulfated glycopeptide involved in cellular proliferation and expansion in *Arabidopsis*. Proceedings of the National Academy of Sciences, 104(46), 18333–18338.

Anjum, N. A., Sofo, A., Scopa, A., Roychoudhury, A., Gill, S. S., Iqbal, M., et al. (2015). Lipids and proteins—major targets of oxidative modifications in abiotic stressed plants. Environmental Science and Pollution Research, 22, 4099–4121.

Anjum, S. A., Wang, L., Farooq, M., Xue, L., & Ali, S. (2011). Fulvic acid application improves the maize performance under well-watered and drought conditions. Journal of Agronomy and Crop Science, 197(6), 409–417. doi: 10.1111/j.1439-037X.2011.00483.x

Bartels, S., & Boller, T. (2015). Quo vadis, Pep? Plant elicitor peptides at the crossroads of immunity, stress, and development. Journal of Experimental Botany, 66(17), 5183–5193.

Blackburn, M. R., Haruta, M., & Moura, D. S. (2020). Twenty years of progress in physiological and biochemical investigation of RALF peptides. Plant Physiology, 182(4), 1657–1666.

Butenko, M. A., Patterson, S. E., Grini, P. E., Stenvik, G. E., Amundsen, S. S., Mandal, A., & Aalen, R. B. (2003). Inflorescence deficient in abscission controls floral organ abscission in *Arabidopsis* and identifies a novel family of putative ligands in plants. The Plant Cell, 15(10), 2296–2307.

Chen, Y. L., Lee, C. Y., Cheng, K. T., Chang, W. H., Huang, R. N., Nam, H. G., & Chen, Y. R. (2014). Quantitative peptidomics study reveals that a wound-induced peptide from PR-1 regulates immune signaling in tomato. The Plant Cell, 26(10), 4135–4148.

Chen, W., Yao, Q., Patil, G. B., Agarwal, G., Deshmukh, R. K., Lin, L., et al. (2016). Identification and comparative analysis of differential gene expression in soybean leaf tissue under drought and flooding stress revealed by RNA-Seq. Frontiers in Plant Science, 7, 1044.

Chen, Z. Q., Zan, Y., Milesi, P., Zhou, L., Chen, J., Li, L., et al. (2021). Leveraging breeding programs and genomic data in Norway spruce (Picea abies L. Karst) for GWAS analysis. Genome Biology, 22(1), 1–30.

Chien, P. S., Nam, H. G., & Chen, Y. R. (2015). A salt-regulated peptide derived from the CAP superfamily protein negatively regulates salt-stress tolerance in *Arabidopsis*. Journal of Experimental Botany, 66(17), 5301–5313.

Cho, S. K., Larue, C. T., Chevalier, D., Wang, H., Jinn, T. L., Zhang, S., & Walker, J. C. (2008). Regulation of floral organ abscission in *Arabidopsis thaliana*. Proceedings of the National Academy of Sciences, 105(40), 15629–15634.

De Storme, N., & Geelen, D. (2014). The impact of environmental stress on male reproductive development in plants: biological processes and molecular mechanisms. Plant, Cell & Environment, 37(1), 1–18.

Delay, C., Chapman, K., Taleski, M., Wang, Y., Tyagi, S., Xiong, Y., et al. (2019). CEP3 levels affect starvation-related growth responses of the primary root. Journal of Experimental Botany, 70(18), 4763–4774.

EL-Beltagi, H. S., Ahmed, O. K., & Hegazy, A. E. (2016). Protective effect of nitric oxide on high temperature induced oxidative stress in wheat (*Triticum aestivum*) callus culture. Notulae Scientia Biologicae, 8(2), 192–198.

Endo, S., Shinohara, H., Matsubayashi, Y., & Fukuda, H. (2013). A novel pollen-pistil interaction conferring high-temperature tolerance during reproduction via CLE45 signaling. Current Biology, 23(17), 1670–1676.

Fahad, S., Bajwa, A. A., Nazir, U., Anjum, S. A., Farooq, A., Zohaib, A., et al. (2017). Crop production under drought and heat stress: plant responses and management options. Frontiers in Plant Science, 1147.

Farooq M., Wahid A., Kobayashi N., Fujita D., Basra S. M. A. (2009). Plant Drought Stress: Effects, Mechanisms and Management. In: E. Lichtfouse, M. Navarrete, P. Debaeke, S. Véronique, & C. Alberola (eds.), Sustainable Agriculture (pp. 153–188), Springer, Dordrecht. doi: 10.1007/978-90-481-2666-8_12

Feller, U., & Vaseva, I. I. (2014). Extreme climatic events: impacts of drought and high temperature on physiological processes in agronomically important plants. Frontiers in Environmental Science, 2, 39.

Feng, W., Kita, D., Peaucelle, A., Cartwright, H. N., Doan, V., Duan, Q., et al. (2018). The FERONIA receptor kinase maintains cell-wall integrity during salt stress through Ca2+ signaling. Current Biology, 28(5), 666–675.

Fletcher, J. C. (2020). Recent advances in *Arabidopsis* CLE peptide signaling. Trends in Plant Science, 25, 1005–1016.

Fu, J., & Huang, B. (2001). Involvement of antioxidants and lipid peroxidation in the adaptation of two cool-season grasses to localized drought stress. Environmental and Experimental Botany, 45(2), 105–114.

Ge, Z., Bergonci, T., Zhao, Y., Zou, Y., Du, S., Liu, M. C., et al. (2017). *Arabidopsis* pollen tube integrity and sperm release are regulated by RALF-mediated signaling. Science, 358(6370), 1596–1600.

Giorno, F., Wolters-Arts, M., Mariani, C., & Rieu, I. (2013). Ensuring reproduction at high temperatures: the heat stress response during anther and pollen development. Plants, 2, 489–506.

Golldack, D., Li, C., Mohan, H., & Probst, N. (2014). Tolerance to drought and salt stress in plants: unraveling the signaling networks. Frontiers in plant science, 5, 151.

Hasanuzzaman, M., Al Mahmud, J., Nahar, K., Anee, T. I., Inafuku, M., Oku, H., & Fujita, M. (2017). Responses, Adaptation, and ROS Metabolism in Plants Exposed to Waterlogging Stress. In: M. Khan, & N. Khan (eds.), Reactive Oxygen Species and Antioxidant Systems in Plants: Role and Regulation under Abiotic Stress (pp. 257–281), Springer, Singapore. 10.1007/978-981-10-5254-5_10

Hemathilake, D., & Gunathilake, D. (2022). Agricultural Productivity and Food Supply to Meet Increased Demands. In: R. Bhat (ed.), Future Foods (pp 539–553). Elsevier.

Hu, L., Zhang, Y., Xia, H., Fan, S., Song, J., Lv, X., & Kong, L. (2019). Photosynthetic characteristics of non-foliar organs in main C3 cereals. Physiologia Plantarum, 166(1), 226–239.

Jackson, M. B., & Colmer, T. (2005). Response and adaptation by plants to flooding stress. Annals of Botany, 96(4), 501–505.

Jebara, S., Jebara, M., Limam, F., & Aouani, M. E. (2005). Changes in ascorbate peroxidase, catalase, guaiacol peroxidase and superoxide dismutase activities in common bean (*Phaseolus vulgaris*) nodules under salt stress. Journal of Plant Physiology, 162(8), 929–936.

Khan, Z., & Shahwar, D. (2020). Role of heat shock proteins (HSPs) and heat stress tolerance in crop plants. Sustainable agriculture in the era of climate change, 211–234.

Kim, M. J., Jeon, B. W., Oh, E., Seo, P. J., & Kim, J. (2021). Peptide signaling during plant reproduction. Trends in Plant Science, 26(8), 822–835.

Kumpf, R. P., Shi, C. L., Larrieu, A., Stø, I. M., Butenko, M. A., Péret, B., et al. (2013). Floral organ abscission peptide IDA and its HAE/HSL2 receptors control cell separation during lateral root emergence. Proceedings of the National Academy of Sciences, 110(13), 5235–5240.

Ladwig, F., Dahlke, R. I., Stührwohldt, N., Hartmann, J., Harter, K., & Sauter, M. (2015). Phytosulfokine regulates growth in *Arabidopsis* through a response module at the plasma membrane that includes CYCLIC NUCLEOTIDE-GATED CHANNEL17, H+-ATPase, and BAK1. The Plant Cell, 27(6), 1718–1729.

Lang, P., Zhang, C. K., Ebel, R. C., Dane, F., & Dozier, W. A. (2005). Identification of cold acclimated genes in leaves of Citrus unshiu by mRNA differential display. Gene, 359, 111–118.

Lata, C., Yadav, A., & Prasad, M. (2011). Role of plant transcription factors in abiotic stress tolerance. Abiotic Stress Response in Plants, INTECH Open Access Publishers, 10, 269–296.

Liang, C., Meng, Z., Meng, Z., Malik, W., Yan, R., Lwin, K. M., et al. (2016). GhABF2, a bZIP transcription factor, confers drought and salinity tolerance in cotton (*Gossypium hirsutum* L.). Scientific Reports, 6(1), 35040.

Mecchia, M. A., Santos-Fernandez, G., Duss, N. N., Somoza, S. C., Boisson-Dernier, A., Gagliardini, V., et al. (2017). RALF4/19 peptides interact with LRX proteins to control pollen tube growth in *Arabidopsis*. Science, 358(6370), 1600–1603.

Munns, R. (2005). Genes and salt tolerance: bringing them together. New phytologist, 167(3), 645–663.

Nakaminami, K., Okamoto, M., Higuchi-Takeuchi, M., Yoshizumi, T., Yamaguchi, Y., Fukao, Y., et al. (2018). AtPep3 is a hormone-like peptide that plays a role in the salinity stress tolerance of plants. Proceedings of the National Academy of Sciences, 115(22), 5810–5815.

Nanjo, Y., Maruyama, K., Yasue, H., Yamaguchi-Shinozaki, K., Shinozaki, K., & Komatsu, S. (2011). Transcriptional responses to flooding stress in roots including hypocotyl of soybean seedlings. Plant molecular biology, 77, 129–144.

Nir, I. D. O., Moshelion, M., & Weiss, D. (2014). The A rabidopsis GIBBERELLIN METHYL TRANSFERASE 1 suppresses gibberellin activity, reduces whole-plant transpiration and promotes drought tolerance in transgenic tomato. Plant, Cell & Environment, 37(1), 113–123.

Ogilvie, H. A., Imin, N., & Djordjevic, M. A. (2014). Diversification of the C-TERMINALLY ENCODED PEPTIDE (CEP) gene family in angiosperms, and evolution of plant-family specific CEP genes. BMC Genomics, 15(1), 1–15.

Oh, E., Seo, P. J., & Kim, J. (2018). Signaling peptides and receptors coordinating plant root development. Trends in Plant Science, 23, 337–351.

Örvar, B. L., Sangwan, V., Omann, F., & Dhindsa, R. S. (2000). Early steps in cold sensing by plant cells: the role of actin cytoskeleton and membrane fluidity. The Plant Journal, 23(6), 785–794.

Ota, R., Ohkubo, Y., Yamashita, Y., Ogawa-Ohnishi, M., & Matsubayashi, Y. (2020). Shoot-to-root mobile CEPD-like 2 integrates shoot nitrogen status to systemically regulate nitrate uptake in *Arabidopsis*. Nature Communications, 11(1), 641.

Parihar, S. P., Guler, R., Khutlang, R., Lang, D. M., Hurdayal, R., Mhlanga, M. M., et al. (2014). Statin therapy reduces the mycobacterium tuberculosis burden in human macrophages and in mice by enhancing autophagy and phagosome maturation. The Journal of Infectious Diseases, 209(5), 754–763.

Pathak, M. R., Teixeira da Silva, J. A., & Wani, S. H. (2014). Polyamines in response to abiotic stress tolerance through transgenic approaches. GM crops & food, 5(2), 87–96.

Patharkar, O. R., & Walker, J. C. (2016). Core mechanisms regulating developmentally timed and environmentally triggered abscission. Plant Physiology, 172(1), 510–520.

Qin, F., Shinozaki, K., & Yamaguchi-Shinozaki, K. (2011). Achievements and challenges in understanding plant abiotic stress responses and tolerance. Plant and Cell Physiology, 52(9), 1569–1582.

Reddy, K. J. (2006). Nutrient stress. In: K. V. M. Rao, A. S. Raghavendra, & J. S. Reddy (eds.), Physiology and molecular biology of stress tolerance in plants (pp. 187–217). Springer Netherlands, Dordrecht.

Reichardt, S., Piepho, H. P., Stintzi, A., & Schaller, A. (2020). Peptide signaling for drought-induced tomato flower drop. Science, 367(6485), 1482–1485.

Rizwan, M., Ali, S., Ibrahim, M., Farid, M., Adrees, M., Bharwana, S. A., et al. (2015). Mechanisms of silicon-mediated alleviation of drought and salt stress in plants: a review. Environmental Science and Pollution Research, 22, 15416–15431.

Roberts, I., Smith, S., Stes, E., De Rybel, B., Staes, A., Van De Cotte, B., et al. (2016). CEP5 and XIP1/CEPR1 regulate lateral root initiation in *Arabidopsis*. Journal of Experimental Botany, 67(16), 4889–4899.

Sahoo, S., Kumar, A., & Upadhyay, A. (2023). How do green knowledge management and green technology innovation impact corporate environmental performance? Understanding the role of green knowledge acquisition. Business Strategy and the Environment, 32(1), 551–569.

Sanghera, G. S., Wani, S. H., Hussain, W., & Singh, N. B. (2011). Engineering cold stress tolerance in crop plants. Current genomics, 12(1), 30.

Sangwan, V., Örvar, B. L., Beyerly, J., Hirt, H., & Dhindsa, R. S. (2002). Opposite changes in membrane fluidity mimic cold and heat stress activation of distinct plant MAP kinase pathways. The Plant Journal, 31(5), 629–638.

Smith, S., Zhu, S., Joos, L., Roberts, I., Nikonorova, N., Dai Vu, L., et al. (2020). The CEP5 peptide promotes abiotic stress tolerance, as revealed by quantitative proteomics, and attenuates the AUX/IAA equilibrium in *Arabidopsis*. Molecular & Cellular Proteomics, 19(8), 1248–1262.

Stenvik, G. E., Butenko, M. A., Urbanowicz, B. R., Rose, J. K., & Aalen, R. B. (2006). Overexpression of INFLORESCENCE DEFICIENT IN ABSCISSION activates cell separation in vestigial abscission zones in *Arabidopsis*. The Plant Cell, 18(6), 1467–1476.

Stührwohldt, N., Bühler, E., Sauter, M., & Schaller, A. (2020). Precursor processing by SBT3. 8 and phytosulfokine signaling contribute to drought stress tolerance in Arabidopsis. bioRxiv, 2020-10.

Tabata, R., Sumida, K., Yoshii, T., et al. (2014). Perception of root-derived peptides by shoot LRR-RKs mediates systemic N-demand signaling. Science, 346 (80), 343–346.

Takahashi, F., Suzuki, T., Osakabe, Y., Betsuyaku, S., Kondo, Y., Dohmae, N., et al. (2018). A small peptide modulates stomatal control via abscisic acid in long-distance signalling. Nature, 556(7700), 235–238.

Wang, J., Li, H., Han, Z., Zhang, H., Wang, T., Lin, G., … & Chai, J. (2015). Allosteric receptor activation by the plant peptide hormone phytosulfokine. Nature, 525(7568), 265–268.

Wani, S. H., Singh, N. B., Haribhushan, A., & Iqbal Mir, J. (2013). Compatible solute engineering in plants for abiotic stress tolerance-role of glycine betaine. Current Genomics, 14(3), 157–165.

Xiao, F., & Zhou, H. (2023). Plant salt response: Perception, signaling, and tolerance. Frontiers in Plant Science, 13, 1053699.

Xu, X., Ji, J., Ma, X., Xu, Q., Qi, X., & Chen, X. (2016). Comparative proteomic analysis provides insight into the key proteins involved in cucumber (*Cucumis sativus* L.) adventitious root emergence under waterlogging stress. Frontiers in Plant Science, 7, 1515.

Yadav, S. K. (2010). Cold stress tolerance mechanisms in plants. Agronomy Sustainable Development, 57, 515–527.

Yang, F., Zhang, Q., Yao, Q., Chen, G., Tong, H., Zhang, J., et al. (2020). Direct and indirect plant defenses induced by (Z)-3-hexenol in tomato against whitefly attack. Journal of Pest Science, 93, 1243–1254.

Yu, L. H., Wu, S. J., Peng, Y. S., Liu, R. N., Chen, X., Zhao, P., et al. (2016). *Arabidopsis* EDT 1/HDG 11 improves drought and salt tolerance in cotton and poplar and increases cotton yield in the field. Plant Biotechnology Journal, 14(1), 72–84.

Zhang, L., Shi, X., Zhang, Y., Wang, J., Yang, J., Ishida, T., et al. (2019). CLE9 peptide-induced stomatal closure is mediated by abscisic acid, hydrogen peroxide, and nitric oxide in *Arabidopsis thaliana*. Plant, Cell & Environment, 42(3), 1033–1044.

Zhao, C., Zayed, O., Yu, Z., Jiang, W., Zhu, P., Hsu, C. C., … & Zhu, J. K. (2018). Leucine-rich repeat extensin proteins regulate plant salt tolerance in *Arabidopsis*. Proceedings of the National Academy of Sciences, 115(51), 13123–13128.

Zhu, Q., Shao, Y., Ge, S., Zhang, M., Zhang, T., Hu, X., et al. (2019). A MAPK cascade downstream of IDA–HAE/HSL2 ligand–receptor pair in lateral root emergence. Nature Plants, 5(4), 414–423.

Zhu, J. K. (2001). Plant salt tolerance. Trends in Plant Science, 6(2), 66–71.

Zhu, J. K. (2016). Abiotic stress signaling and responses in plants. Cell, 167, 313–324.

17 Role of Salicylic Acid in Abiotic Stresses

Prashant Shaw, Riya Pakhre, Rajkumari Sanayaima Devi, and Sachchidanand Tripathi

17.1 INTRODUCTION

Climate change is foreseen as a major liability to global food and economic security. Rapid urbanization, rampant industrialization, forest degradation and an increase in CO_2 emission levels contribute to climate change in broad categories. An estimated jump of around 45% has been reported in atmospheric carbon dioxide levels since the Industrial Revolution (Ciais et al., 2013). A rapid rise in global temperatures has intensified diverse kinds of stress on plants, which is threatening the world's food security. The report of IPCC 2022 emphasizes reduced crop productivity by one-third since 1961. It has been estimated that adverse growth conditions induced due to abiotic stresses negatively impact 70% of the yield of commercial food crops (Zhang et al., 2020). Such abiotic stress hampers the regular functioning of biological, physiological and other metabolic processes. Phytohormones have been found to significantly reduce the impact of stress through resilient mechanisms. The vital role of salicylic acid has been widely acknowledged for conferring tolerance to abiotic stresses.

Salicylic acid (SA) is a naturally occurring phenol synthesized within plants through two metabolic pathways. First is the phenylalanine ammonia-lyase pathway and the second is the isochorismate synthase pathway (Khan et al., 2015). Contribution and significance of these pathways differ with plant species. For example, both pathways are of comparable significance in soybeans. The latter dominates SA biosynthesis in *Arabidopsis thaliana* (L.) Heynh., whereas the former dictates major synthesis in rice (Lefevere et al., 2020).

17.2 UNDERSTANDING SA'S CONTRIBUTION TO ABIOTIC STRESS RESPONSE: AN OVERVIEW

SA communicates a diverse range of plant responses at physiological, metabolic and molecular levels when associated with non-ideal growth conditions (Khan et al., 2015). It directly or indirectly contributes to alleviation from salinity (Xu et al., 2022), waterlogging (Singh et al., 2017), heat (Nosheen et al., 2021), cold (Wang et al., 2021), drought (Bijanzadeh et al., 2019) and ozone stress. During the vegetative stages, SA is well known for inducing basal levels of thermotolerance, while also enhancing reproductive functions and mitigating high-temperature impacts on plant reproduction (Ku et al., 2003; Oshino et al., 2007). Moreover, it promotes tolerance to extreme temperatures by activating antioxidants, osmolytes and stress-responsive proteins, such as late embryogenesis-abundant (LEA), heat shock proteins (HSP), dehydrins and pathogenesis-related (PR) (Saleem et al., 2021). Ozone exposure drastically increases SA production in Bel-W3 tobacco cultivar leaves (Drzewiecka et al., 2012). Moreover, SA combats drought by regulating stomatal opening/closure, transpiration rates and preventing unchallenged entry of pathogens through stomatal pores (Melotto et al., 2006). SA exhibits stress-relieving utilities as growth regulators due to its ability to neutralize reactive oxygen species (ROS) and

DOI: 10.1201/9781003335788-21

restore membrane integrity (Moustafa-Farag et al., 2020). Therefore, SA is widely recognized as a prominent growth hormone in mitigating abiotic stresses.

17.3 SALICYLIC ACID: COMBATING DIVERSE ABIOTIC STRESSES

17.3.1 Heavy Metal Toxicity

Trace amounts of heavy metals exist naturally in the Earth's crust, but their current environmental concentrations exceed normal levels due to anthropogenic contaminants. However, a direct relation between SA and its role in heavy metal tolerance is yet to be established, as they are genetically controlled and dose-dependent. Many members of distantly related plant families are heavy metal "hyperaccumulators," which are known to accumulate heavy metals in surplus amounts in their aerial parts without suffering any phytotoxic effects (Kovács et al., 2014). These hyperaccumulators constitute about 450 plant species (0.2% of angiosperms) of all known species and are resistant to phytotoxic damage, even when accumulating more than 10 milligrams (mg) per gram of Mg and Zn; 1 mg per gram of Pb, Ni, Co and As; and 0.1 mg per gram of Cd in their aerial organs (Verbruggen et al., 2009). Cd treatment in different wheat genotypes promoted Cd tolerance in seedlings via enhancement of the glutathione cycle (Kovács et al., 2014). Additionally, enhanced root length and increased dry mass were observed in SA-primed wheat seedlings compared to solely Cd-treated ones. Pb negatively affects plants' antioxidant capabilities by suppressing the activities of catalase, ascorbic acid, peroxide etc., whereas endogenous SA assists in antagonizing the oxidative damage effects of Pb by upregulating the known enzymic (catalase, peroxidase, Fe-superoxide dismutase) and non-enzymic (ascorbate, glutathione etc.) antioxidant compounds (Chen et al., 2007; López et al., 2014; Wani et al., 2017). Under heavy metal stress, SA participates in alleviating photosynthetic attributes of *Brassica juncea* (L.) Czern (Indian mustard). Chlorophyll (chl) fluorescence, which decreased by 5% and 22% in 500 mg and 1,000 mg Pb treatment, respectively, observed slight improvement under SA treatment than Pb-alone counterparts. Also, a combined treatment with SA and Pb resulted in an increment of 25% chl a and 112% chl b content in 2,000 mg and 1,000 mg Pb treatments, respectively (Agnihotri et al., 2018).

17.3.2 Soil Salinity

Soil salinization is caused by the dissolution of excess minerals and salts in water. Approximately 50% of the arable land will be under the influence of salinity by the year 2050 (Kumar & Sharma, 2020). Two major impacts in crops that result due to high salt content are ionic toxicity and osmotic stress. As a result, increased salt concentration beyond the optimum level imposes a serious threat to cell growth, ionic uptake, photosynthetic attributes, cytosolic metabolism and ROS production (Yadav et al., 2020), which creates imbalances that restrict the plant's minerals (K^+, Ca^{2+}) and water-absorbing capacity. Excess Na^+ and Cl^- accumulation further acts to prompt secondary osmotic stress, leading to depolarization of cell membranes. The consequences of salinity on root growth are much more prominent than on shoot growth (Souri & Tohidloo, 2019). In an experiment, aimed at investigating the impact of salinity treatments on plant growth, a noticeable decrease in the fresh weight of *Dianthus superbus* L. seedlings was observed. The reductions were recorded as 10.7%, 28.13% and 38.5% on treatments with 0.3%, 0.6% and 0.9% NaCl, respectively (Ma et al., 2017). SA influences many physiological traits of the roots that mitigate protection against salinity stress via osmotic adjustments, minimizing the Na^+/K^+ ratio, reducing ion (Na^+, Cl^-) uptake, upregulating H^+- ATPases and preventing K^+ ion leakage (Jayakannan et al., 2015; Souri & Tohidloo, 2019). The SA-deficient *sid2* mutants show a substantial delay in seed germination under high salinity growth conditions, whereas exogenous SA application recovers seed germination. On the contrary, *NahG* transgenic lines with lower SA accumulation than the wild-type plants show no significant effects on germination under different salt concentrations (Lee & Park, 2010). *Triticum aestivum* cv. Samma under growth conditions shows a downward trend when supplemented with NaCl in the growth medium. Proline accumulation, impaired chl a and chl b contents and suppressed carbonic anhydrase (CA) activity were observed under NaCl treatments relative to the control plants. Treatment with SA alone alleviates the cultivar from salt stress by enhancing chl a, CA activity, chl b and proline contents. Upregulated SA signaling genes like non-expresser of pathogenesis-related gene 1 (*NPR1*), support augmented stress responses. Typically found at low levels in plants, its concentration increases nearly three folds after SA treatment. *NPR1* functions as a high-affinity SA-binding protein, highlighting the importance of its SA-binding activity in the context of SA-induced immunity. A collaborative process involving *NPR1*-dependent transcriptional regulation and biosynthesis of biochemicals, counter salinity effects and sustain plant cellular redox homeostasis (Khalid et al., 2021).

17.3.3 Roles in Heat and Cold Stresses

Heat stress beyond the plant's optimum temperatures negatively impacts the plant's development and growth dynamics (Lamaoui et al., 2018). Upon a rise in the environmental temperature, plants display protective responses via changes in morphological, physiological (photosynthesis, respiration, transpiration, membrane thermostability) (Zhao et al., 2020) and biochemical attributes. Plants acquire heat tolerance under a signalling cascade in which re-establishment of the ROS detoxification, enzyme reactivation and cellular homeostasis is achieved via the activation of heat-induced stress-alleviating genes (Hasanuzzaman et al., 2013). In *Arabidopsis*, components of cyclic nucleotide-gated (CNG) Ca^+ channels are encoded by the CNGC2 gene (Bita & Gerats, 2013). These membrane CNGs serve as a primary thermosensor in triggering an ideal thermotolerance response. An initial response to heat stress induces Ca^{2+} influx and cytoskeletal restructuring, which causes activation of the Ca-dependent signalling network (Saidi et al., 2009), leading to the production of antioxidants, osmoprotectants and expression of HSPs (Lamaoui et al., 2018). Photosynthetic attributes and sexual reproduction of the plants are most vulnerable to increasing temperature. An increase in temperature beyond the optimum levels damaged the thylakoid membrane, reduced photosynthetic rates and inhibited membrane-bound electron carriers, especially photosystem II (Zhao et al., 2020; Camejo et al., 2005). Under non-heat-stressed conditions, SA did not lead to a noteworthy enhancement in the net photosynthesis rate (Pn) of grapevine leaves. However, pre-treatment with SA expedited the recovery of Pn under heat-stressed conditions by maintaining elevated levels of RuBisCo and Hsp21 (Wang et al., 2010).

The plants show different sets of responses following chilling stress. At extremely low temperatures, ice crystals are formed in the cell liquids, leading to cell death (Miura & Tada, 2014). Janda et al. (1999) first demonstrated the vital contribution of SA in conferring tolerance against low or chilling temperatures. Both endogenous as well as exogenous SA mitigate the severity of cold stress by participating in various metabolic pathways, mainly Ca^+ signalling, mitogen-activated protein kinase, reactive nitrogen species and others (Saleem et al., 2021). Enhancement of alternative oxidase when supplemented with MESA and MEJA aids in providing relief from chilling injuries (Fung et al., 2004). The efficacy of exogenously applied SA may render either positive or negative effects on plant health, which is concentration-dependent. A low SA concentration (0.1 to 0.5 mM) invoked tolerance in beans and tomatoes, whereas a high concentration SA (1 mM) failed to do so (Senaratna et al., 2000). Cold stress also influences the activation of disease resistance mechanisms, which show significant dependency on SA (Wu et al., 2019). Hence, all these assertions indicate an interlinked and tissue-specific function of SA in relieving cold stress.

17.3.4 Ozone Stress

Rapid industrialization has increased environmental pollutants and tropospheric ozone (O_3) levels. O_3 concentration exceeding 60 parts per billion (ppb) hampers crop productivity and vegetation (Fiscus et al., 2005). The symptoms associated with ozone-induced damage are dependent on concentration and period of exposure. Transient exposure to an intensive concentration of O_3 (>150 ppb) results in visible damage, whereas long exposure to low O_3 concentrations causes growth reduction without any visible damage (Sharma et al., 1996). Ozone exposure leads to ozone-induced effects such as necrotic lesions, chlorosis (Sharma et al., 2017), early flowering and a decrease in pollinator abundance (Duque et al., 2021). O_3-induced leaf damage was estimated by measuring ion leakage in O_3-exposed *Nicotiana* leaves, which resulted in a 57% and 73% increase in 4 hours and 8 hours, respectively. SA accumulation was reported to occur concomitantly, with an increase in ion leakage from leaves (Ogawa et al., 2005). Exposure to O_3 activates membrane-localized oxidases capable of generating activated oxygen species, which is an SA-mediated mitigation response. In three different *Arabidopsis* ecotypes, a highly O_3-sensitive ecotype (Cvi-0) has been found to hyper-accumulate SA upon ozone exposure (Rao & Davis, 1999). SA interacts with other phytohormones like jasmonic acid, ethylene and abscisic acid and elicits a combined response to develop tolerance against the injurious effects of ozone stress (Vahala et al., 2003).

17.3.5 Combating Drought and Waterlogging Conditions

Plants experienced significant impacts on their physical characteristics and physiological parameters due to drought stress. SA alleviates its detrimental effects such as decreased relative water content, reduced biomass both above and below the ground, lowered chlorophyll content, impaired gas-exchange properties, membrane damage, diminished photosynthesis and a decline in total proteins. However, foliar SA application retains membrane integrity, improves photosynthetic parameters and counterbalances ROS effects (Tayyab et al., 2020; Shemi et al., 2021). Exogenous treatment with 100 mM SA significantly intensified the leaf NPK contents in two wheat cultivars, namely Anaj-17 and Barani-17 (Ahmad et al., 2021). Under water-insufficiency conditions, plant phenolic compounds immensely contribute to neutralizing ROS (Kumar et al., 2020). Foliar SA application enhances the accumulation of phenolics as well as total soluble proteins in maize (Latif et al, 2016). Pre-treated muskmelon seeds containing different concentrations of acetylsalicylic acid (ASA) protects them under drought-stressed conditions. Also, ASA pre-treatment displayed the most effective results when soaked at lower concentrations (Korkmaz et al., 2007). Drought-induced endogenous production of abscisic acid, known to promote stomata closure, is also induced by the interplay of SA and methyl jasmonate (MeJA). SA-accumulating mutants, namely *cpr5* and *acd6*, exhibited stomatal closure and increased tolerance to drought in *Arabidopsis*. These effects were attributed to the PR gene expression mediated by SA (Liu et al., 2013).

The adverse impacts of waterlogging and SA-induced tolerance are more likely similar to other stresses (Singh et al., 2017), but varies to some extent. Exogenous SA promoted rapid formations of aerenchyma, exile roots and adventitious roots. Axile and adventitious root development were independently attributed as a function of SA, but aerenchyma development required ethylene in addition to SA (Koramutla et al., 2022). Response mechanisms to waterlogging involve complex crosstalks of endogenous hormones and are not understood completely. In the waterlogging-tolerant line (WTL), there was a superior development of adventitious roots compared to the waterlogging-susceptible line (WSL). Additionally, significant variations in endogenous hormones were observed between WTL and WSL. The WTL exhibited a higher ratio of ethylene production and methionine levels compared to WSL. Moreover, the contents of endogenous abscisic acid were lower in WTL compared to the WSL (Kim et al., 2015). SA and kinetin enhance ROS metabolism and upregulate the glyoxalase system in soybeans by reducing leaf electrolyte leakage, H_2O_2, MDA and proline contents under waterlogging (Hasanuzzaman et al., 2022).

17.4 METHYL SALICYLATE (MESA) IN CONFERRING ABIOTIC STRESS TOLERANCE VIA PLANT-PLANT COMMUNICATION

Plants produce a variety of volatile organic compounds (VOCs) to establish communications with diverse organisms (Loreto & Schnitzler, 2010). A volatile derivative of SA, MeSA is released from a plant in reply to biotic plus abiotic stresses (Lee & Seo, 2014), which also acts as an airborne signalling molecule to alert nearby plants of the incoming stress. Hence, MeSA establishes a plant-plant communication network, where defence genes are expressed upon the reception of airborne signal molecules by the neighbouring plants (Shulaev et al., 1997). Neighbouring plants absorb high MeSA concentrations and convert them to SA, which is utilized to confer SA-mediated defence responses (Chen et al., 2003). On perceiving stress, the stressed "elicitor" plant releases MeSA and other VOCs, triggering non-stressed "recipient" plants to activate defence mechanisms in anticipation of such stressors. VOC emissions from a non-stressed recipient plant are dose-dependent on the elicitor's MeSA emissions. A higher MeSA concentration from a stressed "elicitor" plant is expected to induce higher VOCs from non-stressed "recipient" plants (Liu et al., 2018).

17.5 CONCLUDING REMARKS AND FUTURE DIRECTIONS

This chapter highlights the critical role of salicylic acid in enhancing abiotic stress tolerance and facilitating cellular communication in plants. Numerous studies have investigated the involvement of SA in inducible defence mechanisms and the production of protective compounds. SA exerts metabolic influences and impacts plant-water relations, making it a key factor in plant responses to stress. To gain a deeper understanding of SA's signalling mechanism, a thorough exploration of the pathways and interconnections with other growth hormones is necessary. Additional research at the genetic level is necessary to evaluate the stress tolerance capacities across various cultivars of crops. Since it renders its effectiveness in a dosage-dependent manner, it becomes necessary to dissect the molecular basis of this phenomenon. Additional research is needed to elucidate its role in inter-plant communication, which necessitates a comprehensive understanding of SA-based receptors and signalling molecules. Moreover, SA exhibits significant potential for biotechnological applications, thereby playing a vital role in the development of sustainable crop varieties capable of thriving under future environmental conditions. Given the current climate change scenario and increasing biotic and abiotic stressors, a comprehensive comprehension of SA's role in plant physiology is imperative for inducing stress tolerance in plants.

REFERENCES

Agnihotri, A., Gupta, P., Dwivedi, A., & Seth, C. S. (2018). Counteractive mechanism (S) of salicylic acid in response to lead toxicity in *Brassica juncea* (L.) Czern. cv. Varuna. Planta, 248 (1), 49–68.

Ahmad, A., Aslam, Z., Naz, M., Hussain, S., Javed, T., Aslam, S., et al. (2021). Exogenous salicylic acid-induced drought stress tolerance in wheat (*Triticum aestivum* L.) grown under hydroponic culture. PLoS One, 16(12), e0260556.

Bijanzadeh, E., Naderi, R., & Egan, T. P. (2019). Exogenous application of humic acid and salicylic acid to alleviate seedling drought stress in two corn (*Zea mays* L.) hybrids. Journal of Plant Nutrition, 42(13), 1483–1495.

Bita, C. E., & Gerats, T. (2013). Plant tolerance to high temperature in a changing environment: scientific fundamentals and production of heat stress-tolerant crops. Frontiers in Plant Science, 4, 273.

Camejo, D., Rodríguez, P., Morales, M. A., Dell'Amico, J. M., Torrecillas, A., & Alarcón, J. J. (2005). High temperature effects on photosynthetic activity of two tomato cultivars with different heat susceptibility. Journal of Plant Physiology, 162(3), 281–289.

Chen, F., D'Auria, J. C., Tholl, D., Ross, J. R., Gershenzon, J., Noel, J. P., & Pichersky, E. (2003). An *Arabidopsis thaliana* gene for methylsalicylate biosynthesis, identified by a biochemical genomics approach, has a role in defense. The Plant Journal, 36(5), 577–588.

Chen, J., Zhu, C., Li, L., Sun, Z., & Pan, X. (2007). Effects of exogenous salicylic acid on growth and H2O2-metabolizing enzymes in rice seedlings under lead stress. Journal of Environmental Sciences, 19(1), 44–49. 10.1016/S1001-0742(07)60007-2

Ciais, P., Sabine, C., Bala, G., Bopp, L., Brovkin, V., Canadell, J., Chhabra, A., DeFries, R., Galloway, J., Heimann, M., Jones, C., Le Quéré, C., Myneni, R. B., Piao, S., & Thornton, P. (2013). Carbon and Other Biogeochemical Cycles. In: T. F. Stocker, D. Qin, G.-K. Plattner, M. Tignor, S. K. Allen, J. Boschung, A. Nauels, Y. Xia, V. Bex, & P. M. Midgley (eds.), Climate Change 2013: The Physical Science Basis. Contribution of Working Group I to the Fifth Assessment Report of the Intergovernmental Panel on Climate Change. Cambridge University Press, Cambridge, United Kingdom and New York, NY, USA.

Drzewiecka, K., Borowiak, K., Bandurska, H., & Golinski, P. (2012). Salicylic acid—a potential biomarker of tobacco Bel-W3 cell death developed as a response to ground level ozone under ambient conditions. Acta Biologica Hungarica, 63(2), 231–249.

Duque, L., Poelman, E. H., & Steffan-Dewenter, I. (2021). Effects of ozone stress on flowering phenology, plant-pollinator interactions and plant reproductive success. Environmental Pollution, 272, 115953.

Fiscus, E. L., Booker, F. L., & Burkey, K. O. (2005). Crop responses to ozone: uptake, modes of action, carbon assimilation and partitioning. Plant, Cell & Environment, 28(8), 997–1011.

Fung, R. W., Wang, C. Y., Smith, D. L., Gross, K. C., & Tian, M. (2004). MeSA and MeJA increase steady-state transcript levels of alternative oxidase and resistance against chilling injury in sweet peppers (*Capsicum annuum* L.). Plant Science, 166(3), 711–719.

Hasanuzzaman, M., Ahmed, N., Saha, T., Rahman, M., Rahman, K., Alam, M. M., … & Nahar, K. (2022). Exogenous salicylic acid and kinetin modulate reactive oxygen species metabolism and glyoxalase system to confer waterlogging stress tolerance in soybean (*Glycine max* L.). Plant Stress, 3, 100057.

Hasanuzzaman, M., Nahar, K., Alam, M. M., Roychowdhury, R., & Fujita, M. (2013). Physiological, biochemical, and molecular mechanisms of heat stress tolerance in plants. International Journal of Molecular Sciences, 14(5), 9643–9684.

Janda, T., Szalai, G., Tari, I., & Paldi, E. (1999). Hydroponic treatment with salicylic acid decreases the effects of chilling injury in maize (*Zea mays* L.) plants. Planta, 208, 175–180.

Jayakannan, M., Bose, J., Babourina, O., Rengel, Z., & Shabala, S. (2015). Salicylic acid in plant salinity stress signalling and tolerance. Plant Growth Regulation, 76, 25–40.

Khalid, M., Ali, M., Hassani, D., Rauf, A., Jan, F., & Hui, N. (2021). Salicylic acid mediated protection of *Brassica campestris* sp. chinensis from saline stress via SA receptor NPR1 dependent transcriptional regulation and biosynthesis of related biochemicals. Environmental Technology & Innovation, 24, 101950.

Khan, M. I. R., Fatma, M., Per, T. S., Anjum, N. A., & Khan, N. A. (2015). Salicylic acid-induced abiotic stress tolerance and underlying mechanisms in plants. Frontiers in Plant Science, 6, 462.

Kim, Y. H., Hwang, S. J., Waqas, M., Khan, A. L., Lee, J. H., Lee, J. D., … & Lee, I. J. (2015). Comparative analysis of endogenous hormones level in two soybean (*Glycine max* L.) lines differing in waterlogging tolerance. Frontiers in Plant Science, 6, 714.

Koramutla, M. K., Tuan, P. A., & Ayele, B. T. (2022). Salicylic acid enhances adventitious root and aerenchyma formation in wheat under waterlogged conditions. International Journal of Molecular Sciences, 23(3), 1243.

Korkmaz, A., Uzunlu, M., & Demirkiran, A. R. (2007). Treatment with acetyl salicylic acid protects muskmelon seedlings against drought stress. Acta Physiologiae Plantarum, 29, 503–508.

Kovács, V., Gondor, O. K., Szalai, G., Darkó, É., Majláth, I., Janda, T., & Pál, M. (2014). Synthesis and role of salicylic acid in wheat varieties with different levels of cadmium tolerance. Journal of Hazardous Materials, 280, 12–19.

Ku, S., Yoon, H., Suh, H. S., & Chung, Y. Y. (2003). Male-sterility of thermosensitive genic male-sterile rice is associated with premature programmed cell death of the tapetum. Planta, 217, 559–565.

Kumar, P., & Sharma, P. K. (2020). Soil salinity and food security in India. Frontiers in Sustainable Food Systems, 4, 533781.

Kumar, S., Bhushan, B., Wakchaure, G. C., Meena, K. K., Kumar, M., Meena, N. L., & Rane, J. (2020). Plant Phenolics under Water-deficit Conditions: Biosynthesis, Accumulation, and Physiological Roles in Water Stress Alleviation. In: R. Lone, R. Shuab, A. Kamili (eds.), Plant Phenolics in Sustainable Agriculture (Volume 1, pp. 451–465), Springer, Singapore.

Lamaoui, M., Jemo, M., Datla, R., & Bekkaoui, F. (2018). Heat and drought stresses in crops and approaches for their mitigation. Frontiers in Chemistry, 6, 26.

Latif, F., Ullah, F., Mehmood, S., Khattak, A., Khan, A. U., Khan, S., & Husain, I. (2016). Effects of salicylic acid on growth and accumulation of phenolics in *Zea mays* L. under drought stress. Acta Agriculturae Scandinavica, Section B—Soil & Plant Science, 66(4), 325–332.

Lee, K., & Seo, P. J. (2014). Airborne signals from salt-stressed *Arabidopsis* plants trigger salinity tolerance in neighboring plants. Plant Signaling & Behavior, 9(3), e28392.

Lee, S., & Park, C. M. (2010). Modulation of reactive oxygen species by salicylic acid in *Arabidopsis* seed germination under high salinity. Plant Signaling & Behavior, 5(12), 1534–1536.

Lefevere, H., Bauters, L., & Gheysen, G. (2020). Salicylic acid biosynthesis in plants. Frontiers in Plant Science, 11, 338.

Liu, B., Kaurilind, E., Jiang, Y., & Niinemets, Ü. (2018). Methyl salicylate differently affects benzenoid and terpenoid volatile emissions in *Betula pendula*. Tree Physiology, 38(10), 1513–1525.

Liu, P., Xu, Z. S., Pan-Pan, L., Hu, D., Chen, M., Li, L. C., & Ma, Y. Z. (2013). A wheat PI4K gene whose product possesses threonine autophophorylation activity confers tolerance to drought and salt in *Arabidopsis*. Journal of Experimental Botany, 64(10), 2915–2927.

López-Orenes, A., Martínez-Pérez, A., Calderón, A. A., & Ferrer, M. A. (2014). Pb-induced responses in *Zygophyllum fabago* plants are organ-dependent and modulated by salicylic acid. Plant Physiology and Biochemistry, 84, 57–66.

Loreto, F., & Schnitzler, J. P. (2010). Abiotic stresses and induced BVOCs. Trends in Plant Science, 15(3), 154–166.

Ma, X., Zheng, J., Zhang, X., Hu, Q., & Qian, R. (2017). Salicylic acid alleviates the adverse effects of salt stress on *Dianthus superbus* (Caryophyllaceae) by activating photosynthesis, protecting morphological structure, and enhancing the antioxidant system. Frontiers in Plant Science, 8, 600.

Melotto, M., Underwood, W., Koczan, J., Nomura, K., & He, S. Y. (2006). Plant stomata function in innate immunity against bacterial invasion. Cell, 126(5), 969–980.

Miura, K., & Tada, Y. (2014). Regulation of water, salinity, and cold stress responses by salicylic acid. Frontiers in Plant Science, 5, 4.

Moustafa-Farag, M., Mohamed, H. I., Mahmoud, A., Elkelish, A., Misra, A. N., Guy, K. M., Kamran, M., Ai, S., & Zhang, M. (2020). Salicylic acid stimulates antioxidant defense and osmolyte metabolism to alleviate oxidative stress in watermelons under excess boron. Plants, 9(6), 724. 10.3390/plants9060724

Nosheen, K., Imran, K., Shehla, S., Muhammad, L., & Muhammad, J. (2021). Salicylic Acid: Its Role in Temperature Stress Tolerance. In: S. Fahad, O. Sonmez, S. S. D. Wang, C. Wu, M. Adnan, & Y. Turan (eds.), Plant Growth Regulators for Climate-Smart Agriculture (pp. 117–132). CRC Press.

Ogawa, D., Nakajima, N., Sano, T., Tamaoki, M., Aono, M., Kubo, A., … & Saji, H. (2005). Salicylic acid accumulation under O_3 exposure is regulated by ethylene in tobacco plants. Plant and Cell Physiology, 46(7), 1062–1072.

Oshino, T., Abiko, M., Saito, R., Ichiishi, E., Endo, M., Kawagishi-Kobayashi, M., & Higashitani, A. (2007). Premature progression of anther early developmental programs accompanied by comprehensive alterations in transcription during high-temperature injury in barley plants. Molecular Genetics and Genomics, 278, 31–42.

Rao, M. V., & Davis, K. R. (1999). Ozone-induced cell death occurs via two distinct mechanisms in *Arabidopsis*: the role of salicylic acid. The Plant Journal, 17(6), 603–614.

Saidi, Y., Finka, A., Muriset, M., Bromberg, Z., Weiss, Y. G., Maathuis, F. J., & Goloubinoff, P. (2009). The heat shock response in moss plants is regulated by specific calcium-permeable channels in the plasma membrane. The Plant Cell, 21(9), 2829–2843.

Saleem, M., Fariduddin, Q., & Janda, T. (2021). Multifaceted role of salicylic acid in combating cold stress in plants: A review. Journal of Plant Growth Regulation, 40, 464–485.

Senaratna, T., Touchell, D., Bunn, E., & Dixon, K. (2000). Acetyl salicylic acid (Aspirin) and salicylic acid induce multiple stress tolerance in bean and tomato plants. Plant Growth Regulation, 30(2), 157–161.

Sharma, S., Chatterjee, S., Kataria, S., Joshi, J., Datta, S., Vairale, M. G., & Veer, V. (2017). A Review on Responses of Plants to UV-B Radiation Related Stress: From Environmental Stressor to Regulator of Plant Growth. In V. P. Singh, S. Singh, S. M. Prasad, & P. Parihar (eds.), UV-B Radiation: From Environmental Stressor to Regulator of Plant Growth (pp. 75–97). John Wiley & Sons Ltd.

Sharma, Y. K., León, J., Raskin, I., & Davis, K. R. (1996). Ozone-induced responses in *Arabidopsis thaliana*: the role of salicylic acid in the accumulation of defense-related transcripts and induced resistance. Proceedings of the National Academy of Sciences, 93(10), 5099–5104.

Shemi, R., Wang, R., Gheith, E. S., Hussain, H. A., Hussain, S., Irfan, M., … & Wang, L. (2021). Effects of salicylic acid, zinc and glycine betaine on morpho-physiological growth and yield of maize under drought stress. Scientific Reports, 11(1), 1–14.

Shulaev, V., Silverman, P., & Raskin, I. (1997). Airborne signalling by methyl salicylate in plant pathogen resistance. Nature, 385(6618), 718–721.

Singh, S. K., Singh, A. K., & Dwivedi, P. (2017). Modulating effect of salicylic acid in tomato plants in response to waterlogging stress. International Journal of Agriculture, Environment and Biotechnology, 10(1), 31. 10.5958/2230-732x.2017.00009.2

Souri, M. K., & Tohidloo, G. (2019). Effectiveness of different methods of salicylic acid application on growth characteristics of tomato seedlings under salinity. Chemical and Biological Technologies in Agriculture, 6(1), 1–7.

Tayyab, N., Naz, R., Yasmin, H., Nosheen, A., Keyani, R., Sajjad, M., … & Roberts, T. H. (2020). Combined seed and foliar pre-treatments with exogenous methyl jasmonate and salicylic acid mitigate drought-induced stress in maize. PLoS One, 15(5), e0232269.

Vahala, J., Keinanen, M., Schutzendubel, A., Polle, A., & Kangasjarvi, J. (2003). Differential effects of elevated ozone on two hybrid aspen genotypes predisposed to chronic ozone fumigation. Role of ethylene and salicylic acid. Plant Physiology, 132(1), 196–205.

Verbruggen, N., Hermans, C., & Schat, H. (2009). Molecular mechanisms of metal hyperaccumulation in plants. New Phytologist, 181(4), 759–776.

Wang, L. J., Fan, L., Loescher, W., Duan, W., Liu, G. J., Cheng, J. S., … & Li, S. H. (2010). Salicylic acid alleviates decreases in photosynthesis under heat stress and accelerates recovery in grapevine leaves. BMC Plant Biology, 10, 1–10.

Wang, W., Wang, X., Huang, M., Cai, J., Zhou, Q., Dai, T., & Jiang, D. (2021). Alleviation of field low-temperature stress in winter wheat by exogenous application of salicylic acid. Journal of Plant Growth Regulation, 40, 811–823.

Wani, A. B., Chadar, H., Wani, A. H., Singh, S., & Upadhyay, N. (2017). Salicylic acid to decrease plant stress. Environmental Chemistry Letters, 15, 101–123.

Wu, Z., Han, S., Zhou, H., Tuang, Z. K., Wang, Y., Jin, Y., Shi, H., & Yang, W. (2019). Cold stress activates disease resistance in *Arabidopsis thaliana* through a salicylic acid dependent pathway. Plant, Cell & Environment, 42(9), 2645–2663. 10.1111/pce.13579

Xu, L., Chen, H., Zhang, T., Deng, Y., Yan, J., & Wang, L. (2022). Salicylic acid improves the salt tolerance capacity of *Saponaria officinalis* by modulating its photosynthetic rate, osmoprotectants, antioxidant levels, and ion homeostasis. Agronomy, 12(6), 1443.

Yadav, S., Modi, P., Dave, A., Vijapura, A., Patel, D., & Patel, M. (2020). Effect of Abiotic Stress on Crops. In: M. Hasanuzzaman, M. C. M. T. Filho, M. Fujita, & T. A. R. Nogueira (eds.), Sustainable Crop Production. IntechOpen.

Zhang, H., Zhao, Y., & Zhu, J. K. (2020). Thriving under stress: how plants balance growth and the stress response. Developmental Cell, 55(5), 529–543.

Zhao, J., Lu, Z., Wang, L., & Jin, B. (2020). Plant responses to heat stress: physiology, transcription, noncoding RNAs, and epigenetics. International Journal of Molecular Sciences, 22(1), 117.

18 Strigolactones

Mediator in Abiotic Stress Responses

Ayyagari Ramlal, Amooru Harika, Dhandapani Raju, Ambika Rajendran, and S K Lal*

18.1 INTRODUCTION

Plants, being sessile, encounter a range of environmental stresses throughout their life cycle. Stress is defined as any external factor, like environmental factors (abiotic and biotic) that prevent plants from attaining their full ability to produce yield or genetic potential, thereby affecting their growth and development. For survival, plants have developed various molecular, biochemical and physio-chemical mechanisms as a part of their defensive strategies to combat different stresses. The abiotic factors include drought, salt and heat, along with physical and chemical conditions. Stresses directly affect the productivity of plants, thereby declining the yield, leading to a threat to food security and sustainable agriculture (Pandey et al., 2016; Mostofa et al., 2018; Bhatla & Lal, 2018). Desiccation is the ability of the plant cell to withstand dry conditions until equilibration with ambient air and have the ability to be revived upon hydration (Vicre et al., 2004). Bewley (1979) proposed three properties associated with protoplasm, which helps the plants to withstand desiccation: limit damage from desiccation/rehydration to a minimum, maintain cellular integrity in the desiccated state and activate and mobilize repair mechanisms upon rehydration (Oliver et al., 2005). Heat is the most severe environmental stress of all as it limits the productivity and growth of plants. It has been reportedly used in crop-based modelling and empirical studies in tropical and sub-tropical regions, and every 1 rise in the temperature, around 2.5–16% yield is lost (Wang et al., 2017). Water is the most crucial factor that helps in the sustenance of all organisms, including both plants and animals. Any of the factors from salinity, cold, drought etc. may lead to deprivation (strive) of water for the plants, leading to the cessation of fundamental processes. Abiotic stresses have a devastating effect on signalling system pathways, leading to concentration variations of solutes, denaturation of proteins, turgor loss, damage to membrane etc. (Bhatla & Lal, 2018).

Phytohormones are known to be involved during different stresses. The identification of receptors and downstream cascade crosstalks mediated by the hormones have revealed the underlying molecular mechanisms during any stress (Mostofa et al., 2018). The traditional phytohormones, like abscisic acid (ABA), auxins, gibberellins, cytokinins, ethylene and brassinosteroids (BAs), are known to be involved in regulating stress responses by acting either distantly or locally at a very low concentration in a range from $10^{-6} - 10^{-5}$ mL^{-1}. The non-traditional plant growth regulators include jasmonic acid (JA), salicylic acid (SA), polyamines and nitric oxide. Strigolactones are considered to be the new addition to this list of non-traditional hormones (Banerjee and Roychoudhary 2018). At first instance, strigolactones were thought to be responsible for the germination of parasitic weeds from the genera *Orobanche* and *Striga*. Soon

* Contributed equally as first author.

DOI: 10.1201/9781003335788-22

after, it was discovered that they are involved in the detection of hosts and branching of fungal hyphal for the arbuscular mycorrhizal fungi (AMF). Furthermore, some reports showed that they act as inhibitors of branching (Saeed et al., 2017). Mutants of *Arabidopsis thaliana MORE AXILLARY GROWTH* (*MAX*) genes, *DECREASED APICAL DOMINANCE* (*DAD*) genes in *Petunia hybrida*, *DWARF* (*D*) genes of *Oryza sativa* and *RAMOSUS* (*RMS*) genes of *Pisum sativum*, including several other mutants (loss-of-function), have shown the roles of SLs during plant development, emphasizing their functioning in branching, regulation of root architecture, photomorphogenesis, early seedling development, secondary growth and leaf senescence (Brewer et al., 2013; Kapulnik et al., 2011; Toh et al., 2012; Ueda & Kusaba, 2015; Umehara et al., 2008; Urquhart et al., 2015; Mostofa et al., 2018). The genetic analysis of biosynthetic pathways of SL show that it is regulated by many abiotic stresses like nutrient deprivation, salinity and drought (Du et al., 2018; Ha et al., 2014; Lv et al., 2018; Sun et al., 2014; Umehara et al., 2010; Visentin et al., 2016; Zhuang et al., 2017; Mostofa et al., 2018;). It has been shown that SLs act as secondary messengers, which are activated by auxin, but higher concentrations also cause impairments in auxin transport and lead to suppression of axillary buds (Brewer et al., 2009; Ferguson & Beveridge, 2009; Hayward et al., 2009; Marzec et al., 2013).

In the chapter, different roles of strigolactones during abiotic stress via nutrient-deficiency, osmotic, drought, heat and salinity in plants have been discussed. The applications of strigolactones in agriculture have also been discussed.

18.2 HISTORY OF STRIGOLACTONES

Strigolactones were first identified around 55 years ago from the root exudates of cotton. Also, they were known to mediate between the root parasitic weeds (*Orobanche*, *Alectra*, *Phelipanche* and *Striga*) and their hosts (Zwanenburg et al., 2016; Mostofa et al., 2018).

18.2.1 Timeline of Identification of Strigolactones

An insight on the timeline from the identification to the elucidation of structural details of both canonical and non-canonical strigolactones (a new phytohormone) is tabulated in Table 18.1.

TABLE 18.1
Events in the isolation and structural elucidation of strigolactones

Year	Major Event	References
1966	First isolated from root exudates of *Gossypium hirsutum* L. and named as strigol	Cook et al., 1966
1972	• Structure of strigol • Natural occurrence of strigol was reported in *Striga lutea* Lour. (synonym of *S. asiatica* (L.) Kuntze), a parasitic weed as a germination stimulant	Cook et al., 1966, 1972
1985	Detailed structural analysis was done using X-ray diffraction analysis (structure of strigol)	Brooks et al., 1985
1993	Isolation of strigol from maize, proso millet, sorghum and cowpea	Siame et al., 1993
1998	Orobanchol was isolated from *Orobanche minor* Sm., a parasitic weed from its host root exudate *Trifolium pratense* L. (red clover)	Yokota et al., 1998
2005	Hyphal branching was induced from SLs in AM fungi	Akiyama et al., 2005

(*Continued*)

TABLE 18.1 ***(Continued)***
Events in the isolation and structural elucidation of strigolactones

Year	Major Event	References
2008	Branching was regulated by SL (shoot inhibition by strigolactone)	Gomez-Roldan et al., 2008; Umehara et al., 2008
2009	D14 was involved in the SL pathway and functions by inhibiting branching (rice tillering was inhibited by D14)	Arite et al., 2009
2011	The structure of orobanchol was elucidated from root exudates of *Vigna unguiculata* (L.) Walp. and *Striga gesnerioides* (Willd.) Vatke through circular dichroism, ^1H – NMR and HPLC	Ueno et al., 2011
2011	Root growth-promoting and root hair elongation functions of SL was observed	Koltai, 2011
2012	Carlactone was discovered	Alder et al., 2012
2012	DAD2/D14 act as receptors for SLs	Hamiaux et al., 2012
2013	DWARF14 is an essential component in SL signalling. Surface change and interaction of D14 is induced by the free D ring	Nakamura et al., 2013
2013	SL hydrolysis intermediate on D14 active site Serine	Zhao et al., 2013
2014	Increased secondary growth by extending the internodal regions by interacting with auxin	Yamada et al., 2014
2014	Carlactone was an endogenous biosynthetic precursor for SLs	Seto et al., 2014
2014	Avenaol was isolated	Kim et al., 2014
2014	Heliolactone was isolated	Ueno et al., 2014
2015	Highly sensitive receptor for SL was identified	Toh et al., 2015
2015	Intact SL molecule causes the surface change of D14 and interaction with F-box protein	Zhao et al., 2015
2016	His-butenolide complex on *Pisum* RMS3, single-turnover model	de Saint Germain et al., 2016
2016	Crystal structure of AtD14-D3-ASK1, CLIM model	Yao et al., 2016
2017	Structure of Avenaol	Yasui et al., 2017
2018	Structural basis for ligand specificity in *Striga* sp.	Xu et al., 2018
2018	Structural plasticity in D3	Shabek et al., 2018
2018	Zeapyranolactone was isolated	Charnikhova et al., 2018
2019	Structural basis for ligand specificity in *Physcomitrella* sp.	Bürger et al., 2019
2019	Zealactones were isolated	Xie et al., 2017; Charnikhova et al., 2017
2019	Intact SL molecule triggers signalling	Seto et al., 2019
2019	Lotus lactone (non-canonical SL) was isolated from *Lotus japonicus* (Regel) K. Larsen	Xie et al., 2019
2019	Structure of heliolactone	Woo & McErlean, 2019; Yoshimura et al., 2019

18.3 BIOSYNTHESIS OF STRIGOLACTONES

Strigolactones are a group of apocarotenoid compounds that are synthesized serially with the help of various enzymatic reactions occurring in the plastids and cytosol. The carlactone (CL) is the precursor of strigolactones, which is synthesized from all-trans-β-carotene by the serial enzymatic action of the enzymes namely DWARF27 (D27), D17, and D10 in rice, similarly *A. thaliana* D27

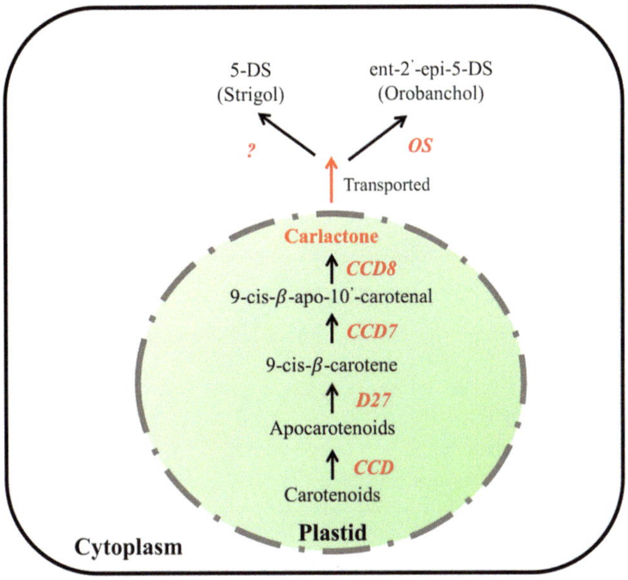

FIGURE 18.1 Formation of strigolactones from the carotenoids (CCD: carotenoid cleavage dioxygenase (CCD, CCD7 and CCD8); D27: DWARF27, OS: Orobanchol synthase and an unknown enzyme).

(*AtD27*), *MORE AXILLARY GROWTH3* (*MAX3*) and *MAX4* in *Arabidopsis. More axillary growth* (*MAX*) genes, including *MAX1/AT2G26170, MAX3/AT2G44990* and *MAX4/ AT4G32810*, have been identified, which are involved in the biosynthetic pathway SL from the non-mycotrophic *Arabidopsis*. Additionally, two more genes, namely *MAX2/AT2G42620* and *BRC1/AT3G18550* (*branched 1*), have been reported to play roles during SL responses. *MAX3* and encodes carotenoid cleavage dioxygenase 7 (CCD7) while *MAX4* encodes CCD8. In the cytosol, carlactone oxidase and orobanchol synthase convert carlactone into strigolactones in rice, whereas in *Arabidopsis*, the consecutive actions of *cytochrome P450 MAX1* act on the *MAX3* and *MAX4* downstream and *LATERAL BRANCHING OXIDOREDUCTASE* (*LBO*), convert the CL into carlactonic acid (CLA) and methyl carlactonic acid (MeCLA) and finally, an unidentified SL-like compound (MeCLA+16 Da). Figure 18.1 represents the synthesis of strigolactones via a sequential series of enzymatic reactions taking place in both plastids and cytosol. SLs are formed by the carotenoid cleavage dioxygenase (CCD)–mediated cleavage of carotenoid molecules, thereby forming apocarotenoids in the plastids. The three enzymes localized in the plastids, namely DWARF 27 (D27), CCD7 and CCD8, sequentially convert all-*trans*-β-carotene into carlactone (CL). The formation of 9-*cis* -β-carotene is catalyzed by the carotenoid isomerase D27. The formed product is then converted to 9-*cis*-β-apo-10′-carotenal by the CCD7. The CCD8 finally catalyzes the conversion of the product to yield CL (Ha et al., 2014; Mostofa et al., 2018).

18.3.1 Natural Occurrence and Synthetic Homologues of SLs

There are around 25 naturally occurring canonical and non-canonical strigolactones that have been identified across the plant kingdom (Figure 18.2). The first naturally occurring strigolactone was strigol and its acetate forms. The natural SLs can be categorized into two main classes, namely orobanchol-type SLs with an α-oriented C ring and strigol-type with a β-oriented C-ring and all possess a 2′(R) configuration (Khetkam et al., 2014). SLs have been found from the root exudates of proso-millet, sorghum and maize (Saime et al., 1993; Xie, 2016). Apart from the strigol, there are other related hormones that were extracted from root exudates of sorghum and cowpea namely sorgolactone and alectrol, respectively (Hauck et al., 1992; Müller et al., 1992; Xie, 2016).

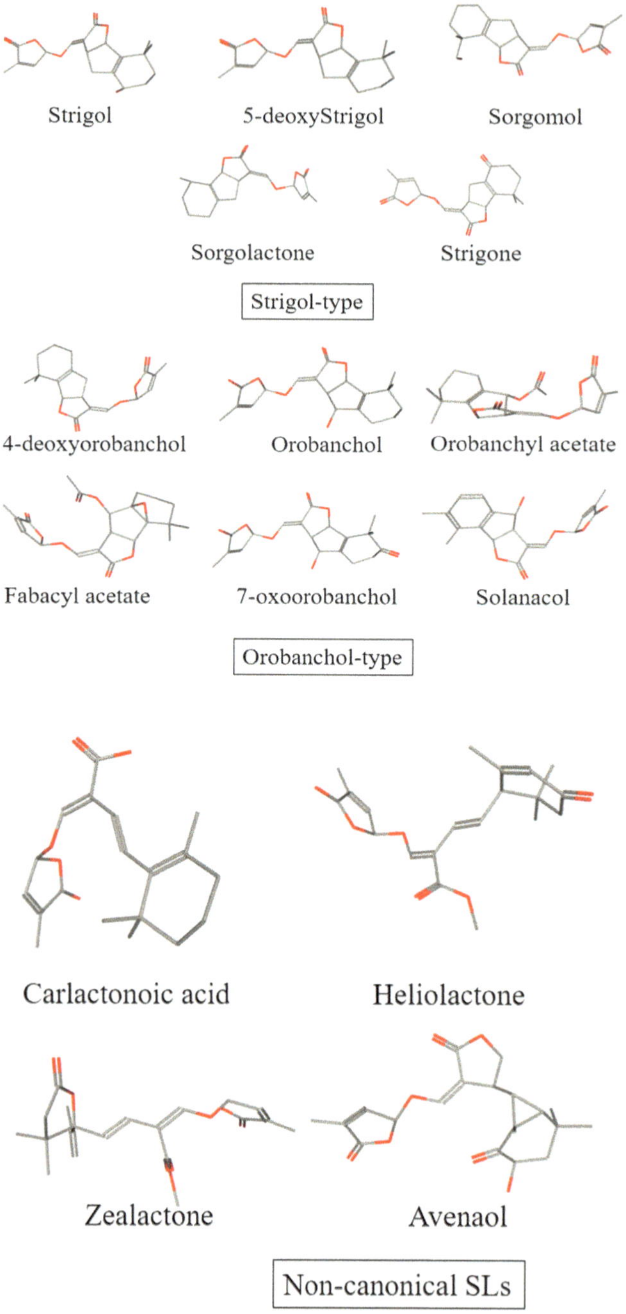

FIGURE 18.2 Shows canonical strigol-type SLs: 5-deoxystrigol (PCID 15102684), Strigol (PCID 5281396), Sorgomol (PCID 24860211), Sorgolactone (PCID 5281395) and Strigone (PCID 23256101) and Orobanchol-type SLs, namely Orobanchol (PCID 10665247), 4-deoxyorobanchol (PCID 102129999), Orobanchyl acetate (PCID 54754035), Fabacyl acetate (PCID 42605181), 7-oxoorobanchol (PCID 71537873) and Solanacol (PCID 51041887) and non-canonical forms of SLs, which include Carlactonoic acid (PCID 90658468), Heliolactone (PCID 101882533), Zealactone (PCID 122389028) and Avenaol (PCID 102236949).

Along with monocots, dicots like white lupin and Chinese milk vetch (*Astragalus sinicus* L.) also produce SLs like sorgomol. The SLs 7-oxoorobanchol, 7-oxoorobanchyl acetate and 7-hydroxyorobanchyl acetate were identified from the root exudates of flax (*Linum usitatissimum* L.) and were detected in root exudates from various plant species, including cucumber (*Cucumis sativus* L.). Members of the Solanaceae family, like tobacco (*Nicotiana tabacum* L.) and tomato (*Solanum lycopersicum* L.), produce strigolactones known as solanacol and solanacyl acetate. These are unique SLs containing a benzene ring. Fabacol and fabacyl acetate are found in root exudates from Fabaceae members like peas (*Pisum sativum* L.), which are epoxides. Medicaol (an isomer of didehydro-orobanchol) contains a seven-membered A ring and is identified from root exudates of barrel medic (*Medicago truncatula* Gaertn.). Recently, root exudates from red clover, tomato and tobacco contain didehydro-orobanchol isomers, which are different from medicaol. Putative desmethyl-orobanchyl acetate isomers, desmethyl-7-hydroxyorobanchyl acetate isomers, dihydro-orobanchol isomers and their derivatives have been detected from various plant species (Xie, 2016).

18.3.2 Synthetic Strigolactones

GR24 is a synthetic strigolactone. When phosphate is available, the application of GR24 to the roots reduces the potential for the formation of lateral roots by suppressing the lateral root primordia. When auxin levels were increased by the exogenous application of synthetic homolog, GR24 showed a stimulatory effect on the lateral roots and promoted the development of the lateral roots (Ruyter-Spira et al., 2011).

18.4 ROLES OF STRIGOLACTONES DURING ABIOTIC STRESSES

There are many kinds of abiotic stresses, which include drought, salinity, extremes of temperatures, unavailability of nutrients and many more. The stresses arise as a result of anthropogenic activities. These stresses orchestrating together pose a serious threat to many aspects, including food security, decreasing economy by reducing the crop yield and decreasing plant growth (production). These conditions are directly responsible for crop losses and critically affect agricultural practices and production. Therefore, to cater to the needs of an exploding population, it is crucial to formulate new strategies to improve crop production. It has been observed that under many conditions, strigolactones can confer abiotic stress tolerance across plant species (Banerjee and Roychoudhary 2018).

18.4.1 Drought

Several pathways were found to be activated during biotic and abiotic stresses, like abscisic acid (ABA), salicylic acid (SA), jasmonic acid (JA) and others. When the wildtype (WT) plants were treated with the 9-cis-epoxycarotenoid dioxygenase (NCED) inhibitor abamineSG and the untreated deficient mutants of ABA, they exhibited reduced ABA and SL levels. The expression analysis of ABA-deficient mutants revealed the down-regulation of *Lycopersicum esculentum* carotenoid cleavage dioxygenase (*LeCCD7* and *LeCCD8*) transcripts. The regulatory role of SLs was revealed using *Arabidopsis* under the drought stress using the loss-of-function approach, which was later confirmed by the rescue of the germination phenotype by SL treatment in biosynthesis mutants (*max3* and *max4*), although it was not observed in the SL response mutant (*max2*). Furthermore, SLs regulate drought stress response partially through ABA signalling, indicated by the lower sensitivity of all the max mutants to ABA, as compared to WT during germination under drought. Also, it was observed that the increased rate of transpiration and closure of ABA-dependent stomata and density of stomata changed in *Arabidopsis* due to ABA-SL signalling (Pandey et al., 2016). In a similar experiment, the role of SL was determined in *Arabidopsis* in which all max mutant and wild-type (WT) plants displayed similar plant size and growth rate without the application of drought stress. WT and mutants were subjected to abiotic stress. SL-signalling mutants, *max3* and *max4* mutants along with *max2* were given drought stress to 3-week-old plants and it was found that a low number of mutants survived compared to the normal types (WT) (Ha et al., 2014). The microarray analysis carried out using WT and mutant (*max2*) reveals the involvement of SL in providing tolerance against the abiotic stresses through the abiotic stress-responsive genes and hormones (abscisic acid and cytokinin). The SL response mutant (*max2*) showed down-regulation of cytokinin catabolism genes, namely *CKX1*, *CKX2*, *CKX3* and *CKX5;* ABA import genes *ABCG22* and *ABCG40;* osmotic genes *CIPK1;* positive regulators of ABA and abiotic responsive genes like *AtNAC2*. These mutants have a large aperture of stomata and a thin cuticle (Bu et al., 2014; Pandey et al., 2016). Inducible ABA marker genes, namely *RD29B* (*RESPONSIVE TODEHYDRATION29B*), *COR47* (*COLD- REGULATEDPROT-EIN47*) and *KIN1* (*COLDINDUCIBLE*) and *Responsive to ABA*, *RD29A* (*RESPONSIVE TO DEHYDRATION29A*), along with genes involved in biosynthesis, catabolism, transport and signalling of ABA like *NCED3*, *ABCG22*, *ABA insensitive1*, *Cytochrome P450*, *707A3* and *Hypersensitive to ABA1,* their expression was found to decrease (Pandey et al., 2016). The role of SL in drought was also demonstrated using the water deficit hypersensitive responses of the *Arabidopsis* SL-biosynthetic *max1* mutants, and the SL-depleted *L. japonicus LjCCD7*-silenced and tomato *SlCCD7*-silenced transgenic plants, which confirmed that SLs are involved in drought stress and also proved that *max3* and *max4* mutants (SL-biosynthetic) and *max2* mutants (SL-response) were susceptible to drought, as compared to the WTs (Liu et al., 2015; Visentin et al., 2016; Zhang et al., 2018; Mostofa et al., 2018). The MAX2, an SL receptor, F-box component of this plays an important role in drought responses by providing resistance to the plant while there is a decrease in the levels of exogenous ABA by the guard cells (Bu et al., 2014; Ha et al., 2014; Liu et al., 2015). SLs are also involved in developmental and functional modification of stomata under drought and salinity stresses (Waadt et al., 2022).

18.4.2 Salinity

Plants show salt-sensitive phenotypes at the germination and vegetative stages when exposed to salt stress. Ha and co-workers (2014) studied the tolerance for salinity using 3-week-old plants with wild-types (WT) and max mutants (*max3*, *max4* and *max2*), which were grown in 200 mM NaCl without water for 6 days and followed by water for 4 days; thereby, the effect of salt stress on plants was then analyzed. A significantly higher number of SL-deficient and SL-signalling max mutant plants died compared to the WT plants in response to salinity stress. To confirm the results, they were germinated on a medium (GM) containing 100 mM NaCl, which also resulted in reduced germination for the max mutants than the wild-types. Thus, the max mutants are much sensitive to salt, which is evident by the reduced growth in both SL-deficient and SL-signalling mutants (Ha et al., 2014). *max2* mutants showed a time- and concentration-dependent decrease in germinative greening under salinity stress, as compared to the WTs (Mostofa et al., 2018). In the seedlings of *Sesbania cannabina*, SL was shown to be interacting with hydrogen peroxide in association with arbuscular mycorrhizal fungi (AMF) and thereby involved in increasing the salt stress (Kong et al., 2017; Mostofa et al., 2018). GR24 acts as a positive regulator in salinity and drought stresses; its exogenous application can enhance resistance in such extreme conditions (Ha et al., 2014; Kapulnik & Koltai, 2014). In *Brassica rapa*, the salinity stress during its germination limits growth and development. The morphological, physiological and biochemical responses to the application of GR24 in rapeseed supports a strong root–shoot interaction. As it was shown that when GR24 is applied to the roots, they are transported through the shoots, and they play an important role in stress (Koltai 2011; Khosla & Nelson 2016; Ma et al., 2017).

18.4.3 Osmotic Stress

It was observed in *Lotus* that the osmotic stress decreases the levels of SL in roots and tissues by suppressing the transcription factors that are involved in the SL biosynthesis. While the plants were pre-treated with SL, osmotic stress was inhibited with the increase in the levels of ABA in roots by down-regulating the *LjNCED2* gene of ABA biosynthesis (Pandey et al., 2016). On the other hand, under the osmotic stress in *Lotus japonicus*, ABA was significantly reduced in both roots and shoots (Liu et al., 2015). The production of SL in roots is negatively regulated in plants like tomato, lotus and lettuce under salinity and drought (osmotic-related stress) (Aroca et al., 2013; Liu et al., 2015; Ruiz-Lozano et al., 2016; Mostofa et al., 2018). It has been reported that phosphorus starvation leads to increases in the effects of osmotic stress (and vice versa). This increase in P starvation enhances the production of SL (Liu et al., 2015).

18.4.4 Heat Stress

Temperature is an important primary factor affecting plant growth and development. Extreme fluctuations in temperature may lead to an impact in productivity and growth of a plant. Pollination is mainly affected by changes in temperature (Hatfield & Prueger, 2015). Germination of seeds requires optimum temperatures for proper growth. For instance, high temperatures inhibit seed germination in *Arabidopsis*. SLs induce seed germination in both root parasitic weeds and other plants. Cytokinin and GA have a positive effect on seed germination, while ABA shows a negative effect. It has been observed that SL decreases the GA and ABA levels, thereby increasing cytokinin levels and thus promotes seed germination. With the application of GR24, germination in SL-defective *Arabidopsis* mutants under high-temperature conditions was stimulated. The seed dormancy of broomrape is overcome by the release of SLs, thereby reducing the levels of ABA during warm conditions (Tsuchiya et al., 2010; Lechat et al., 2015; Pandey et al., 2016; Mishra et al., 2017).

18.4.5 Nutrient Deficiency Stress

During the conditions of nutrient stress (deficiency), plants produce more SL to suppress branching and promote symbiosis (Gomez-Roldan et al., 2008; Umehara et al., 2008; Mishra et al., 2017). SLs play a role in nitrogen (N) and phosphorous (P) deficiencies. It has been observed that the synthesis of SLs increased in N and P deprivation conditions (Marzec et al., 2013). Through the modifications in their root system, they employ symbiotic bacteria and AM fungi. The association of AM fungi ensures the supply of water and minerals like P and N through the extension of their hyphae. Low phosphate and high strigolactones with reduced bud outgrowth were reported in *Arabidopsis* and rice (Umehara et al., 2008; Kohlen et al., 2011).

18.5 APPLICATIONS OF STRIGOLACTONES

The scope for strigolactones and their analogs and mimics have immense potential applications in the field of agriculture. The area of current interest and research is controlling parasitic weeds using SLs. One of the approaches is the suicidal germination approach. A germination stimulant, which is an easily accessible synthetic analog, is applied to the field before the host presence. The weed seeds will germinate, but due to the scarcity of nutrients, they eventually die. After that, the host crop is planted, which is free from side effects caused by weeds (Zwanenburg et al., 2009, 2016). There are many applications of SLs in agriculture as they can be used to improve abiotic stress tolerant varieties (drought, salinity, nutrient deficiency, temperature and osmotic), soil remediation (controlling of weeds and heavy metal sequestration), disease resistance, recruit beneficial microbial communities that can be helpful in symbiosis (therefore, can be used against the pathogens or provide defence against biotic agents), nutrient uptake and soil stability and promote plant architectural modifications, including boosting yield, uniform ripening, elongated shelf life and efficient rooting (Aliche et al., 2020). SLs act as natural stimulants. They are potential molecules for the investigation in basic science about plant-rhizosphere interactions and in applied science for their use in modern agriculture in the fields of crop productivity, resilience challenges and promoting more sustainable agronomical practices (Mostofa et al., 2018; De Mesmaeker et al., 2019; Bouwmeester et al., 2019; Aliche et al., 2020; Borghi et al., 2021).

18.6 CONCLUSION AND FUTURE PROSPECT

The environment has various factors, like drought and high salinity, that may have a negative effect on the growth and development of plants, thereby affecting their normal functioning and metabolic processes. These factors could be either living (biotic) or non-living (abiotic) in nature. Plant stress responses are found to be regulated by various phytohormones, for instance, strigolactones. The roles of strigolactones (SLs) remain elusive. With the help of different molecular and physiological methods carried out in *Arabidopsis*, it was reported and showed that SL acts as a positive regulator of plant responses to drought and salt stress, which was associated with the shoot- rather than root-related traits. Comparative transcriptome analysis suggests that plants integrate multiple hormone-response pathways of SL, abscisic acid and cytokinin pathways for adaptation to environmental stresses. Therefore, it is important to know and elucidate the roles and functions of SL, which could provide a new approach for the development of crops with improved stress tolerance.

REFERENCES

Akiyama, K., Matsuzaki, K. I., & Hayashi, H. (2005). Plant sesquiterpenes induce hyphal branching in arbuscular mycorrhizal fungi. Nature, 435(7043), 824–827.

Alder, A., Jamil, M., Marzorati, M., Bruno, M., Vermathen, M., Bigler, P., et al. (2012). The path from β-carotene to carlactone, a strigolactone-like plant hormone. Science, 335(6074), 1348–1351.

Aliche, E. B., Screpanti, C., De Mesmaeker, A., Munnik, T., & Bouwmeester, H. J. (2020). Science and application of strigolactones. New Phytologist, 227(4), 1001–1011.

Arite, T., Umehara, M., Ishikawa, S., Hanada, A., Maekawa, M., Yamaguchi, S., & Kyozuka, J. (2009). d14, a strigolactone-insensitive mutant of rice, shows an accelerated outgrowth of tillers. Plant and Cell Physiology, 50(8), 1416–1424.

Aroca, R., Ruiz-Lozano, J. M., Zamarreño, Á. M., Paz, J. A., García-Mina, J. M., Pozo, M. J., & López-Ráez, J. A. (2013). Arbuscuflar mycorrhizal symbiosis influences strigolactone production under salinity and alleviates salt stress in lettuce plants. Journal of Plant Physiology, 170, 47–55.

Banerjee, A., & Roychoudhury, A. (2018). Strigolactones: Multi-level regulation of biosynthesis and diverse responses in plant abiotic stresses. Acta Physiologiae Plantarum, 40. 10.1007/s11738-018-2660-5.

Bewley, J. D. 1979. Physiological aspects of desiccation tolerance. Annal Review of Plant Physiology, 30, 195–238.

Bhatla, S. C. & Lal, M. A. (2018). Plant Physiology, Development and Metabolism. Springer, Singapore.

Borghi, L., Screpanti, C., Lumbroso, A., Lachia, M., Gübeli, C., & De Mesmaeker, A. (2021). Efficiency and bioavailability of new synthetic strigolactone mimics with potential for sustainable agronomical applications. Plant and Soil, 1–15.

Bouwmeester, H. J., Fonne-Pfister, R., Screpanti, C., & De Mesmaeker, A. (2019). Strigolactones: plant hormones with promising features. Angewandte Chemie International Edition, 58(37), 12778–12786.

Brewer, P. B., Dun, E. A., Ferguson, B. J., Rameau, C., & Beveridge, C. A. (2009). Strigolactone acts downstream of auxin to regulate bud outgrowth in pea and *Arabidopsis*. Plant Physiology, 150(1), 482–493.

Brewer, P. B., Koltai, H., & Beveridge, C. A. (2013). Diverse roles of strigolactones in plant development. Molecular Plant, 6, 18–28.

Brooks, D. W., Bevinakatti, H. S., & Powell, D. R. (1985). The absolute structure of (+)-strigol. The Journal of Organic Chemistry, 50(20), 3779–3781.

Bu, Q., Lv, T., Shen, H., Luong, P., Wang, J., Wang, Z., et al. (2014). Regulation of drought tolerance by the F-box protein MAX2 in *Arabidopsis*. Plant Physiology, 164(1), 424–439.

Bürger, M., Mashiguchi, K., Lee, H. J., Nakano, M., Takemoto, K., Seto, Y., et al. (2019). Structural basis of karrikin and non-natural strigolactone perception in *Physcomitrella patens*. Cell Reports, 26(4), 855–865.

Charnikhova, T. V., Gaus, K., Lumbroso, A., Sanders, M., Vincken, J. P., De Mesmaeker, A., et al. (2017). Zealactones. Novel natural strigolactones from maize. Phytochemistry, 137, 123–131.
Charnikhova, T. V., Gaus, K., Lumbroso, A., Sanders, M., Vincken, J. P., De Mesmaeker, A., et al. (2018). Zeapyranolactone– A novel strigolactone from maize. Phytochemistry Letters, 24, 172–178.
Cook, C. E., Whichard, L. P., Turner, B., Wall, M. E., & Egley, G. H. (1966). Germination of witchweed (*Striga lutea* Lour.): isolation and properties of a potent stimulant. Science, 154(3753), 1189–1190.
Cook, C. E., Whichard, L. P., Wall, M., Egley, G. H., Coggon, P., Luhan, P. A., & McPhail, A. T. (1972). Germination stimulants. II. Structure of strigol, a potent seed germination stimulant for witchweed (*Striga lutea*). Journal of the American Chemical Society, 94(17), 6198–6199.
De Mesmaeker, A., Screpanti, C., Fonné-Pfister, R., Lachia, M., Lumbroso, A., & Bouwmeester, H. (2019). Design, synthesis and biological evaluation of strigolactone and strigolactam derivatives for potential crop enhancement applications in modern agriculture. CHIMIA International Journal for Chemistry, 73(7-8), 549–560.
de Saint Germain, A., Clavé, G., Badet-Denisot, M. A., Pillot, J. P., Cornu, D., Le Caer, J. P., et al. (2016). An histidine covalent receptor and butenolide complex mediates strigolactone perception. Nature Chemical Biology, 12(10), 787–794.
Du, H., Huang, F., Wu, N., Li, X., Hu, H., & Xiong, L. (2018). Integrative regulation of drought escape through ABA dependent and independent pathways in rice. Molecular Plant, 11, 584–597.
Ferguson, B. J., & Beveridge, C. A. (2009). Roles for auxin, cytokinin, and strigolactone in regulating shoot branching. Plant Physiology, 149(4), 1929–1944.
Gomez-Roldan, V., Fermas, S., Brewer, P. B., Puech-Pagès, V., Dun, E. A., Pillot, J. P., et al. (2008). Strigolactone inhibition of shoot branching. Nature, 455(7210), 189–194.
Ha, C. V., Leyva-Gonzalez, M. A., Osakabe, Y., Tran, U. T., Nishiyama, R., Watanabe, Y., et al. (2014). Positive regulatory role of strigolactone in plant responses to drought and salt stress. Proceedings of the National Academy of Sciences of the United States of America, 111, 851–856.
Hamiaux, C., Drummond, R. S., Janssen, B. J., Ledger, S. E., Cooney, J. M., Newcomb, R. D., & Snowden, K. C. (2012). DAD2 is an α/β hydrolase likely to be involved in the perception of the plant branching hormone, strigolactone. Current Biology, 22(21), 2032–2036.
Hatfield, J. L., & Prueger, J. H. (2015). Temperature extremes: Effect on plant growth and development. Weather and Climate Extremes, 10, 4–10.
Hauck, C., Müller, S., & Schildknecht, H. (1992). A germination stimulant for parasitic flowering plants from *Sorghum bicolor*, a genuine host plant. Journal of Plant Physiology, 139(4), 474–478.
Hayward, A., Stirnberg, P., Beveridge, C., & Leyser, O. (2009). Interactions between auxin and strigolactone in shoot branching control. Plant Physiology, 151(1), 400–412.
Kapulnik, Y., and Koltai, H. (2014). Strigolactone involvement in root development, response to abiotic stress, and interactions with the biotic soil environment. Plant Physiology, 166, 560–569. doi: 10.1104/pp.114.244939
Kapulnik, Y., Delaux, P.-M., Resnick, N., Mayzlish-Gati, E., Wininger, S., Bhattacharya, C., et al. (2011). Strigolactones affect lateral root formation and root-hair elongation in Arabidopsis. Planta, 233, 209–216.
Khetkam, P., Xie, X., Kisugi, T., Kim, H. I., Yoneyama, K., Uchida, K., et al. (2014). 7α-and 7β-Hydroxyorobanchyl acetate as germination stimulants for root parasitic weeds produced by cucumber. Journal of Pesticide Science, D14-038.
Khosla, A., and Nelson, D. C. (2016). Strigolactones, super hormones in the fight against *Striga*. Current Opinon in Plant Biology, 33, 57–63. doi: 10.1016/j.pbi.2016.06.001
Kim, H. I., Kisugi, T., Khetkam, P., Xie, X., Yoneyama, K., Uchida, K., et al. (2014). Avenaol, a germination stimulant for root parasitic plants from *Avena strigosa*. Phytochemistry, 103, 85–88.
Kohlen, W., Charnikhova, T., Liu, Q., Bours, R., Domagalska, M. A., Beguerie, S., et al. (2011). Strigolactones are transported through the xylem and play a key role in shoot architectural response to phosphate deficiency in nonarbuscular mycorrhizal host *Arabidopsis*. Plant Physiology, 155(2), 974–987.
Koltai, H. (2011). Strigolactones are regulators of root development. New Phytologist, 190, 545–549. doi: 10.1111/j.1469-8137.2011.03678.x
Kong, C. C., Ren, C. G., Li, R. Z., Xie, Z. H., & Wang, J. P. (2017). Hydrogen peroxide and strigolactones signaling are involved in alleviation of salt stress induced by arbuscular mycorrhizal fungus in *Sesbania cannabina* seedlings. Journal of Plant Growth Regulation, 36(3), 734–742.
Lechat, M. M., Brun, G., Montiel, G., Véronési, C., Simier, P., Thoiron, S., et al. (2015). Seed response to strigolactone is controlled by abscisic acid-independent DNA methylation in obligate root parasitic plant, *Phelipanche ramose* L. Pomel. Journal of Experimental Botany 66, 3129–3140. doi:10.1093/jxb/erv119

Liu, J., He, H., Vitali, M., Visentin, I., Charnikhova, T., Haider, I., et al. (2015). Osmotic stress represses strigolactone biosynthesis in *Lotus japonicus* roots: exploring the interaction between strigolactones and ABA under abiotic stress. Planta, 241(6), 1435–1451.

Lv, S., Zhang, Y., Li, C., Liu, Z., Yang, N., Pan, L., et al. (2018). Strigolactone-triggered stomatal closure requires hydrogen peroxide synthesis and nitric oxide production in an abscisic acid-independent manner. New Phytologist, 217, 290–304.

Ma, N., Hu, C., Wan, L., Hu, Q., Xiong, J., & Zhang, C. (2017). Strigolactones improve plant growth, photosynthesis, and alleviate oxidative stress under salinity in rapeseed (*Brassica napus* L.) by regulating gene expression. Frontiers in Plant Science, 8, 1671.

Marzec, M., Muszynska, A., & Gruszka, D. (2013). The role of strigolactones in nutrient-stress responses in plants. International Journal of Molecular Sciences, 14(5), 9286–9304.

Mishra, S., Upadhyay, S., & Shukla, R. K. (2017). The role of strigolactones and their potential cross-talk under hostile ecological conditions in plants. Frontiers in Physiology, 7, 691.

Mostofa, M. G., Li, W., Nguyen, K. H., Fujita, M., & Tran, L. S. P. (2018). Strigolactones in plant adaptation to abiotic stresses: An emerging avenue of plant research. Plant, Cell & Environment, 41(10), 2227–2243.

Müller, S., Hauck, C., & Schildknecht, H. (1992). Germination stimulants produced by *Vigna unguiculata* Walp *cv* Saunders Upright. Journal of Plant Growth Regulation, 11(2), 77–84.

Nakamura, H., Xue, Y. L., Miyakawa, T., Hou, F., Qin, H. M., Fukui, K., ... & Asami, T. (2013). Molecular mechanism of strigolactone perception by DWARF14. Nature Communications, 4(1), 1–10.

Oliver, M. J., Velten, J., & Mishler, B. D. (2005). Desiccation Tolerance in Bryophytes: A Reflection of the Primitive Strategy for Plant Survival in Dehydrating Habitats? Integrative and Comparative Biology, 45, 788–799.

Pandey, G. K., Pandey, A., Prasad, M., & Böhmer, M. (2016). Abiotic stress signaling in plants: functional genomic intervention. Frontiers in Plant Science, 7, 681.

Ruiz-Lozano, J. M., Aroca, R., Zamarreño, Á. M., Molina, S., Andreo-Jiménez, B., Porcel, R., ...López-Ráez, J. A. (2016). Arbuscular mycorrhizal symbiosis induces strigolactone biosynthesis under drought and improves drought tolerance in lettuce and tomato. Plant, Cell & Environment, 39, 441–452.

Ruyter-Spira, C., Kohlen, W., Charnikhova, T., van Zeijl, A., van Bezouwen, L., de Ruijter, N., ... & Bouwmeester, H. (2011). Physiological effects of the synthetic strigolactone analog GR24 on root system architecture in Arabidopsis: another belowground role for strigolactones?. Plant Physiology, 155(2), 721–734.

Saeed, W., Naseem, S., & Ali, Z. (2017). Strigolactones biosynthesis and their role in abiotic stress resilience in plants: A critical review. Frontiers in Plant Science, 8. 10.3389/fpls.2017.01487.

Seto, Y., Sado, A., Asami, K., Hanada, A., Umehara, M., Akiyama, K., & Yamaguchi, S. (2014). Carlactone is an endogenous biosynthetic precursor for strigolactones. Proceedings of the National Academy of Sciences, 111(4), 1640–1645.

Seto, Y., Yasui, R., Kameoka, H., Tamiru, M., Cao, M., Terauchi, R., ... & Yamaguchi, S. (2019). Strigolactone perception and deactivation by a hydrolase receptor DWARF14. Nature Communications, 10(1), 1–10.

Shabek, N., Ticchiarelli, F., Mao, H., Hinds, T. R., Leyser, O., & Zheng, N. (2018). Structural plasticity of D3–D14 ubiquitin ligase in strigolactone signalling. Nature, 563(7733), 652–656.

Siame, B. A., Weerasuriya, Y., Wood, K., Ejeta, G., & Butler, L. G. (1993). Isolation of strigol, a germination stimulant for *Striga asiatica*, from host plants. Journal of Agricultural and Food Chemistry, 41(9), 1486–1491.

Sun, H., Tao, J., Liu, S., Huang, S., Chen, S., Xie, X., ...Xu, G. (2014). Strigolactones are involved in phosphate- and nitrate-deficiency induced root development and auxin transport in rice. Journal of Experimental Botany, 65, 6735–6746.

Toh, S., Holbrook-Smith, D., Stogios, P. J., Onopriyenko, O., Lumba, S., Tsuchiya, Y., ... & McCourt, P. (2015). Structure-function analysis identifies highly sensitive strigolactone receptors in *Striga*. Science, 350(6257), 203–207.

Toh, S., Kamiya, Y., Kawakami, N., Nambara, E., McCourt, P., & Tsuchiya, Y. (2012). Thermoinhibition uncovers a role for strigolactones in *Arabidopsis* seed germination. Plant & Cell Physiology, 53, 107–117.

Tsuchiya, Y., Vidaurre, D., Toh, S., Hanada, A., Nambara, E., Kamiya,Y., et al.. (2010). A small-molecule screen identifies new functions for the plant hormone strigolactone. Nat. Chem. Biol., 6, 741–749. doi:10.1038/nchembio.435

Ueda, H., & Kusaba, M. (2015). Strigolactone regulates leaf senescence in concert with ethylene in Arabidopsis. Plant Physiology, 169, 138–147.

Ueno, K., Furumoto, T., Umeda, S., Mizutani, M., Takikawa, H., Batchvarova, R., & Sugimoto, Y. (2014). Heliolactone, a non-sesquiterpene lactone germination stimulant for root parasitic weeds from sunflower. Phytochemistry, 108, 122–128.

Ueno, K., Nomura, S., Muranaka, S., Mizutani, M., Takikawa, H., & Sugimoto, Y. (2011). Ent-2′-epi-orobanchol and its acetate, as germination stimulants for Striga gesnerioides seeds isolated from cowpea and red clover. Journal of Agricultural and Food Chemistry, 59(19), 10485–10490.

Umehara, M., Hanada, A., Magome, H., Takeda-Kamiya, N., & Yamaguchi, S. (2010). Contribution of strigolactones to the inhibition of tiller bud outgrowth under phosphate deficiency in rice. Plant & Cell Physiology, 51, 1118–1126.

Umehara, M., Hanada, A., Yoshida, S., Akiyama, K., Arite, T., Takeda-Kamiya, N., ... & Yamaguchi, S. (2008). Inhibition of shoot branching by new terpenoid plant hormones. Nature, 455(7210), 195–200.

Urquhart, S., Foo, E., & Reid, J. B. (2015). The role of strigolactones in photomorphogenesis of pea is limited to adventitious rooting. Physiologia Plantarum, 153, 392–402.

Vicre, M., Farrant, J. M., & Driouich A. (2004). Insights into the cellular mechanisms of desiccation tolerance among angiosperm resurrection plant species. Plant, Cell and Environment, 27, 1329–1340.

Visentin, I., Vitali, M., Ferrero, M., Zhang, Y., Ruyter-Spira, C., Novák, O., et al. (2016). Low levels of strigolactones in roots as a component of the systemic signal of drought stress in tomato. New Phytologist, 212(4), 954–963.

Waadt, R., Seller, C. A., Hsu, P. K., Takahashi, Y., Munemasa, S., & Schroeder, J. I. (2022). Plant hormone regulation of abiotic stress responses. Nature Reviews Molecular Cell Biology, 23(10), 680–694.

Wang, X., Xu, C., Cai, X., Wang, Q., & Dai, S. (2017). Heat-responsive photosynthetic and signaling pathways in plants: insight from proteomics. International Journal of Molecular Sciences, 18(10), 2191.

Woo, S., & McErlean, C. S. (2019). Total synthesis and stereochemical confirmation of heliolactone. Organic Letters, 21(11), 4215–4218.

Xie, X. (2016). Structural diversity of strigolactones and their distribution in the plant kingdom. Journal of Pesticide Science, J16-02.

Xie, X., Kisugi, T., Yoneyama, K., Nomura, T., Akiyama, K., Uchida, K., ... & Yoneyama, K. (2017). Methyl zealactonoate, a novel germination stimulant for root parasitic weeds produced by maize. Journal of Pesticide Science, 42(2), 58–61.

Xie, X., Mori, N., Yoneyama, K., Nomura, T., Uchida, K., Yoneyama, K., & Akiyama, K. (2019). Lotuslactone, a non-canonical strigolactone from *Lotus japonicus*. Phytochemistry, 157, 200–205.

Xu, Y., Miyakawa, T., Nosaki, S., Nakamura, A., Lyu, Y., Nakamura, H., et al. (2018). Structural analysis of HTL and D14 proteins reveals the basis for ligand selectivity in Striga. Nature Communications, 9(1), 1–11.

Yamada, Y., Furusawa, S., Nagasaka, S., Shimomura, K., Yamaguchi, S., & Umehara, M. (2014). Strigolactone signaling regulates rice leaf senescence in response to a phosphate deficiency. Planta, 240, 399–408.

Yao, R., Ming, Z., Yan, L., Li, S., Wang, F., Ma, S., et al. (2016). DWARF14 is a non-canonical hormone receptor for strigolactone. Nature, 536(7617), 469–473.

Yasui, M., Ota, R., Tsukano, C., & Takemoto, Y. (2017). Total synthesis of avenaol. Nature Communications, 8(1), 1–9.

Yokota, T., Sakai, H., Okuno, K., Yoneyama, K., & Takeuchi, Y. (1998). Alectrol and orobanchol, germination stimulants for Orobanche minor, from its host red clover. Phytochemistry, 49(7), 1967–1973.

Yoshimura, M., Fonné-Pfister, R., Screpanti, C., Hermann, K., Rendine, S., Dieckmann, M., ... & De Mesmaeker, A. (2019). Total synthesis and biological evaluation of heliolactone. Helvetica Chimica Acta, 102(11), e1900211.

Zhang, Y., Lv, S., & Wang, G. (2018). Strigolactones are common regulators in induction of stomatal closure in planta. Plant Signaling & Behavior, 13(3), e1444322.

Zhao, L. H., Zhou, X. E., Wu, Z. S., Yi, W., Xu, Y., Li, S., et al. (2013). Crystal structures of two phytohormone signal-transducing α/β hydrolases: karrikin-signaling KAI2 and strigolactone-signaling DWARF14. Cell research, 23(3), 436–439.

Zhao, L. H., Zhou, X. E., Yi, W., Wu, Z., Liu, Y., Kang, Y., et al. (2015). Destabilization of strigolactone receptor DWARF14 by binding of ligand and E3-ligase signaling effector DWARF3. Cell Research, 25(11), 1219–1236.

Zhuang, L., Wang, J., & Huang, B. (2017). Drought inhibition of tillering in *Festuca arundinacea* associated with axillary bud development and strigolactone signaling. Environmental and Experimental Botany, 142, 15–23.

Zwanenburg, B., Mwakaboko, A. S., Reizelman, A., Anilkumar, G., & Sethumadhavan, D. (2009). Structure and function of natural and synthetic signalling molecules in parasitic weed germination. Pest Management Science: Formerly Pesticide Science, 65(5), 478–491.

Zwanenburg, B., Pospíšil, T., & Zeljković, S. Ć. (2016). Strigolactones: new plant hormones in action. Planta, 243(6), 1311–1326.

Unit III

Molecular Interactions and Future Prospects

19 Signaling Transduction, Molecular Interactions and Crosstalk of Hormones during Abiotic Stress

Aravindan Shanmugam, Raman Pushpa, Chandrasekaran Deepika, Nallamuthu Ramya Selvi, Kamaraj Keerthana, Sakthivel Viswabharathy, and Ramalingam Suresh

19.1 INTRODUCTION

Climate change accounts for around 60% of yield variability. It is thought to be a key factor affecting crop productivity and farmers' income (Aryal et al., 2020). Abiotic stresses, such as drought, heat, salinity, flooding and cold, have profound impacts on plant growth and survival. Due to the intricacy of the tolerance-related features, there are several limitations in the standard breeding methods used to deal with these stresses. The receptors found in the cell membrane are the first to detect the stress signals. A number of stress-responsive genes are then activated as the signal information is transduced downstream (Tuteja and Sopory, 2008). Different chemicals are involved in a signal transduction pathway. Among them, the role of phytohormones in reducing the negative impact of abiotic stress has undergone a radical change as a result of advances in plant molecular biology (Raza et al., 2019). Plant hormones are a group of small, diffusible molecules with distinct structural characteristics that help to enhance various plant growth and developmental processes and have notable effects against various abiotic stresses. Cytokinin, auxin, ethylene, abscisic acid, gibberellin, salicylic acid, jasmonic acid and brassinosteroid are some of the major classical hormones that fall under the category of plant growth regulators; nevertheless, there are still a great deal of undiscovered hormones. For the plants to adapt and tolerate such stresses, this requires complex sensing, signaling and stress mechanisms. It is now clear that physiological processes are controlled in a complicated fashion by the crosstalk of multiple hormones, despite prior studies considerably advancing our knowledge of how hormones affect plant growth and development and stress responses by focusing on a single chemical (Munné-Bosch & Müller, 2013). It would be much easier for scientists to improve food crops that can flourish in decreasing environmental conditions and even improve crop productivity if they understood the combined effect of various signaling pathways of different hormones in plants.

19.2 DROUGHT

Drought is one of the most significant abiotic stimuli that impairs plant function and has a detrimental impact on grain and straw yield (Fahad et al., 2017). Drought stress results from a lack of water reaching the roots and/or increased transpiration rates brought on by high atmospheric temperatures. The osmotic equilibrium is changed, pigment concentration and photosynthetic

DOI: 10.1201/9781003335788-24

activity are decreased and the membrane integrity is damaged (Praba et al., 2009). Plants have extensive stress-responsive mechanisms that help them resist stress created due to water deficiency (Jogawat et al., 2021). To advance transgenic and logical crop breeding strategies to increase the stress tolerance of crops with major economic value, it is crucial to comprehend the molecular underpinnings of the signal transduction system related to drought stress in agricultural plants (Lata et al., 2015). This chapter aims to review several responses of plants to drought stress and also some general drought stress signal transduction traits.

19.2.1 Signal Transduction Pathway

Drought stress is perceived by receptors such as plasma membrane proteins and are bound to extracellular proteins such as ligands or elicitors in the signal transduction pathway. Secondary messengers like ROS (reactive oxygen species) and inositol phosphates are generated as the stress signal gets transmitted that can modify intercellular Ca^{2+} levels. Calcium sensors recognize this modification and initiate a chain of phosphorylation that ultimately targets genes that respond to stress after changing conformation. These genes that respond to stress may function as transcription factors (TFs), which control the expression of downstream stress-regulatory genes or they may directly contribute to cellular defence (Mahajan & Tuteja, 2005). In addition to aiding plants in adapting to and surviving harsh environmental conditions, the by-products of genes that respond to stress are also engaged in the production of plant hormones like ethylene, salicylic acid and ABA. The primary stress-induced gene products can be roughly grouped into two categories: the first category contains proteins that are directly connected to stress tolerance, such as aquaporins and late embryogenesis abundant (LEA) proteins. The second category consists of protein kinases, phosphatases, cis-regulatory elements and transcription factors (TFs), which control signaling within cells and the stress gene expression (Lata et al., 2015).

19.2.2 Expression of ABA Biosynthesis Genes

ABA, an important plant growth hormone is involved in the regulation of various plant physiology and development. (Lata & Prasad, 2011). The function of ABA goes beyond controlling developmental pathways to activating genes related to osmotic adaptation, ion partitioning, root permeability coefficient, control of shoot growth and root growth, limiting transpiration and thereby wilting, and eventually lowering plant water loss (Verslues & Zhu, 2005). ABA is also involved in upregulating various stress-responsive genes by altering gene expression during osmotic imbalances (Ingram & Bartels, 1996). In contrast, by following exogenous ABA treatment, many genes express themselves in anticipation of drought stress, while some other genes do not. This shows that crop plants have both ABA-dependent and ABA-independent signaling cascades as possible signaling pathways (Lata & Prasad, 2011). The mechanism for ABA-dependent signaling activates two distinct regulons: (1) AREB/ABF (ABA-responsive element-binding protein/ABA-binding factor) and (2) MYC/MYB regulons to help plants respond to stress. Cold binding factor/dehydration-responsive element binding (CBF/DREB) regulon and NAC and ZF-HD (zinc-finger homeodomain) regulon are the ABA-independent regulons, respectively (Lata & Prasad, 2011). Although stress-response pathways typically work independently of one another, there can occasionally be some level of interaction between signaling pathways. RD20A, RD29B, RD22 and other genes that are induced by stress are regulated by the endogenous ABA that gets accumulated in response to drought stress (Figure 19.1).

19.2.3 Crosstalk of Phytohormones during Drought

Drought stress regulates numerous genes involved in jasmonic acid (JA) signaling (Huang et al., 2008). CPK6 (Ca^{2+} Dependent Protein Kinase 6), a component of ABA signaling, increases JA

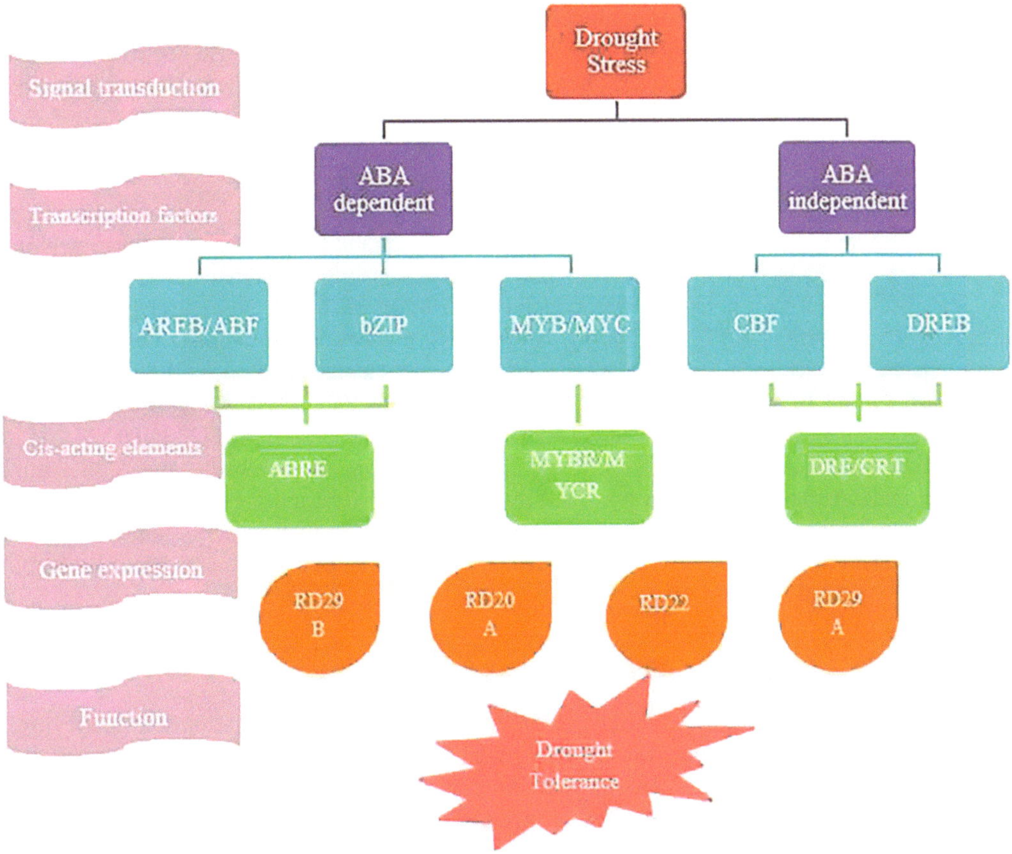

FIGURE 19.1 Signal transduction pathway during drought stress.

signallng in guard cells (Munemasa et al., 2011). At the genomic level, a number of elements of the ABA-salicylic acid (SA) network have been found. In contrast to wild-type plants, mutants exhibit hypersensitivity to ABA and enhanced SA-induced gene expression, as seen by the activation of AHG2, which encrypts a poly (A)-specific ribonuclease (Nishimura et al., 2009). Based on a recent study, during drought stress, crosstalk between ABA and strigolactone (SL) affects the function and development of stomata by connecting several stress signals. The participation of SL-mediated network pathways that control responses due to drought via a variety of hormonal response pathways, including via cytokinins and ABA, has been further identified by microarray (Ha et al., 2014).

19.3 SALINITY

The saline condition in soil is characterized by the prevalence of ions that are soluble, which adversely reduces the growth of the plant and metabolic activities (Pan et al., 2021). Saline soil also enhances the stress condition by decreasing the potential of soil osmosis (Arif et al., 2020), thereby causing disturbances to dividing and expanding stages of cells, which leads to the formation of smaller and thicker leaves, roots and stems (Zorb et al., 2019). Salinity also causes structural disruption to various membrane-bound organelles (El Ghazali, 2020) and affects the physiological process pertaining to developmental stages of the plant. It causes ionic imbalance by increasing the uptake of Na^+ and Cl^- ions and decreasing K^+ and Ca^{2+} accumulation (Joshi et al., 2017) as Ca^{2+} maintains ion homeostasis by regulating the Na^+ and K^+ transporters. In order to overcome salinity, plants regulate multicomponent signal transduction pathways that are further subdivided into a Ca^{2+}-independent mitogen-activated protein kinase (MAPK) pathway and Ca^{2+}-dependent pathway such as salt overly sensitive (SOS), ABA and Ca^{2+}-dependent protein kinase (CDPK) pathway (Hao et al., 2021).

19.3.1 Location and Perception of Saline Stress Response in Plants

The first and foremost organ exposed to salt stress is a root and the root hairs. Primary salt stress sensors are located in roots and the mechanism behind sensing and signal transduction remains unclear, but to a certain extent, it was concluded to be localized in the central portion of root meristem (Nakamura et al., 2021). During saline stress, higher sodium interacts with the negative components of the cell wall, which causes conformational changes and opens the ion channels to increase the influx of calcium ions in the cytosol (Wang et al., 2022). In addition, ROS also induces Ca^{2+} inflow to the cytoplasm by activating REDUCED HYPEROSMOLALITY-INDUCED [Ca^{2+}] INCREASE1 (OSCA1) (Zhai et al., 2020). The MAPK cascades pathway and SOS pathway interacts and crosstalk exists between these two pathways (Rodriguez et al., 2010). Generally, adaptation to salinity is regulated by signal perception, which further activates multiple signaling pathways, enabling salt stress resistance in plants. Potential salt stress sensors and their localization have been compiled in Table 19.1.

19.3.2 Salinity-Responsive Signal Transduction Pathways

When the plants are exposed to saline stress, an irregular Na^+ concentration elevates cytosolic calcium, which further activates downstream signal transduction pathways either through a SOS signal transduction pathway or through a calcineurin pathway (CBL-CIPK signalling complex) (Seifikalhor et al., 2019). Signal perception in roots and shoots is done by SOS3 (a Ca^{2+}-binding protein) and SCaBP8/CBL10, which further activates a kinase SOS2 (or PKSs/CIPKs). The SOS3-SOS2 complex further phosphorylates SOS1/NHX 7, which is localized in a plasma membrane and functions in transport and sodium ion exclusion (Nongpiur et al., 2020). The complex further activates NHX1, which helps in sodium compartmentalization in the vacuole. ABA, which is a

TABLE 19.1
Putative sensors for saline response and their location in plants

Location	Salt Sensors	References
Cell wall	Glycine rich proteins (GRP)	Tenhaken, 2015
Endoplasmic reticulum (ER)	ER chaperone binding immunoglobulin protein (BIP) and the INOSITOL-REQUIRING PROTEIN 1 (IRE1) protein kinase	Zhu, 2016
Cell wall and plasma membrane	*Pisum sativum* lectin receptor-like kinase (PsLecRLKs)	Ye et al., 2017
Plasma membrane	FERONIA (FER)	Feng et al., 2018
Plasma membrane lipid	Glycosyl inositol phosphorylceramide (GIPC)	Jiang et al., 2019
Cell wall	*Solanum lycopersicum* wall-associated receptor-like kinases (SlWAK1)	Meco et al., 2020
Cell wall and plasma membrane	*Oryza sativa* lectin receptor-like kinase (OsLecRLKs)	Passricha et al., 2020
Root cell wall	Leucine-rich repeat extensins (LRXs) such as LRX3, LRX4 and LRX5	Zhao et al., 2020

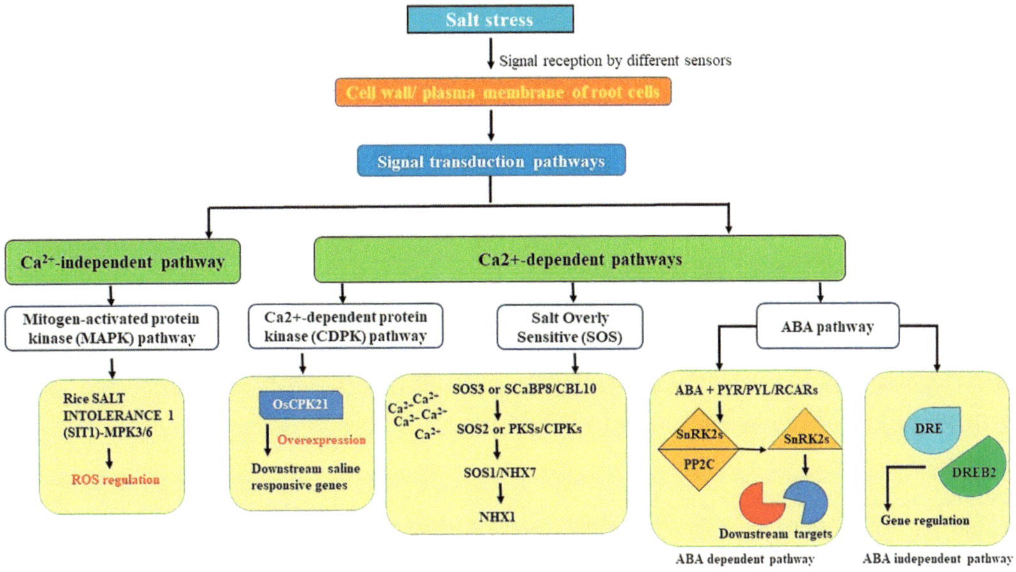

FIGURE 19.2 Different signal transduction pathways for salinity stress resistance in plants.

major phytohormone, regulates the stress signaling pathway by binding to its receptors, such as PYR/PYL/RCARs, which causes the inactivation of PP2C. The SnRK2s-PP2C complex dissociates and the free SnRK2s phosphorylates the downstream targets, which further activates the stress-induced responses at the physiological and molecular level (Dong et al., 2015). Considering the ABA independent pathway, the dehydration responsive element (DRE) and its binding protein 2 (DREB2) are the major factors needed for the regulation of the stress gene (Yoshida et al., 2014).

In addition, calcium-dependent protein kinases (CDPKs) also regulate the salinity response by overexpressing OsCPK21, which improves salt tolerance by activating downstream saline responsive genes (Chen et al., 2021). In the absence of calcium signaling during high saline conditions, the MAPK signaling cascade accumulates phospholipase D in *Asterochloris erici*. In rice, a signaling cascade called the RICE SALT INTOLERANCE 1 (SIT1)-MPK3/6 regulates the ROS response, the hormone ethylene production and signaling (Zhang et al., 2022). Besides ABA, the key hormone of a saline stress response is ethylene. An ethylene-based signal transduction model was developed in *Arabidopsis,* which includes five major receptors *viz.,* CTR1 (a negative regulator), EIN2 (a positive regulator), EIN3/EILs (primary transcription factors) and many other downstream factors. The *Arabidopsis* model was different from rice, where different ethylene response mutants (*maohuzi*) had been isolated and studied (Ma et al., 2010) (Figure 19.2).

19.4 TEMPERATURE STRESS

Global agriculture is greatly relies upon climatic conditions and is highly affected by temperature fluctuations. The average temperature of the world is constantly increasing due to undesired human activities and pollution. Earth's temperature has risen by 0.32 °F per decade since 1981 is more than twice that of previous decades (NOAA, 2022), as shown in Figure 19.3. The average temperature is expected to rise by 2.6–4.8°C between 2081–2100, according to the Intergovernmental Panel on Climate Change (IPCC, 2014). Therefore, heavy yield reduction in food crops are seen. To cope up with rising temperature, heat- or cold-resilient crops need be developed by understanding molecular mechanisms and plant responses under heat stress. Plants can receive and emit signals according to

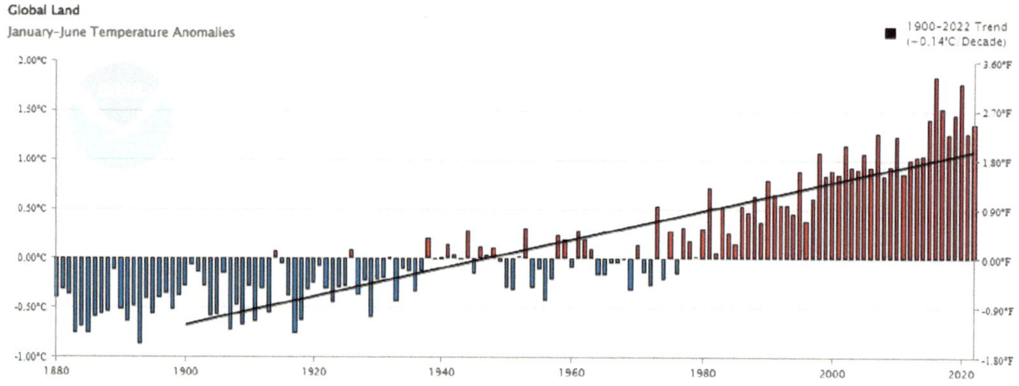

FIGURE 19.3 Global temperature from 1900–2022 (NOAA, 2022).

internal and external environments. Exposing plants to extended periods of temperature above or below the threshold level could change the growth pattern and their life cycle.

The plant hormones play a role in defense-related pathways, growth metabolisms and stress responses. Generally, 40 plant growth regulators (present as endogenous or exogenous) with active ingredients have been reported so far. All are balancing the intrinsic hormone levels within plants by altering the hormone-induced signal transduction pathways. Auxin (AUX), cytokinins (CKs), ethylene (ET), abscisic acid (ABA), salicylic acid (SA), jasmonic acid (JA), strigolactones (SLs) and brassinosteroids (BRs) are natural hormones with defensive responses from environmental stimuli for different stresses, including temperature/heat stress. Further information pinpoints the role, crosstalk or interactions of hormones during signal transduction pathways under abnormal temperatures based on studies undertaken in the model organism *Arabidopsis,* followed by some other crops.

19.4.1 Role of Phytohormones in Signal Transduction Pathways

Stress-mediated interactions of hormones in living systems remain to be learned about under molecular levels *viz.,* transcription factors, mitogen activated protein kinase (MAPKs), signal transduction components, upstream regulators and ROS systems. Crosstalk between different biochemical pathways involves tens and thousands of molecular proteins, including hormones that can help the plants to sustain themselves under stress. The functions and interactions of many hormones will be addressed below and Table 19.2 shows the genes and proteins involved in various pathways.

19.4.2 Auxin

Auxin, a phytohormone, contributes to thermomorphogenesis by way of leaf hyponasty and linear hypocotyl elongation (Kupers et al., 2020). The level of auxin is greatly elevated in growing seedlings under heat stress. ARFs (auxin response factors) play a major contributing transcription factor for activating auxin-responsive genes since ARF-lacking plants showed lower responses to maximum temperatures (Reed et al., 2018). Apart from this, phytochrome interacting factors (PIFs) are the main actors in auxin-mediated thermo-morphogenesis by regulating auxin biosynthesis genes like YUC8. When temperatures are high, especially in epidermal cells, the amount of endogenous PIF4 protein gets increased (Kim et al., 2020). Auxins might also be involved in the opening of stomata by reducing the production of hydrogen peroxide in guard cells (Song et al., 2006), in which it maintains the coolest plant atmospheric conditions by means of transpiration. In addition to this, PIN-LIKES (PILS) proteins are recognized auxin carriers in ER, where they work for the distribution of auxin intercellularly to maintain homeostasis under stress conditions (Sauer & Kleine-Vehn, 2019).

TABLE 19.2
Functions of sensory factors involved in signal transduction pathways under heat stress

S.No.	Hormones/Metabolic Process under Which Genes Are Activated	Genes/Transcription Factors/Proteins	Functions	References
1.	Auxin	bHLH	Basic helix loop helix transcription regulators responsible for leaf hyponasty, regulator for PIFs	Fiorucci et al., 2020
		CRY1	Blue-light receptor cryptochrome 1 acts as a negative regulator by which it inhibits PIF4 activity	Ma et al., 2016
		phyB	Negative regulator for PIF 4 since it degrades PIF 4 by 26 S proteosome pathway	Kim et al., 2020
		SAUR	Small auxin up regulator that negatively regulates auxin production	Zhang et al., 2012
2.	Cytokinin	Ipt (isopentyl transferase) genes -IPT3, IPT4, IPT7	Cytokinin biosynthesis-related enzymes are repressed during heat stress	Dobra et al., 2015
		CKX3, CKX6 - CK oxidase/dehydrogenase3	Cytokinin degradation enzymes are upregulated in the leaves under heat stress	
3.	Ethylene	ETR1, ETR2, ERS1, ERS2 and EIN2, EIN3, EIN4 (Ethylene insensitive)	Ethylene response receptors/sensors receive signals from ethylene during heat stress	Abdelrahman et al., 2017
		NAC families, AP2/ERF, SAGs (senescence associated genes) and WRKY53	Ethylene downstream responsive genes that activate the network of regulatory proteins to control premature leaf senescence	Zhang et al., 2012 Xie et al., 2014
		PA (Phosphatidic Acid)	Regulates ethylene signaling	Jakubowicz et al., 2010
		GCC box	Involved in ethylene-induced gene expression	Zhang et al., 2012
4.	Abscisic acid	NAC like protein OsNAP	Induced by ABA and down regulation of OsNAP induces delayed leaf senescence in rice	Lv et al., 2011
		PLD (Phospholipase D), PA (Phosphatidic Acid)	Stimulate ABA synthesis by activating SAPK and downstream components	Li et al., 2021
		CE3, ABRE	Involved in ABA signaling pathway	Zhang et al., 2012
		COR (cold responsive) genes	Induced by ABA which encodes LEA like stress proteins, responsible for cold acclimation	Xiong et al., 2002
5.	Brassinosteroids	BZR 1	Transcription factor for BR signaling. BZR1 could be activated PIF4 under heat stress	Oh et al., 2012

(Continued)

TABLE 19.2 *(Continued)*
Functions of sensory factors involved in signal transduction pathways under heat stress

S.No.	Hormones/Metabolic Process under Which Genes Are Activated	Genes/Transcription Factors/Proteins	Functions	References
6.	ROS system	Rboh B	Homologue of NADP oxidase on which Rac1 of GTPase binding could result in increase of ROS	Wong et al., 2007
7.	Cytoprotective signaling pathway	bZIP60	Splicing of this mRNA will induce the UPR	Neill et al., 2019
		bZIP28	ER membrane associated TF mobilized from ER to nucleus in response to stress	
8.	Cellular homeostasis	OsNIP2, OsNIP4, OsTIP1	Upregulated in rice upon heat stress	Zhang et al., 2012
		FtsH gene	Protease involved in thermotolerance in rice and *Arabidopsis*	Chen et al., 2006

19.4.3 Cytokinin

CKs are involved from the germination of seeds to senescence, including stress-regulated responses under hot ambient temperatures. They also enhance transpiration through stomatal openings as a protective mechanism for heat stress. Most of the studies reported limited levels of bioactive CKs under temperature stress; for example, Dobra et al. (2015) reported a notable increase of bioactive CK content 45 minutes after imposing heat, followed by gradual reduction of CK content in *Arabidopsis*. The suppression of CK levels under heat stress is found to have the advantage of reduced shoot growth, which enables longer root growth for uptake water from deeper soil zones. In comparison, higher CK in leaves acts as a protective mechanism for emerging flowers under heat stress; increased antioxidant metabolism by the activities of catalase (CAT), ascorbate peroxidase (APX), superoxide dismutase (SOD) and guaiacol peroxidase (GPX) in roots; osmolyte accumulation; and cell membrane integrity, followed by recovery of chloroplast function and photosynthetic ability (Li et al., 2021). The heat stress significantly reduces the rate of cytokinin transpiration and sap flow in xylem. Treatment with CK oxidase inhibitors prevents the reduction of CKs on tolerance during heat stress tolerance in *Arabidopsis* (Prerostova et al., 2020).

19.4.4 Ethylene

Gaseous hormone ethylene (ET) acts as an anti-regulator during heat stress in *Arabidopsis* since it reduces photosynthesis, sucrose content, antioxidants and leaf senescence at the pre-maturation stage (Maduraimuthu & Prasad, 2010). Leaf senescence acquired by heat-induced ethylene production has been studied extensively. Under heat stress, repression of stress-induced ethylene production/signaling would be employed to delay grain filling time (in the case of rice) and slow down leaf senescence. By creating transgenic plants with the appropriate changes and applying exogenous ethylene inhibitors, it might be accomplished.

19.4.5 Abscisic Acid

The stress hormone ABA significantly increased during heat stress, in which a higher accumulation of ABA was seen in heat-sensitive plants than heat-tolerant plants. As with other hormones, it also upregulates the antioxidant enzymes and ROS scavenging enzymes. According to Lv et al. (2011), ABA might be involved in the removal of negative effects by excess proline production during stress conditions.

19.4.6 Salicylic Acid

SA takes part in stress regulations by activating different signaling pathways. In this context, SA controls a heat-induced response by the production of HSPs (Li et al., 2021). The accumulation of ROS scavengers and proline can alleviate the detrimental effects of heat/temperature stress. SA highly regulates the chloroplast HSP21 proteins by which it facilitates recovery of photosynthesis by increasing antioxidant-related enzymes and maintains the relative water content after heat stress.

19.4.7 Jasmonic Acid

Oxylipin-based hormone JAs accumulate as stress hormones during high-temperature stress. By the view of reported studies, exogenous application of JAs before heat stress will reduce the adverse effects (Clarke et al., 2009) and also arrest the stomatal growth, followed by induction of stomatal closure. Studies in *Arabidopsis thaliana* reported the role of JA in cold stress response by controlling the CBF pathway (C-repeat binding factor) (Hu et al., 2017).

19.4.8 Brassinosteroids

Brassinosteroids (BRs) are interplayed in alleviation of adverse effects of maximum and minimum temperature stress by regulating the biochemical and physiological metabolisms by crosstalk with other phytohormones. BRs regulate the mitogen-activated protein kinases (MAPK1 and MAPK3) cascade upon signal transduction; in addition, they control plant growth and developmental stages. Exogenous application of BRs on crops could activate oxidative enzymes, antioxidant enzymes, Calvin cycle–related enzymes and proline, by which they improve the osmoregulation and photosynthesis; trihydroxylated spirotane (Analogue of BR) applied to bananas at 34°C greatly enhances the growth (Gonzalez-Olmedo et al., 2005). As the view of some studies, BRs can also induce the expression of HSPs in plants under stress. Dhaubhadel et al. (1999) found that different kinds of HSPs viz., HSP70, HSP90 and HSP100 and some other low molecular weight HSPs were involved in heat stress regulation in *Brassica napus* and tomato plants. In addition to this, Mt-shsp (mitochondrial small hsp) was also expressed at 38°C when an exogenous application of BRs was activated.

19.4.9 Crosstalk and Molecular Interactions of Phytohormones during Temperature Stress

Plants exposed to stresses are accompanied by various metabolic and osmotic imbalances. One of the most common signals is an increase in the level of ABA, which in turn triggers a variety of hormones, including ET, JA, SA and CKs (Li et al., 2021). They often have either synergistic or antagonistic effects with ABA. Hence, ABA act as a molecular switch for managing the stresses.

19.4.10 Synergistic Effect

The genes participated in sucrose transport *viz*., sucrose synthase, sucrose transporters and invertase and are promoted by ABA upon heat stress. Among them, sucrose alone is majorly involved in the thermal regulation of plants. Hence, sucrose and ABA show synergistic effects on sustaining plants under stress (Wind et al., 2010). According to O'Brien & Benkova, (2013), reduced CK levels promoted apical dominance with ABA-regulated stomatal aperture changes that helped to withstand the plants under heat stress. Against this view, increased levels of IPT (CK biosynthesis gene) correlate with an assimilation of C and N in sinks through increased levels of hexokinase and nitrogen reductase (NR). Hence, CKs play a complicated role in hormonal homeostasis under stress conditions. Under heat stress, the BZR1–PIF4 interaction expresses the positive correlation between brassinosteroids and auxins, which regulate a transcriptional network and are involved in plant morphological development (Oh et al., 2012).

19.4.11 Antagonistic Effect

According to Yasuda et al., (2008), application of SA analogue 1,2-benzisothiazol-3(2 H)-one1,1-dioxide (BIT) would activate SA synthesis; in contrast to this repression of ABA, synthesis occurred in *Arabidopsis* during stress conditions. The enzyme NADPH oxidase induces the production of the ROS system; in turn, it will further stimulate ABA signaling by PLD, which negatively regulates the ethylene signaling pathway. Li et al., (2021) reported that ET and JA are accumulated and act antagonistically in a heat stress response. The positive interaction between ABA and nitric oxide (NO) was seen during an elevation of heat stress; these molecules associated with anti-oxidant enzymes, osmoprotectants (proline, glycine, betaine), secondary messengers (Ca^2+) and HSPs/HSFs (heat stress transcription factors) lead to tolerance for heat stress (Iqbal et al., 2021). As supported by the view of O'Brien & Benkova, (2013), they showed that auxin and CKs might act in the opposite role in a plant defense response. A greater CK/ABA ratio was noticed while plants were exposed to heat stress; it implied an antagonistic interaction between the two phytohormones (Cerny et al., 2014).

TABLE 19.3
Functions of three important heat shock proteins under temperature stress

Heat Shock Proteins	Functions
Hsp 21 (Li et al., 2021)	Chloroplast HSP involved in the improvement of photosynthetic efficiency under heat stress
Hsp70 (Song et al., 2001)	Regulation of Bag 1 function, which was involved in key functions such as regulation of cell death, anti-apoptotic protein Bcl-2 and growth regulator Raf-1
Hsp90 (Song et al., 2001)	It can be regulated by co-chaperones Hop and p23

The role of different heat shock proteins can interact with different molecular components during a signal transduction pathway under heat stress (Table 19.3). Apart from this, "STAY-GREEN" is a major trait that keeps the plants with an active photosynthetic state (by limiting the chlorophyll degradation) under heat stress (Abdelrahman et al., 2017). In conclusion, phytohormones play a complex function in hormonal homeostasis during stress conditions; hence, a complete understanding of hormonal functions and interactions in plant thermotolerance must await further studies.

19.5 ETHYLENE RESPONSE DURING SUBMERGENCE

Plants struggle to obtain adequate oxygen when they are submerged. By causing the internode growth of stems to increase, deep-water rice is able to endure periodic floods and the resulting lack of oxygen. During submerged conditions, the leaves of some terrestrial plants like *A. thaliana* and tomatoes turned pale yellow and their development were stunted (Tang et al., 2021). Other species, like *R. palustris*, have extended leaves and reduced leaf thickness, which aid the plant in obtaining a comparatively greater gas exchange when submerged. In contrast to submerged leaves, which are usually thin, narrow, or split and have reduced stomata, many aquatic, dimorphic plant species have thick, broad and complete aerial leaves.

Aquatic plants have developed adaptive defenses to endure in water. When rice is subjected to flooding, high levels of ethylene buildup, activating the *SEMIDWARF 1 (SD1)* gene, promoting GA-dependent elongation thereby making an alternative mechanism ('escape' strategy) establishing contact with air again (Kuroha et al., 2018). According to recent research, GA accumulation caused by submergence activates *ACE1 (ACCELERATOR OF INTERNODE ELONGATION 1*), which gives cells in the intercalary meristematic area the capacity to divide. This results in internode elongation when GA is present. Contrarily, *DEC1 (DECELERATOR OF INTERNODE ELONGATION 1)* down-regulation promotes internode elongation while strong GA repression inhibits it (Nagai et al., 2020). During submerged conditions, ethylene also stimulates the expression of two ethylene response factors (ERFs), *SNORKEL1 (SK1)* and *SK2*, to cause considerable internode elongation through GA (Hattori et al., 2009). When terrestrial plants like *A. thaliana* are submerged, the constrained gas transport results in passive ethylene buildup, which increases the production of the nitric oxide (NO) scavenger PHYTOGLOBIN 1 and signals dependent on ETHYLENE INSENSITIVE 2 (EIN2) and EIN3/EIN3-like 1 (EIL1) (PGB1). The increased PGB1 levels cause a reduction in NO, which improves the stability of group VII ethylene response factor (ERFVII) (Gibbs et al., 2014).

19.6 CONCLUSION AND FUTURE PERSPECTIVES

A changing climate causes several abiotic stresses, which have severe impacts on plant growth and metabolism. Cell surface receptors detect the stress signals and transduce them to the nucleus, where the stress-responsive gene has been transcribed and translated to produce the concerned proteins and hormones to combat various stresses. There exists crosstalk between phytohormones and enzymes that may synergistically or antagonistically regulate plant defense responses. To comprehend the genetic mechanism behind stress responses in various agricultural plants, it is necessary to further elucidate the signal transduction pathways that have been activated in response to various abiotic stresses. It will remain crucial to analyze these intricate pathways using both forward and reverse genetics methods. Despite the fact that conventional genetic assays have provided useful information about stress signal transduction, this strategy may eventually have certain limitations due to the functional redundancy of components within signaling pathways. To firmly establish the specificity or crosstalk of the signaling pathways, it will be necessary to integrate cell biology study of temporal as well as spatial expression patterns with biochemical characterization of the components, especially identification of signaling complexes. This will further assist in the development of stress-resistant cultivars.

REFERENCES

Abdelrahman, M., Sayed, M. E., Jogaiah, S., Burritt, D. J., & Phan Tran, L. S. (2017). The "STAY-GREEN" trait and phytohormone signaling networks in plants under heat stress. Plant Cell Reports, 36(7), 1009–1025.

Arif, Y., Singh, P., Siddiqui, H., Bajguz, A., & Hayat, S. (2020). Salinity induced physiological and biochemical changes in plants: an omic approach towards salt stress tolerance. Plant Physiology and Biochemistry, 156, 64–77.

Aryal, J. P., Sapkota, T. B., Khurana, R., Khatri-Chhetri, A., Rahut, D. B., & Jat, M. L. (2020). Climate change and agriculture in South Asia: Adaptation options in smallholder production systems. Environment, Development and Sustainability, 22(6), 5045–5075.

Cerny, M., Jedelsky, P. L., Novak, J. A. N., Schlosser, A., & Brzobohaty, B. (2014). Cytokinin modulates proteomic, transcriptomic and growth responses to temperature shocks in *Arabidopsis*. Plant, Cell & Environment, 37(7), 1641–1655. doi: 10.1111/pce.12270.

Chen, J., Burke, J. J., Velten, J., & Xin, Z. (2006). FtsH11 protease plays a critical role in *Arabidopsis* thermotolerance. The Plant Journal, 48(1), 73–84.

Chen, T., Shabala, S., Niu, Y., Chen, Z. H., Shabala, L., Meinke, H. et al. (2021). Molecular mechanisms of salinity tolerance in rice. The Crop Journal, 9(3), 506–520.

Clarke, S. M., Cristescu, S. M., Miersch, O., M Harren, F. J., Wasternack, C., & Mur, L. A. J. (2009). Jasmonates act with salicylic acid to confer basal thermotolerance in *Arabidopsis thaliana*. New Phytologist, 182(1), 175–187. doi: 10.1111/j.1469-8137.2008.02735.x.

Dhaubhadel, S., Chaudhary, S., Dobinson, K. F., & Krishna, P. (1999). Treatment with 24-epibrassinolide, a brassinosteroid, increases the basic thermotolerance of *Brassica napus* and tomato seedlings. Plant Molecular Biology, 40(2), 333–342.

Dobra, J., Cerny, M., Storchova, H., Dobrev, P., Skalak, J., Jedelsky, P. L., Luksanoa, H., et al. (2015). The impact of heat stress targeting on the hormonal and transcriptomic response in *Arabidopsis*. Plant Science, 231, 52–61.

Dong, T., Park, Y., & Hwang, I. (2015). Abscisic acid: Biosynthesis, inactivation, homoeostasis and signalling. Essays Biochemistry, 58, 29–48.

El Ghazali, & Gamal, E. B. (2020). *Suaeda vermiculata* Forssk. ex JF Gmel.: structural characteristics and adaptations to salinity and drought: a review. International Journal of Scientific, 9, 28–33.

Fahad, S., Bajwa, A. A., Nazir, U., Anjum, S. A., Farooq, A., Zohaib, A., et al. (2017). Crop production under drought and heat stress: plant responses and management options. Frontiers in Plant Science, 1147.

Feng, W., Kita, D., Peaucelle, A., Cartwright, H. N., Doan, V., Duan, Q., et al. (2018). The FERONIA receptor kinase maintains cell-wall integrity during salt stress through Ca^{2+} Signaling. Current Biology, 28, 666–675.

Fiorucci, A. S., Galvão, V. C., Ince, Y. Ç., Boccaccini, A., Goyal, A., Petrolati, L. A., Trevisan, M., & Fankhauser, C. (2020). PHYTOCHROME INTERACTING FACTOR 7 is important for early responses to elevated temperature in *Arabidopsis* seedlings. New Phytologist, 226(1), 50–58. doi: 10.1111/nph.16316.

Gibbs, D. J., Isa, N. M., Movahedi, M., Lozano-Juste, J., Mendiondo, G. M., Berckhan, S., & Holdsworth, M. J. (2014). Nitric oxide sensing in plants is mediated by proteolytic control of group VII ERF transcription factors. Molecular Cell, 53(3), 369–379.

Gonzalez-Olmedo, J. L., Cordova, A., Aragon, C. E., Pina, D., Rivas, M., & Rodríguez, R. (2005). Effect of an analogue of brassinosteroid on FHIA-18 plantlets exposed to thermal stress. InfoMusa, 14(1), 18–20.

Ha, C. V., Leyva-Gonzalez, M. A., Osakabe, Y., Tran, U. T., Nishiyama, R., Watanabe, Y., et al. (2014). Positive regulatory role of strigolactone in plant responses to drought and salt stress. Proceedings of the National Academy of Sciences of the United States of America, 111, 851–856.

Hao, S., Wang, Y., Yan, Y., Liu, Y., Wang, J., & S. Chen. (2021). A review on plant responses to salt stress and their mechanisms of salt resistance. Horticulturae, 7, 132.

Hattori, Y., Nagai, K., Furukawa, S., Song, X. J., Kawano, R., Sakakibara, H., & Ashikari, M. (2009). The ethylene response factors SNORKEL1 and SNORKEL2 allow rice to adapt to deep water. Nature, 460(7258), 1026–1030.

Hu, Y., Jiang, Y., Han, X., Wang, H., Pan, J., & Yu, D. (2017). Jasmonate regulates leaf senescence and tolerance to cold stress: crosstalk with other phytohormones. Journal of Experimental Botany, 68(6), 1361–1369. doi: 10.1093/jxb/erx004.

Huang, D., Wu, W., Abrams, S. R., Adrian, J., & Cutler, A. J. (2008). The relationship of drought-related gene expression in *Arabidopsis thaliana* to hormonal and environmental factors. Journal of Experimental Botany, 59, 2991–3007.

Ingram, J., & Bartels, D. (1996). The molecular basis of dehydration tolerance in plants. Annual Review of Plant Physiology and Plant Molecular Biology, 47, 377–403.

IPCC. (2014). Climate Change (2014): Synthesis Report. Contribution of Working Groups I, II and III to the Fifth Assessment Report of the Intergovernmental Panel on Climate Change [Core Writing Team, R.K. Pachauri and L.A. Meyer (eds.)]. IPCC, Geneva, Switzerland, 151 pp.

Iqbal, N., Umar, S., Khan, N. A., & Corpas, F. J. (2021). Crosstalk between abscisic acid and nitric oxide under heat stress: Exploring new vantage points. Plant Cell Reports, 40(8), 1429–1450.

Jakubowicz, M., Gałgańska, H., Nowak, W., & Sadowski, J. (2010). Exogenously induced expression of ethylene biosynthesis, ethylene perception, phospholipase D, and Rboh-oxidase genes in broccoli seedlings. Journal of Experimental Botany, 61(12), 3475–3491.

Jiang, Z., Zhou, X., Tao, M., Yuan, F., Liu, L., Wu, F., et al. (2019). Plant cell-surface GIPC sphingolipids sense salt to trigger Ca^{2+} influx. Nature, 572, 341–346.

Jogawat, A., Yadav, B., Lakra, N., Singh, A. K., & Narayan, O. P. (2021). Crosstalk between phytohormones and secondary metabolites in the drought stress tolerance of crop plants: A review. Physiologia Plantarum, 172, 1106–1132.

Joshi, R., Anwar, K., Das, P., Singla-Pareek, S. L., & Pareek, A. (2017). Overview of Methods for Assessing Salinity and Drought Tolerance of Transgenic Wheat Lines. In: P. Bhalla, & M. Singh (eds.), Wheat Biotechnology. Methods in Molecular Biology, vol 1679. Humana Press, New York, NY. 10.1007/978-1-4939-7337-8_5

Kim, S., Hwang, G., Kim, S., Thi, T. N., Kim, H., Jeong, J., Kim, J., Kim, J., Choi, G., & Oh, E. (2020). The epidermis coordinates thermoresponsive growth through the phyB-PIF4-auxin pathway. Nature Communications, 11(1), 1–13. doi: 10.1038/s41467-020-14905-w.

Kupers, J. J., Oskam, L., & Pierik, R. (2020). Photoreceptors regulate plant developmental plasticity through auxin. Plants, 9(8), 940. doi: 10.3390/plants9080940.

Kuroha, T., Nagai, K., Gamuyao, R., Wang, D. R., Furuta, T., Nakamori, M., & Ashikari, M. (2018). Ethylene-gibberellin signaling underlies adaptation of rice to periodic flooding. Science, 361(6398), 181–186.

Lata, C., & Prasad, M. (2011). Role of DREBs in regulation of abiotic stress responses in plants. Journal of Experimental Botany, 62, 4731–4748.

Lata, C., Muthamilarasan, M., & Prasad, M. (2015). Drought stress responses and signal transduction in plants in Elucidation of abiotic stress signaling in plants, Springer, 195–225.

Li, N., Euring, D., Cha, J. Y., Lin, Z., Lu, M., Huang, L. J., & Kim, W. Y. (2021). Plant hormone-mediated regulation of heat tolerance in response to global climate change. Frontiers in Plant Science, 11, 627969. doi: 10.3389/fpls.2020.627969.

Lv, W. T., Lin, B., Zhang, M., & Hua, X. J. (2011). Proline accumulation is inhibitory to *Arabidopsis* seedlings during heat stress. Plant Physiology, 156(4), 1921–1933. doi: 10.1104/pp.111.175810

Ma, B., Chen, S. Y., & Zhang, J. S. (2010). Ethylene signaling in rice. Chinese Science Bulletin, 55, 2204–2210.

Ma, D., Li, X., Guo, Y., Chu, J., Fang, S., Yan, C., Noel, J. P., & Liu, H. (2016). Cryptochrome 1 interacts with PIF4 to regulate high temperature-mediated hypocotyl elongation in response to blue light. Proceedings of the National Academy of Sciences, 113(1), 224–229. doi: 10.1073/pnas.1511437113.

Maduraimuthu, D., & Prasad, P. V. V. (2010). Ethylene production under high temperature stress causes premature leaf senescence in soybean. Functional Plant Biology, 37(11), 1071–1084.

Mahajan, S., & Tuteja, N. (2005). Cold, salinity and drought stresses: an overview. Archives of biochemistry and biophysics, 444, 139–158.

Meco, V., Egea, I., Ortíz-Atienza, A., Drevensek, S., Esch, E., Yuste-Lisbona, F. J., et al. (2020). The salt sensitivity induced by disruption of cell wall-associated kinase 1 (SlWAK1) tomato gene is linked to altered osmotic and metabolic homeostasis. International Journal of Molecular Sciences, 21,6308.

Munemasa, S., Hossain, M. A., Nakamura, Y., Mori, I. C., & Murata, Y. (2011). The *Arabidopsis* calcium dependent protein kinase, CPK6, functions as a positive regulator of methyl jasmonate signaling in guard cells. Plant Physiology, 155, 553–561.

Munné-Bosch, S., & Müller, M. (2013). Hormonal cross-talk in plant development and stress responses. Frontiers in Plant Science, 4, 529.

Nagai, K., Mori, Y., Ishikawa, S., Furuta, T., Gamuyao, R., Niimi, Y., Hobo, T. et al. (2020). Antagonistic regulation of the gibberellic acid response during stem growth in rice. Nature, 584(7819), 109–114.

Nakamura, C., Takenaka, S., Nitta, M., Yamamoto, M., Kawazoe, T., Ono, S., et al. (2021). High sensitivity of roots to salt stress as revealed by novel tip bioassay in wheat seedlings. Biotechnology and Biotechnological Equipment, 35, 246–254.

Neill, E. M., Byrd, M. C. R., Billman, T., Brandizzi, F., & Stapleton, A. E. (2019). Plant growth regulators interact with elevated temperature to alter heat stress signaling via the Unfolded Protein Response in maize. Scientific Reports, 9(1), 1–10.

Nishimura, N., Okamoto, M., Narusaka, M., Yasuda, M., Nakashita, H., Shinozaki, K., Narusaka, Y., & Hirayama, T. (2009). ABA hypersensitive germination 2-1 causes the activation of both abscisic acid and salicylic acid responses in *Arabidopsis*. Plant Cell Physiology, 50, 2112–2122.

NOAA. (2022). National Centers for Environmental Information, Monthly Global Climate Report for Annual 2022, published online January 2023, retrieved on September 30, 2023 fromhttps://www.ncei.noaa.gov/access/monitoring/monthly-report/global/202213.

Nongpiur, R. C., Singla-Pareek, S. L., & Pareek, A. (2020). The quest for osmosensors in plants. Journal of Experimental Botany, 71, 595–607.

O'Brien, J. A., & E. Benkova. (2013). Cytokinin cross-talking during biotic and abiotic stress responses. Frontiers in Plant Science, 4, 451. doi: 10.3389/fpls.2013.00451.

Oh, E., Zhu, J. Y., & Wang, Z. Y. (2012). Interaction between BZR1 and PIF4 integrates brassinosteroid and environmental responses. Nature Cell Biology, 14 (8), 802–809. doi: 10.1038/ncb2545.

Pan, T., Liu, M., Kreslavski, V. D., Zharmukhamedov, S. K., Nie, C., Yu, M. et al. (2021). Non-stomatal limitation of photosynthesis by soil salinity. Critical Reviews in Environmental Science and Technology, 51, 791–825.

Passricha, N., Saifi, S. K., Kharb, P., & Tuteja, N. (2020). Rice lectin receptor-like kinase provides salinity tolerance by ion homeostasis. Biotechnology and Bioengineering, 117, 498–510.

Praba, M. L., Cairns, J. E., Babu, R. C., & Lafitte, H. R. (2009). Identification of physiological traits underlying cultivar differences in drought tolerance in rice and wheat. Journal of Agronomy and Crop Science, 195, 30–46.

Prerostova, S., Dobrev, P. I., Kramna, B., Gaudinova, A., Knirsch, V., Spichal, L., Zatloukal, M., & Vankova, R. (2020). Heat acclimation and inhibition of cytokinin degradation positively affect heat stress tolerance of *Arabidopsis*. Frontiers in Plant Science, 11(87). doi: 10.3389/fpls.2020.00087.

Raza, A., Razzaq, A., Mehmood, S., Zou, X., Zhang, X., Lv, Y., & Xu, J. (2019). Impact of climate change on crops adaptation and strategies to tackle its outcome: A review. Plants, 8(2), 34.

Reed, J. W., Wu, M. F., Reeves, P. H., Hodgens, C., Yadav, V., Hayes, S., & Pierik, R. (2018). Three auxin response factors promote hypocotyl elongation. Plant Physiology, 178(2), 864–875. doi: 10.1104/pp.18.00718.

Rodriguez, M. C. S., Petersen, M., & Mundy, J. (2010). Mitogen-activated protein kinase signaling in plants. Annual Review of Plant Biology, 61, 621–649.

Sauer, M., & Kleine-Vehn, J. (2019). PIN-FORMED and PIN-LIKES auxin transport facilitators. Development, 146(1), dev168088.

Seifikalhor, M., Aliniaeifard, S., Shomali, A., Azad, N., Hassani, B., Lastochkina, O. et al. (2019). Calcium signaling and salt tolerance are diversely entwined in plants. Plant Signaling & Behavior, 14, 1665455

Song, J., Takeda, M., & Morimoto, R. I. (2001). Bag1–Hsp70 mediates a physiological stress signalling pathway that regulates Raf-1/ERK and cell growth. Nature Cell Biology, 3(3), 276–282.

Song, X. G., She, X. P., He, J. M., Huang, C., & Song, T. (2006). Cytokinin-and auxin-induced stomatal opening involves a decrease in levels of hydrogen peroxide in guard cells of *Vicia faba*. Functional Plant Biology, 33 (6), 573–583.

Tang, H., Bi, H., Liu, B., Lou, S., Song, Y., Tong, S., Chen, N., Jiang, Y., Liu, J., & Liu, H. (2021). WRKY33 interacts with WRKY12 protein to up-regulate RAP2. 2 during submergence induced hypoxia response in *Arabidopsis thaliana*. New Phytologist, 229, 106–125.

Tenhaken, R. (2015). Cell wall remodeling under abiotic stress. Frontiers in Plant Science, 5, 771.

Tuteja, N., & Sopory, S. K. (2008). Chemical signaling under abiotic stress environment in plants. Plant Signaling & Behavior, 3(8), 525–536.

Verslues, P. E., &. Zhu, J. K. (2005). Before and beyond ABA, upstream sensing and internal signals that determine ABA accumulation and response under abiotic stress. Biochemical Society Transaction, 33, 375–379.

Wang, C. F., Han, G. L., Yang, Z. R., Li, Y. X., & Wang, B. S. (2022). Plant Salinity Sensors: Current Understanding and Future Directions. Frontiers in plant science, 13, 1–13.

Wind, J., Smeekens, S., & Hanson, J. (2010). Sucrose: metabolite and signaling molecule. Phytochemistry, 71 (14-15), 1610–1614.

Wong, H. L., Pinontoan, R., Hayashi, K., Tabata, R., Yaeno, T., Hasegawa, K., Kojima, C. et al. (2007). Regulation of rice NADPH oxidase by binding of Rac GTPase to its N-terminal extension. The Plant Cell, 19(12), 4022–4034.

Xie, Y., Huhn, K., Brandt, R., Potschin, M., Bieker, S., Straub, D., Doll, J., Drechsler, T., Zentgraf, U., & Wenkel, S. (2014). REVOLUTA and WRKY53 connect early and late leaf development in *Arabidopsis*. Development, 141 (24), 4772–4783.

Xiong, L., Schumaker, K. S., & Zhu, J. K. (2002). Cell signaling during cold, drought, and salt stress. The plant cell, 14, S165–S183.

Yasuda, M., Ishikawa, A., Jikumaru, Y., Seki, M., Umezawa, T., Asami, T., Nakashita, A. M. et al. (2008). Antagonistic interaction between systemic acquired resistance and the abscisic acid–mediated abiotic stress response in *Arabidopsis*. The Plant Cell, 20 (6), 1678–1692.

Ye, Y., Ding, Y., Jiang, Q., Wang, F., Sun, J., & Zhu, C. (2017). The role of receptor-like protein kinases (RLKs) in abiotic stress response in plants. Plant Cell Reports, 36, 235–242.

Yoshida, T., Mogami, J., & Yamaguchi-Shinozaki, K. (2014). ABA-dependent and ABA-independent signaling in response to osmotic stress in plants. Current Opinion in Plant Biology, 21, 133–139.

Zhai, Y., Wen, Z., Han, Y., Zhuo, W., Wang, F., Xi, C., et al. (2020). Heterogeneous expression of plasma-membrane-localised OsOSCA1.4 complements osmotic sensing based on hyperosmolality and salt stress in *Arabidopsis* osca1 mutant. Cell Calcium, 91,102–261.

Zhang, H., Zhu, J., Gong, Z., & Zhu, J. K. (2022). Abiotic stress responses in plants. Nature Reviews Genetics, 23, 104–119.

Zhang, X., Li, J., Liu, A., Zou, J., Zhou, X., Xiang, J., Rerksiri, W., Peng, Y., Xiong, X., & Chen, X. (2012). Expression profile in rice panicle: insights into heat response mechanism at reproductive stage. PloS One, 7 (11), e49652.

Zhao, C., Zhang, H., Song, C., Zhu, J.-K., & Shabala. S. (2020). Mechanisms of plant responses and adaptation to soil salinity. Innovation, 1, 100017.

Zhu, J. K. (2016). Abiotic stress signaling and responses in plants. Cell, 167, 313–324.

Zorb, C., Geilfus, C. M., & Dietz, K. J. (2019). Salinity and crop yield. Plant Biology, 21, 31–38.

20 Engineering Phytohormones for the Development of Tolerant Varieties

Ankita Rajendra Parab, Ayyagari Ramlal, and Sreeramanan Subramaniam

20.1 INTRODUCTION

The morphology and physiology of plants are entirely dependent on chemical messengers produced by plants called phytohormones. These are intercellular messenger molecules that control the different characteristics of growth, such as cellular structure and development, reproductive methods, plant metabolism and stress tolerance (Davies, 2010; Small & Degenhardt, 2018). They are required in low concentrations to carry out the entire life cycle of a plant. There are different types of plant hormones and each of them has different functions (Davies, 2010; Oshchepkov et al., 2020; Torun et al., 2022). Based on their application, phytohormones can be endogenous or exogenous. They primarily belong to five classes, namely cytokinin, auxins, gibberellins, ethylene and abscisic acid (ABA). Synthetic biochemicals, such as brassinosteroids, jasmonic acid (JA), salicylic acid and strigolactones, are also reported to perform growth regulatory functions in plants. These are the types of growth regulators that are essential either in combination or independently (Jiang & Asami, 2018). The immobility in plants cause restricted plant responses and, therefore, plants respond to any stimulus in the form of biosynthesis of these regulatory biochemicals (Jiang & Asami, 2018). Phytohormones are the primary responders during biotic and abiotic stress conditions (Ku et al., 2018). The external administration of phytohormones to increase stress tolerance has been used frequently. However, these practices can be labour-intensive and economically demanding. A long-term solution for plant stress tolerance is to tackle it on a cellular and molecular level (Chhaya et al., 2021). The following chapter aims to discuss biological regenerative pathways to produce modified phytohormones that may help to enhance the tolerance levels among plants that experience environmental stress.

20.1.1 PHYTOHORMONES

Biosynthesis of plant hormones may occur at any location in the plant, like stems, roots or leaves. Plants produce these hormones from the stimulus they get. Localized production of hormones also entails that many times the phytohormones may get transported to the desired location (Kieber & Schaller, 2014; Premkumar et al., 2011). A combination of cytokinins and auxins together have positive interactions in plant cells and, therefore, aid in plant tissue development. In tissue culture, cytokinins such as 6-benzylaminopurine (BAP) have proven to be highly successful in the process of organogenesis (Bhatt & Dhar, 2000; Lijalem & Feyissa, 2020; Parab et al., 2021). Cytokinins are responsible for shoot development, whereas auxins are responsible for root development. However, in many cases, a combination treatment has also yielded positive effects on tissue culture optimization and regeneration experiments (Checker et al., 2018; Danial et al., 2014; Faisal et al., 2018).

 DOI: 10.1201/9781003335788-25

20.1.2 Response to Stimuli

Plants generally undergo environmental stress regularly and therefore need to process protective mechanisms for survival. Environmental conditions produce abiotic and biotic stress on plants. Ku et al. (2018) have reported intercellular signal transduction due to environmental stress, producing important secondary metabolites that protect the plants against deteriorating climate change. Their research revealed that Ca^{2+}-signalling pathways constitute the first line of defence against stress in plants. "Calcium waves" are a phenomenon in plants that occurs due to fluctuations in calcium levels due to environmental stress conditions. These fluctuations trigger chain reactions that develop plant hormone responses accordingly. This is followed by ABA, JA and strigolactone interactions and, therefore, create a crossover effect of phytohormones to ensure plant tolerance toward environmental stress.

The issues for production, metabolism, transport and signal transduction have been improved due to molecular biology techniques. With the aid of the CRISPR-Cas9 technology, research on phytohormone synthesis and sequencing has made a significant contribution to the advancement of plant genetic information. To understand the physiological mechanisms of crossover and biological activities of chemical molecules in plants, the field of chemical biology has proven to be at the epitome of discovery. To avoid gene redundancy, small chemical compounds with bioactivities can be applied to plants to examine gene function. For instance, the synthetic antagonist of the ABA receptor, pyrabactin, which activates only in a handful of the receptors among its 14 homologs, is responsible for characterizing ABA (Jiang & Asami, 2018).

20.2 ENVIRONMENTAL STRESS CONDITIONS

Various environmental conditions, such as climate change, drought, floods, soil composition, temperature and heavy metal accumulation, can be considered a class of abiotic stress endured by sessile plants. To sustain these environmental conditions, plants respond by producing secondary metabolites or phytohormones. For example, according to Wani et al. (2016), the salt concentration fluctuations in soil contributes to major abiotic stress on plants. Productivity has been drastically reduced in the dry regions of the world because of the salt stress. The phytohormones that are produced in response to such stresses can be potentially re-engineered to enhance the tolerance in plants.

Primary metabolites are precursors for biosynthetic pathways for producing stress-induced phytohormones or secondary metabolites. Biotic and abiotic stress conditions in plants affect the normal growth and development of the plants. Biotic stress on the plants is caused by pests, such as insects, rodents and other microorganisms that include viruses. Abiotic stress involves climatic conditions, such as drought, floods or even heat fluctuations and global warming (Mahajan et al., 2020).

Pandey et al. (2017) have found biotic stress can be caused in combinations, for example, viruses and bacteria together can have a combined effect called "brown apical necrosis" on the walnut plant. This can be termed "combined biotic stress." Drought and flood conditions can cause a case of combined abiotic stress. Certain microorganisms, including *Macrophomina* sp., *Erysiphe* sp., *Fusarium* sp., *Rhizoctonia* sp. and *Puccinia* sp., have elevated inductive responses toward plants in drought stress conditions. Dry root rot is a very common fungal infection in plants growing in dry conditions and high temperatures. Powdery mildew fungal growth can be observed on sugar beet plants in extended drought conditions with warm mornings and colder nights.

20.3 STRESS TOLERANCE: PHYTOHORMONE RESPONSE TO ENVIRONMENTAL STRESS

Plants naturally have symbiotic and endophytic interactions with a variety of microbes. These interactions result in the active production of numerous biochemicals that are useful for plants.

Microbial-mediated plant protective responses have been beneficial to plants (Egamberdieva et al., 2017). However, the pathogenic cycle is the biggest threat to genetic resistance. Research has yet to understand plant reactions to climate change. Plant interactions with respective pathogens due to environmental stress conditions are a limitation to studying the exact response mechanisms. Generally, phytohormones are the primary defence molecules in plants. However, naturally occurring phytohormones may not be sufficient to provide the necessary antagonistic actions. In many cases, plant species have a genetic modification to produce altered and enhanced phytohormones to survive extreme environmental conditions. Reactive oxygen species (ROS) formation is also a threat to plant growth and development. In the case of cellular destruction due to oxidative stress, ROS is the primary concern. Proline, glycine-betaine, polyamines and various sugars are the secondary metabolite responses produced by plants during such conditions (Sharma et al., 2019). Various transcription factors and genes code for specific tolerance mechanisms in different plants. It is important to identify these genes and their loci to modify the responses on the genetic level (Singh et al., 2020). For example, to learn more about how turfgrass species tolerate salt, the molecular responses to salt stress were examined. Numerous potential unigenes associated with chemical processes to salt stress were found after a thorough analysis of the transcriptome profiles of various turfgrass species. These findings are essential in transcriptomic and proteomic analyses for phytohormone engineering (Fan et al., 2020).

20.3.1 Phytohormone Responses

Polyamines play a crucial role in inducing genetic-level modifications to reduce stress-induced diseases in plants. Polyamines interact with negatively charged molecules in plant nuclei i.e., DNA and RNA. However, the effects vary between plant species and across polyamines. Exogenous administration of polyamines to stress-induced cells has typically resulted in damage reduction and promotion of growth (Liu et al., 2007).

Glutathione is a plant molecule that interacts favourably with plant cells. It prevents the denaturation of proteins under oxidative stress conditions. It is a precursor for the production of phytochelatins. These molecules help to chelate toxic metals through plant vacuoles. Owing to these properties, glutathione becomes an important biomolecule for plant stress tolerance (Hasanuzzaman et al., 2017).

In the case of soil salinity, proline helps to promote cell growth and to increase salt tolerance. Gibberellins also help to improve salt tolerance and tolerance against heavy metal toxicity (Asif et al., 2022; Wani et al., 2016). Wani et al. (2016) have also studied the role of ABA, which is popularly known to be used for abiotic stresses in plants. ABA is responsible for triggering multiple signalling pathways that produce important plant defence-related biomolecules during stress conditions. For example, during drought conditions, ABA will activate the transcription factors, such as bZIPs, MYB/MYC2 and WRKY, that bind to their corresponding receptors. This initiates the blocking of stomata, which prevents transpiration activity through leaves (Muhammad Aslam et al., 2022).

20.3.2 Molecular Changes to Phytohormones

In a study, plant hormones have been compared with the hormones of the mammalian class for structural comparisons to purine bases. Nucleotides and cyclic nucleotides are closely related to phytohormones. Therefore, metabolic and biosynthetic pathways associated with phytohormones may be greatly affected and influenced by such nucleotide bases and their ability to induce molecular-level changes. Secondary messengers, ion transporters and kinases contribute greatly toward plant cellular signalling pathways (Williams, 2021). During drought-induced stress in plants, genetic modifications occur at the transcriptome level. This causes the plant tissues to modify in a way by which the water transport channels, aquaporins, can reduce transpiration. It has also been reported that during environmental stress conditions, such as changes in temperature, light, salinity in soil and

TABLE 20.1
Some plant species, regulatory genes and their functions during stress conditions

Plant Species	Regulatory Gene	Functions
Solanum lycopersicum	*SlILR1, SlILR5* and *SlILR6*	Floral pedicle abscission
	miR167a	Auxin signalling
Nicotiana tabaccum	*NtPT2*	Tolerance to phosphorus stress
	PIN1, PIN3, PIN3b, PIN4 and *PIN9*	Auxin transport during drought stress
	GmERF3	Improve drought tolerance by enhancing proline content
Major crop plants	*TaSAUR75*	Auxin upregulation during water-stress conditions
	Isopentenyl transferase	Cytokinin biosynthetic gene during drought tolerance
Arabidopsis sp.	*SnRK2s* and *IbARF5*	ABA-mediated stress response
	AtNIP5;1	Ethylene and ABA-induced boron uptake cycle

heavy metal accumulation, the auxin biosynthesis pathway is triggered. Auxin synthesis is directly associated with the induction of the signalling pathways of ABA response genes, which is a highly crucial biomolecule during drought-related stress in plants. Table 20.1 shows the list of genes responsible for triggering different biosynthesis pathways for stress tolerance in different plant species (Bozbuga et al., 2022; Ku et al., 2018; Ullah et al., 2018). In temperate and polar climates, plants typically tolerate freezing conditions by moving nutrients from the leaf and shoot regions to storage organs. Dehydrative stress is brought on by freezing stress because it prevents water from being absorbed and transported to aerial plant parts. A signal-cascading pathway is triggered to activate the ROS accumulation and enzyme activities (Raza et al., 2023). Stress-regulated gene interactions and proteomics have a great potential to engineer phytohormones to make plants more resilient to environmental stress conditions (Chhaya et al., 2021).

20.4 VARIATIONS IN MORPHOGENETIC EFFECTS OF PHYTOHORMONES

The field of molecular biology has succeeded in producing genetic-level enhancements in plants. These enhancements are positive in terms of tolerance. Genetic engineering has tremendous potential for improving plant quality and quantity and stress tolerance. Expression of genes that regulate the signalling pathways for secondary metabolite production is the key to a successful morphogenetic variance among the phytohormones (Kumar et al., 2016). Epigenetic mechanisms, such as DNA methylation, RNA actions and histone modifications, are responsible for altering the transcriptional pathways. These mechanisms generally result from stress conditions. For example, salt stress causes salt-tolerance genes to accumulate active histone markers like H3K9K14Ac and H3K4me3 and to lose repressive markers like H3K9me2 and H3K27me3. As a result, the peroxidase gene is activated. An increased peroxidase gene expression has been associated with the activation of the ABA pathway, which reduces the formation of ROS (reactive oxygen species) and boosts levels of osmotic metabolites, improving salt tolerance (Singroha et al., 2022).

Gene expression due to stress triggers is highly dependent on signalling pathways induced by ABA. PYR/PYL/RCARs are ABA receptors that are required to induce signalling pathways of ABA responses. In different plants, different genes encode for these receptor proteins. *Arabidopsis* contains 14 gene codes for ABA receptors, whereas a rice plant contains 11. Protein kinases are also essential for ABA stress responses. Different types of transcription factors control ABA-dependent signalling pathways such as MYB, MYC and NAC. Calcium-dependent protein kinases (CDPKs) are another type of transcription factor that contributes toward a positive ABA stress-signalling response. The two types of genes, CDPK4 and CDPK11, encode for ABA signalling (Ullah et al., 2018, Aslam et al., 2022).

Ullah et al. (2018) also describe jasmonates or JA, which are generally derived from flowers. The proteins JA13/JAZ will bind to transcription factors (such as MYC2) that induce the production of JA signalling pathways. JA signalling pathways can induce the production of essential volatile oils that can fight plant pathogens. Crossover effects can also be observed with JA and other phytohormones, due to dynamic gene regulation by JA. There are occurrences of multiple JA-mediated crossover effects with phytohormones, such as gibberellins, cytokinins and auxins, all of which have plant developmental properties. For example, the JA–GA interactions play harmoniously because under stress conditions, phytohormone crosstalk effects are beneficial. The JA mediating pathways will affect the JAZ and the GA pathway will trigger GA-dependent plant coordination in the form of DELLA mutant genes. Eventually, JA–GA interact antagonistically to induce plant growth and defence mechanisms (Jang et al., 2020).

Auxin-induced response factors (ARFs) for biotic stress help to induce phytohormones that aid in root diseases. Root length in *Arabidopsis* is altered to extend the survival of the plant in drought and high salinity conditions with the help of $TaSAUR_{75}$. In drought and salt-induced stress, there is over-expression of ABA, which causes genetic mutations in the plants. Genetic modifications in plants where mutations have occurred on the O_SSWEET13 and O_SSWEET15 genes increase sugar production to combat salt-induced stress (Bozbuga et al., 2022).

20.5 ROLE OF GENETIC VARIATIONS ON PHYTOHORMONES

There is an urgent need for plants to develop better stress tolerance due to the rapidly deteriorating climatic circumstances. It has been predicted that by the year 2050, if the environmental conditions keep on deteriorating, plant life is highly likely to vanish. Therefore, there is an urgency for phytohormone improvement and development. Plants naturally produce phytohormones for growth and development purposes. Types of hormones are also induced during environmental stress conditions as mechanisms for stress tolerance. According to the external stimuli, clients may alter the genes to trigger a change in the biosynthetic pathways of the naturally occurring phytohormones to enhance their ability to sustain themselves in adverse environmental stress conditions (Ghini et al., 2008; Small & Degenhardt, 2018; Oshchepkov et al., 2020; Chhaya et al., 2021).

Genome editing methodology, which includes enzymes such as meganucleases and endonucleases, has been widely used for more than a decade. Meganucleases can cut larger DNA sequences. Utilizing the host's homologous repair system, meganucleases cause double-stranded DNA (dsDNA) breaks in the host genome at a specific location and then spread throughout the provided genome. However, the re-engineering of meganucleases is a tedious process. Therefore, in modern genomics, endonucleases such as the zinc-finger nucleases (ZFNs) are used. The ZFNs can be engineered to be receptor-specific and can produce a wide range of genetic modifications. ZFNs can be made by various protein engineering techniques to essentially target any unique DNA stretch. Another class of genome editing molecules are transcription activator-like effector nucleases (TALENs). These belong to the restriction endonuclease family and can be synthetically customized as per requirement. Clustered regularly interspaced short palindromic repeats/CRISPR-associated protein (CRISPR/Cas) is considered to be a breakthrough tool in gene modifications in plants. The CRISPR/Cas system is adopted through bacteriophage mechanisms. Essentially, a CRISPR array transcribes a mature crRNA, which is then used as a guide for Cas endonuclease regulation. CRISPR mechanisms are used to enhance plant tolerance against abiotic stresses (Al Aboud & Jialal, 2022; Ghosh & Dey, 2022; Kaur et al., 2022).

Cellular growth and development in the meristematic tissues is highly controlled by *WUSCHEL* (*WUS* genes) and *CLAVATA* (*CLV* genes). Reportedly, any mutations in these genes have resulted in programmed cell death post-organogenesis. The WUS protein is responsible for suppressing the activity of helix-loop-helix (bHLH) gene expression. DEC1 and DEC2 are sub-members of the bHLH protein family, which helps in the activity of repressor genes. Therefore, it can be concluded that the fate of meristematic cells is dependent on the expression of WUS and CLV genes

(Fujimoto et al., 2007; Lee et al., 2019). In the tomato, potato and soybean plants, there are almost 20 StARFs that are induced by the auxin-responsive factors (ARFs) during temperature, salt and osmotic stress, and 9 OsARFs genes are expressed during drought and salt stress. These genes are responsible for producing plant immuno-proteins during environmental stress conditions (Shankaraswamy & Ento, 2022).

YUCCA genes are essential elements in the auxin biosynthesis pathways in plants such as tomato, potato, rice, maize and strawberry. These genes help to maintain the levels of auxins in plants. Auxins are essential biochemicals that regulate cellular functions including apical dominance, differentiation and tolerance. Poplar plants are considered important from an environmental point of view because they are useful for reforestation purposes. A transgenic variety of poplar plant, *P. alba* × *P. glandulosa*, which possesses *Arabisopsia YUCCA6*, shows stress tolerance mediated by auxin-induced signal transduction (Ke et al., 2015). Ke et al., (2016) also studied the same phenomenon on oxidative stress tolerance using the *CodA* gene regulation and were able to successfully modify the transgenic poplar species using *Agrobacterium* transformation.

On many occasions, ABA has proven to be a crucial phytohormone that regulates growth and stress responses in plants. In the case of wheat plants, the heat shock factor, HsfA6f, acts primarily to improve heat tolerance by inducing biosynthesis of ABA. However, there is a difference in gene expression in transgenic plants compared with wild-type plants. This interaction can also be termed a loop, where HsfA6f is triggered by ABA and, in turn, HsfA6f enhances ABA production (Li et al., 2021).

20.6 CONCLUSION AND FUTURE PROSPECTS

Climate change and deteriorating environmental conditions have created a global phenomenon. Primarily, the natural environment expresses stress on plants in such adverse conditions. Drought, floods, soil contamination due to heavy metal accumulation, fluctuations in soil salinity and temperature changes are some of the abiotic stresses plants must endure. Plants also need to sustain biotic stress, which includes parasitic attacks by microorganisms such as bacteria, fungi and viruses. A combined effect of biotic and abiotic stress can cause multiple negative effects on plants. The prime response of plants during such stress conditions is the production of phytohormones or secondary metabolites. Phytohormones are biochemicals that help plants overcome the biotic and abiotic environmental stress conditions. Cytokinins, auxins, gibberellins, jasmonates, salicylic acid, brassinosteroids and abscisic acid are the classifications of phytohormones that help in the overall growth and development of plants. Hormones are voluntarily produced by plants during stress conditions. Because of the combined effects of biotic and abiotic stress, it is essential to analyze the morphogenetic variations among phytohormones, which will improve the plant tolerance level. Intercellular cell signalling pathways are responsible for triggering the production of these biomolecules during stress conditions. Genetic alterations during biosynthesis pathways can cause positive mutations in these biochemicals. This results in improved stress tolerance in plants. Moreover, this field is scarcely studied and, therefore, research based on natural and synthetically directed genetic changes in the plant genome to induce genetically engineered phytohormones is highly anticipated.

REFERENCES

Al Aboud, N. M., & Jialal, I. (2022). Genetics, Epigenetic Mechanism. StatPearls, 1–5. http://www.ncbi.nlm.nih.gov/pubmed/30422591

Asif, R., Yasmin, R., Mustafa, M., Ambreen, A., Mazhar, M., Rehman, A., Umbreen, S., & Ahmad, M. (2022). Phytohormones as Plant Growth Regulators and Safe Protectors against Biotic and Abiotic Stress. Plant Hormones - Recent Advances, New Perspectives and Applications, 1–18. 10.5772/intechopen.102832

Bhatt, I. D., & Dhar, U. (2000). Combined effect of cytokinins on multiple shoot production from cotyledonary node explants of Bauhinia Vahlii. Entomologia Experimentalis et Applicata, 103(3), 239–248. 10.1023/A

Bozbuga, R., Bulent Arpaci, B., Uluisik, S., Gok Guler, P., Nilufer Yildiz, H., & Yalcin Ates, S. (2022). Genetic Modification of Plant Hormones Induced by Parasitic Nematodes, Virus, Viroid, Bacteria, and Phytoplasma in Plant Growing. Plant Hormones - Recent Advances, New Perspectives and Applications, 1–43. 10.5772/intechopen.102721

Checker, V. G., Kushwaha, H. R., Kumari, P., & Yadav, S. (2018). Role of phytohormones in plant defense: Signaling and cross talk. In Molecular Aspects of Plant-Pathogen Interaction. 10.1007/978-981-10-7371-7_7

Chhaya, Yadav, B., Jogawat, A., Gnanasekaran, P., Kumari, P., Lakra, N., Lal, S. K., Pawar, J., & Narayan, O. P. (2021). An overview of recent advancement in phytohormones-mediated stress management and drought tolerance in crop plants. Plant Gene, 25(November 2020), 100264. 10.1016/j.plgene.2020.100264

Danial, G., Ibrahim, D., Brkat, S., & Khalil, B. (2014). Multiple shoots production from shoot tips of fig tree (Ficus carica L.) and callus induction from leaf segments. International Journal of Pure and Applied Sciences and Technology, 20(1), 117–124.

Davies, P. J. (2010). The Plant Hormones: Their Nature, Occurrence, and Functions. In Nature, Occurrence and Functions (Issue October). 10.1007/978-1-4020-2686-7

Egamberdieva, D., Wirth, S. J., Alqarawi, A. A., Abd-Allah, E. F., & Hashem, A. (2017). Phytohormones and beneficial microbes: Essential components for plants to balance stress and fitness. Frontiers in Microbiology, 8(OCT), 1–14. 10.3389/fmicb.2017.02104

Faisal, M., Ahmad, N., Anis, M., Alatar, A. A., & Qahtan, A. A. (2018). Auxin-cytokinin synergism in vitro for producing genetically stable plants of Ruta graveolens using shoot tip meristems. Saudi Journal of Biological Sciences, 25(2), 273–277. 10.1016/j.sjbs.2017.09.009

Fan, J., Zhang, W., Amombo, E., Hu, L., Kjorven, J. O., & Chen, L. (2020). Mechanisms of environmental stress tolerance in turfgrass. Agronomy, 10(4), 1–23. 10.3390/agronomy10040522

Fujimoto, K., Hamaguchi, H., Hashiba, T., Nakamura, T., Kawamoto, T., Sato, F., Noshiro, M., Bhawal, U. K., Suardita, K., & Kato, Y. (2007). Transcriptional repression by the basic helix-loop-helix protein Dec2: Multiple mechanisms through E-box elements. International Journal of Molecular Medicine, 19(6), 925–932. 10.3892/ijmm.19.6.925

Ghini, R., Hamada, E., & Bettiol, W. (2008). Climate change and plant diseases. Scientia Agricola, 65, 98–107.

Ghosh, S., & Dey, G. (2022). Biotic and abiotic stress tolerance through CRISPR-Cas mediated genome editing. Journal of Plant Biochemistry and Biotechnology, 31(2), 227–238. 10.1007/s13562-021-00746-1

Hasanuzzaman, M., Nahar, K., Anee, T. I., & Fujita, M. (2017). Glutathione in plants: biosynthesis and physiological role in environmental stress tolerance. Physiology and Molecular Biology of Plants, 23(2), 249–268. 10.1007/s12298-017-0422-2

Jang, G., Yoon, Y., & Choi, Y. Do. (2020). Crosstalk with jasmonic acid integrates multiple responses in plant development. International Journal of Molecular Sciences, 21(1), 1–20. 10.3390/ijms21010305

Jiang, K., & Asami, T. (2018). Chemical regulators of plant hormones and their applications in basic research and agriculture. Bioscience, Biotechnology and Biochemistry, 82(8), 1265–1300. 10.1080/09168451.2018.1462693

Kaur, N., Sharma, S., Hasanuzzaman, M., & Pati, P. K. (2022). Genome editing: A promising approach for achieving abiotic stress tolerance in plants. International Journal of Genomics, 2022, 1–12. 10.1155/2022/5547231

Ke, Q., Wang, Z., Ji, C. Y., Jeong, J. C., Lee, H. S., Li, H., Xu, B., Deng, X., & Kwak, S. S. (2015). Transgenic poplar expressing *Arabidopsis* YUCCA6 exhibits auxin-overproduction phenotypes and increased tolerance to abiotic stress. Plant Physiology and Biochemistry, 94, 19–27. 10.1016/j.plaphy.2015.05.003

Ke, Q., Wang, Z., Ji, C. Y., Jeong, J. C., Lee, H. S., Li, H., Xu, B., Deng, X., & Kwak, S. S. (2016). Transgenic poplar expressing codA exhibits enhanced growth and abiotic stress tolerance. Plant Physiology and Biochemistry, 100, 75–84. 10.1016/j.plaphy.2016.01.004

Kieber, J. J., & Schaller, G. E. (2014). Cytokinins. The Arabidopsis Book, 12(January), 1–25. 10.1199/tab.0168

Ku, Y. S., Sintaha, M., Cheung, M. Y., & Lam, H. M. (2018). Plant hormone signaling crosstalks between biotic and abiotic stress responses. In International Journal of Molecular Sciences (Vol. 19, Issue 10). 10.3390/ijms19103206

Kumar, V., Sah, S. K., Khare, T., Shriram, V., & Wani, S. H. (2016). Engineering Phytohormones for Abiotic Stress Tolerance. Plant Hormones under Challenging Environmental Factors, 247–266. 10.1007/978-94-017-7758-2

Lee, Z. H., Hirakawa, T., Yamaguchi, N., & Ito, T. (2019). The roles of plant hormones and their interactions with regulatory genes in determining meristem activity. International Journal of Molecular Sciences, 20(16), 4065. 10.3390/ijms20164065

Li, N., Euring, D., Cha, J. Y., Lin, Z., Lu, M., Huang, L. J., & Kim, W. Y. (2021). Plant hormone-mediated regulation of heat tolerance in response to global climate change. Frontiers in Plant Science, 11(February), 1–11. 10.3389/fpls.2020.627969

Lijalem, T., & Feyissa, T. (2020). In vitro propagation of Securidaca longipedunculata (Fresen) from shoot tip: an endangered medicinal plant. Journal of Genetic Engineering and Biotechnology, 18(1), 1–10. 10.1186/s43141-019-0017-0

Liu, J. H., Kitashiba, H., Wang, J., Ban, Y., & Moriguchi, T. (2007). Polyamines and their ability to provide environmental stress tolerance to plants. Plant Biotechnology, 24(1), 117–126. 10.5511/plantbiotechnology.24.117

Mahajan, M., Kuiry, R., & Pal, P. K. (2020). Understanding the consequence of environmental stress for accumulation of secondary metabolites in medicinal and aromatic plants. Journal of Applied Research on Medicinal and Aromatic Plants, 18(September), 100255. 10.1016/j.jarmap.2020.100255

Muhammad Aslam, M., Waseem, M., Jakada, B. H., Okal, E. J., Lei, Z., Saqib, H. S. A., Yuan, W., Xu, W., & Zhang, Q. (2022). Mechanisms of Abscisic Acid-Mediated Drought Stress Responses in Plants. International Journal of Molecular Sciences, 23(3). 10.3390/ijms23031084

Oshchepkov, M. S., Kalistratova, A. V., Savelieva, E. M., Romanov, G. A., Bystrova, N. A., & Kochetkov, K. A. (2020). Natural and synthetic cytokinins and their applications in biotechnology, agrochemistry and medicine. Russian Chemical Reviews, 89(8), 787–810. 10.1070/rcr4921

Pandey, P., Irulappan, V., Bagavathiannan, M. V., & Senthil-Kumar, M. (2017). Impact of combined abiotic and biotic stresses on plant growth and avenues for crop improvement by exploiting physio-morphological traits. Frontiers in Plant Science, 8(April), 1–15. 10.3389/fpls.2017.00537

Parab, A. R., Chew, B. L., Yeow, L. C., & Subramaniam, S. (2021). Organogenesis on apical buds in common fig (Ficus carica) var. Black Jack. Electronic Journal of Biotechnology, 54, 69–76. 10.1016/j.ejbt.2021.10.001

Premkumar, G., Sankaranarayanan, R., Jeeva, S., & Rajarathinam, K. (2011). Cytokinin induced shoot regeneration and flowering of Scoparia dulcis L. (Scrophulariaceae)-an ethnomedicinal herb. Asian Pacific Journal of Tropical Biomedicine, 1(3), 169–172. 10.1016/S2221-1691(11)60020-8

Raza, A., Charagh, S., Najafi-Kakavand, S., Abbas, S., Shoaib, Y., Anwar, S., Sharifi, S., Lu, G., & Siddique, K. H. M. (2023). Role of phytohormones in regulating cold stress tolerance: Physiological and molecular approaches for developing cold-smart crop plants. Plant Stress, 8(March), 100152. 10.1016/j.stress.2023.100152

Shankaraswamy, J., & Ento, S. (2022). Phytohormones: Role In Manipulation Of Genome-Wide Associated Horticultural Traits For Crop Improvement. In Plant Growth Regulators (PGRs) (pp. 118–138).

Sharma, A., Shahzad, B., Kumar, V., Kohli, S. K., Sidhu, G. P. S., Bali, A. S., Handa, N., Kapoor, D., Bhardwaj, R., & Zheng, B. (2019). Phytohormones regulate accumulation of osmolytes under abiotic stress. Biomolecules, 9(7), 285–321. 10.3390/biom9070285

Singh, A. K., Dhanapal, S., & Yadav, B. S. (2020). The dynamic responses of plant physiology and metabolism during environmental stress progression. Molecular Biology Reports, 47(2), 1459–1470. 10.1007/s11033-019-05198-4

Singroha, G., Kumar, S., Gupta, O. P., Singh, G. P., & Sharma, P. (2022). Uncovering the Epigenetic Marks Involved in Mediating Salt Stress Tolerance in Plants. Frontiers in Genetics, 13(April), 1–10. 10.3389/fgene.2022.811732

Small, C. C., & Degenhardt, D. (2018). Plant growth regulators for enhancing revegetation success in reclamation: A review. Ecological Engineering, 118(September 2017), 43–51. 10.1016/j.ecoleng.2018.04.010

Torun, by H., Novák, O., Mikulík, J., Strnad, M., & Ayaz, F. A. (2022). The effects of exogenous salicylic acid on endogenous phytohormone status in hordeum vulgare L. under salt stress. Plants, 11, 618.

Ullah, A., Manghwar, H., Shaban, M., Khan, A. H., Akbar, A., Ali, U., Ali, E., & Fahad, S. (2018). Phytohormones enhanced drought tolerance in plants: a coping strategy. Environmental Science and Pollution Research, 25(33), 33103–33118. 10.1007/s11356-018-3364-5

Wani, S. H., Kumar, V., Shriram, V., & Sah, S. K. (2016). Phytohormones and their metabolic engineering for abiotic stress tolerance in crop plants. Crop Journal, 4(3), 162–176. 10.1016/j.cj.2016.01.010

Williams, W. R. (2021). Phytohormones: structural and functional relationship to purine nucleotides and some pharmacologic agents. Plant Signaling and Behavior, 16(1), 1–6. 10.1080/15592324.2020.1837544

21 Phytohormones

Past, Present and Future

Ayyagari Ramlal, Ambika Rajendran, Dhandapani Raju, and Virendra Pal Singh

21.1 INTRODUCTION

Plant hormones, also known as phytohormones, are signal molecules made by plants that exist in incredibly small amounts. Consequently, phytohormones are chemical mediators that regulate plant cellular processes. Plant hormones regulate every stage of plant growth and development, including embryogenesis, the management of organ size, disease defence, stress tolerance and reproductive development. The occurrence of plant hormones is ubiquitous; they are present in all higher plants and lower plants as well. Synthesis, metabolism, transport to the desired tissue and signal transmission, which govern its actions in the plant, all work to maintain their homeostasis. These tiny compounds, which come from secondary metabolism, are what allow plants to respond to their surroundings. Because their surroundings are constantly shifting, plants need these phytohormones to respond in the right way. Different transport systems are used to regulate the effect of plant hormones at nearby and distant locations. From the initial discovery of auxin as the first phytohormone to the most recent discovery of indole amines, 15 categories of phytohormones have so far been identified, including auxins, cytokinins (CKs), gibberellins (GAs), abscisic acid (ABA), ethylene (ETH), brassinosteroids (BR), salicylates (SAs), jasmonates (JAs), strigolactones (SLs), oxylipins, indoleamines, karrikins, nitric oxide, peptide hormones and polyamines (PAs). Auxin, CK, GA, ABA and ETH are the first five phytohormones, and they are sometimes referred to as the "traditional/classical" phytohormones. The remaining set of hormones, like brassinosteroids (BRs), jasmonates, salicylates (SAs), jasmonates (JAs), strigolactones (SLs), oxylipins, indole amines, karrikins, nitric oxide, peptide hormones and polyamines (PAs) are referred as non-classical/non-traditional phytohormones. Different transport systems control how plant hormones operate at near and far locations. By loading hormones from the source into the xylem or phloem, hormone transport to a remote destination is facilitated. Several proteins that function as hormone transporters at remote locations have been discovered in the last ten years, while symplast, apoplast or transcellular mechanisms are responsible for the movement of hormones locally (Abualia et al., 2018). At one end, cytokinins are carried from the roots to the leaves, where they delay senescence and keep the metabolism active, whereas at the other, the creation of the gas ethylene may cause changes within the same tissue or cell where it is made. Plant hormones are chemically diverse, containing indole, steroids, terpenes, carotenoids, fatty acids and compounds of adenine. This diversity reflects the variety of biochemical roles that these hormones serve (Srivastava et al., 2002). In this chapter, we will briefly describe the progress in the research of phytohormones. Without ignoring the reality that most hormones are active throughout a plant's life, we will introduce each hormone in the context of one of its functions, especially in abiotic stress conditions.

21.2 LESSONS FROM THE PAST

The five traditional plant hormones, which were discovered in the early to mid-20th century, are frequently cited. They are auxin (isolated in 1926 by F. Went), cytokinins (1950s, F. Skoog),

DOI: 10.1201/9781003335788-26

ethylene (1901, D. Neljubow), gibberellins (1926, E. Kurosawa) and abscisic acid (ABA; 1950s, T. Bennett-Clark and N. Kefford).

Even after more than a century of research, the signalling, transport and biosynthetic mechanisms for auxin are still unclear. IAA (indole-3-acetic acid) is one of the most versatile phytohormones and is essential for controlling and/or coordinating plant growth under stressful situations, in addition to supporting plant growth and development (Ke et al., 2015). Growing research suggests that IAA is essential for plants to adjust to salt stress. Auxin is recognized as a key component of defensive responses, due to its ability to regulate a wide range of genes and mediate the communication between abiotic and biotic stress responses (Fahad et al., 2015a,b). Only a few of the several tetracyclic diterpenoid carboxylic acids known as gibberellins (GAs) are capable of acting as growth hormones in higher plants; the most common types are GA1 and GA4. They serve growth-stimulating roles for plants throughout their life cycle. Transitions between developmental phases are also encouraged. It's interesting to note that there is mounting proof of their critical functions in abiotic stress response and adaptability (Colebrook et al., 2014). As master regulators throughout plant growth and development, CKs have a significant impact on a variety of processes related to plant growth and development. Abiotic stress, including salt and drought, is indicated by altered endogenous levels of CKs in response to stress (O'Brien & Benkovā, 2013). Abscisic acid (ABA), which gets its name from its function in the abscission of plant leaves, is referred to as a "stress hormone" and is, therefore, the phytohormone that has been the subject of the most research due to its specific role in plant adaptation to abiotic stressors. Under conditions of drought stress and nitrogen deficiency, ABA is also engaged in strong root development and other architectural alterations. ABA controls the creation of LEA proteins, dehydrins and other protective proteins as well as the regulation of other stress-responsive genes (Versleus et al., 2006). The production of osmoprotectants and antioxidant enzymes, as well as activities involved in cell turgor maintenance, are all enhanced by ABA water-saving anti-transpirant activity in water-deficit conditions. ABA content increased proportionately when plants were exposed to salt, according to research on conferring desiccation tolerance. A crucial regulator of stress reactions, ethylene (ET) is a gaseous phytohormone that plays a role in several stages of plant growth and development, including fruit ripening, floral senescence and leaf and petal abscission (Gamalero & Glick, 2012). Low temperature and salt are abiotic stressors that affect plant endogenous ET levels. Higher ET concentrations were thus successful in enhancing tolerance. Additionally, ET is crucial for plants' defensive mechanisms against heat stress (Shi et al., 2012).

21.3 CURRENT SCENARIO OF PHYTOHORMONES

There have been many other substances discovered during the last 50 years or so that fit the definition of hormones. Brassinosteroids (BRs), jasmonates, salicylates and strigolactones are four of the most recent hormones to be discovered. They are often termed "non-traditional phytohormones." The polyhydroxy steroidal plant hormones known as brassinosteroids (BRs) have a significant potential to promote growth and development. In the rapeseed plant (*Brassica napus*) pollen, they were initially discovered and identified. However, new research suggests that BRs and related chemicals have stress-impact-reducing activities in a variety of plants exposed to different abiotic pressures (Bajguz & Hayat, 2009). Abiotic stressors include extreme heat, cold, soil salinity, light, drought, flooding, metals/metalloids and organic contaminants. JAs are cyclopentanone phytohormones that are produced by the metabolism of membrane fatty acids, chiefly methyl jasmonate (MeJA) and its free acid JA. They are quite common in the plant kingdom. Involved in vital plant growth and survival activities such as reproduction, blooming, fruiting, senescence, secondary metabolism and direct and indirect defensive responses are these multifunctional chemicals (Fahad et al., 2015a,b). JAs are essential signalling molecules that are brought on by a variety of environmental stressors, including salt, drought, and UV radiation. A naturally occurring phenolic substance called SA controls the expression of proteins linked to pathogenesis. It is crucial for the control of plant growth, ripening and development as well as reactions to abiotic stressors in addition to defensive responses

(Hara et al., 2012). As a result of abiotic stressors, including heat, cold, salt and drought, SA affects how plants react. A tiny family of carotenoid-derived chemicals known as strigolactones (SLs) was initially identified more than 45 years ago as seed germination stimulants in root parasite plants such *Striga*, *Orobanche* and *Phelipanche*. SLs are a significant class of signalling molecules that play a critical role in controlling how plants evolve to adapt to shifting environmental circumstances.

A few more hormones have also been identified as phytohormones in addition to these. They are oxylipins, indole amines, karrikins, nitric oxide, peptide hormones and polyamines. These are a part of non-traditional phytohormones. Small aliphatic nitrogenous bases known as polyamines (PAs) are created as a result of cellular metabolism. Although the PAs are not hormones, they have been proposed as a novel class of plant growth regulators due to their participation in controlling many growth and development processes and responses to abiotic stress in plants (Bohra et al., 2015). Serotonin and melatonin are the essential indoleamines present in animals and plants. The existence of melatonin (N-acetyl-5-methoxytryptamine) in vascular plants was originally discovered in 1995 by two separate research teams. A pleiotropic chemical, melatonin has a wide range of effects on plants. It is regarded as an antioxidant that plays a significant role in the regulation of ROS, reactive nitrogen species (RNS), other free radicals and dangerous oxidative compounds found in plant cells (Raza et al., 2020). All aerobic organisms produce oxylipins, signalling molecules made from unsaturated fatty acids by enzymes or spontaneously. Oxylipins control how organisms respond to environmental cues and grow and develop. Plants can adapt to abiotic stress conditions such as wounding, poor light and temperature, dehydration and osmotic stress and the impacts of ozone and heavy metals thanks to the role played by oxylipins (Savchenko et al., 2014). Fire, one of the most hazardous environmental forces, releases chemicals that cause a variety of physiological reactions. The substances, known as karrikins, are present in smoke. Unexpectedly, many abiotic stressors, such as drought, salt, low and high temperatures, heavy metals or nutrition deficit, are lessened by the result of a destructive natural force (Antala, 2022). Therefore, karrikins may be a part of the solution for more resilient agriculture in future.

21.4 CONCLUSION AND FUTURE RESEARCH

In the future, there will be immense development in the research on phytohormones, especially on abiotic stress conditions. Crosstalk between the phytohormones leads to further study of hormones in different paths. There is a path to provide crucial research in the phytohormones and advance knowledge of the mechanism behind microbe-mediated stress tolerance. The OsCSN genes were found to be important in rice development and abiotic stress response by mRNA transcription analysis under different plant phytohormone treatments and abiotic challenges. Their findings thus provide the framework for more research on the role of the CSN genes. Global demand for medicinal plants is increasing, which calls for effective control of the supply of mineral nutrients to enable the best possible production of material of therapeutic value. The application of plant growth promoters and bio-stimulants, such as plant-growth-promoting rhizobium, phytohormones, biochar and nanomaterials, can address the nutrient uptake-related issues of medicinal plants caused by abiotic stress. There will be exponential development in genome-wide analysis of different family genes, proteins and development and identification of newer phytohormones and their derivatives under abiotic stress conditions.

REFERENCES

Abualia, R., Benkova, E., & Lacombe, B. (2018). Transporters and mechanisms of hormone transport in Arabidopsis. In: C. Maurel (eds.), Advances in botanical research, vol. 87, (pp. 115–138). US: Academic Press.

Antala, M. (2022). Physiological roles of karrikins in plants under abiotic stress conditions. In: M. Naeem, & Aftab, T. (eds.), Emerging plant growth regulators in agriculture: Role in stress tolerance (pp. 193–204). Academic Press.

Bajguz, A., & Hayat, S. (2009). Effects of brassinosteroids on the plant responses to environmental stresses. Plant Physiology and Biochemistry, 47(1), 1–8.
Bohra, A., Sanadhya, D., & Bhatia, D. S. (2015). Polyamines: metabolism and role in abiotic stress amelioration. Journal of Plant Science Research, 31(2).
Colebrook, E. H., Thomas, S. G., Phillips, A. L., & Hedden, P. (2014). The role of gibberellin signalling in plant responses to abiotic stress. Journal of Experimental Biology, 217(1), 67–75.
Fahad, S., Hussain, S., Matloob, A., Khan, F. A., Khaliq, A., Saud, S., et al. (2015a). Phytohormones and plant responses to salinity stress: a review. Plant Growth Regulation, 75, 391–404.
Fahad, S., Nie, L., Chen, Y., Wu, C., Xiong, D., Saud, S., et al. (2015b). Crop plant hormones and environmental stress. Sustainable Agriculture Reviews, 15, 371–400.
Gamalero, E., & Glick, B. R. (2012). Ethylene and abiotic stress tolerance in plants. Environmental adaptations and stress tolerance of plants in the era of climate change. In: P. Ahmad, & M. Prasad (eds.), Environmental Adaptations and Stress Tolerance of Plants in the Era of Climate Change (pp. 395–412). Springer, New York, NY. 10.1007/978-1-4614-0815-4_18.
Hara, M., Furukawa, J., Sato, A., Mizoguchi, T., & Miura, K. (2012). Abiotic stress and role of salicylic acid in plants. In: P. Ahmad, & M. Prasad (eds.), Abiotic Stress Responses in Plants (pp. 235–251). Springer, New York, NY. 10.1007/978-1-4614-0634-1_13.
Ke, Q., Wang, Z., Ji, C. Y., Jeong, J. C., Lee, H. S., Li, H., et al. (2015). Transgenic poplar expressing Arabidopsis YUCCA6 exhibits auxin-overproduction phenotypes and increased tolerance to abiotic stress. Plant Physiology and Biochemistry, 94, 19–27.
O'Brien, J. A., & Benková, E. (2013). Cytokinin cross-talking during biotic and abiotic stress responses. Frontiers in Plant Science, 4, 451.
Raza, M. A., van der Werf, W., Ahmed, M., & Yang, W. (2020). Removing top leaves increases yield and nutrient uptake in maize plants. Nutrient Cycling in Agroecosystems, 118, 57–73.
Savchenko, T. V., Zastrijnaja, O. M., & Klimov, V. V. (2014). Oxylipins and plant abiotic stress resistance. Biochemistry (Moscow), 79, 362–375.
Shi, Y., Tian, S., Hou, L., Huang, X., Zhang, X., Guo, H., & Yang, S. (2012). Ethylene signaling negatively regulates freezing tolerance by repressing expression of CBF and type-A ARR genes in *Arabidopsis*. The Plant Cell, 24(6), 2578–2595.
Srivastava, L. M. (2002). Plant growth and development: Hormones and environment. USA: Academic Press.
Verslues, P. E., Guo, Y., Dong, C. H., Ma, W., & Zhu, J. K. (2006). Mutation of SAD2, an importin β-domain protein in Arabidopsis, alters abscisic acid sensitivity. The Plant Journal, 47(5), 776–787.

Index

Note: Locators in *italics* represent figures and **bold** indicate tables in the text.

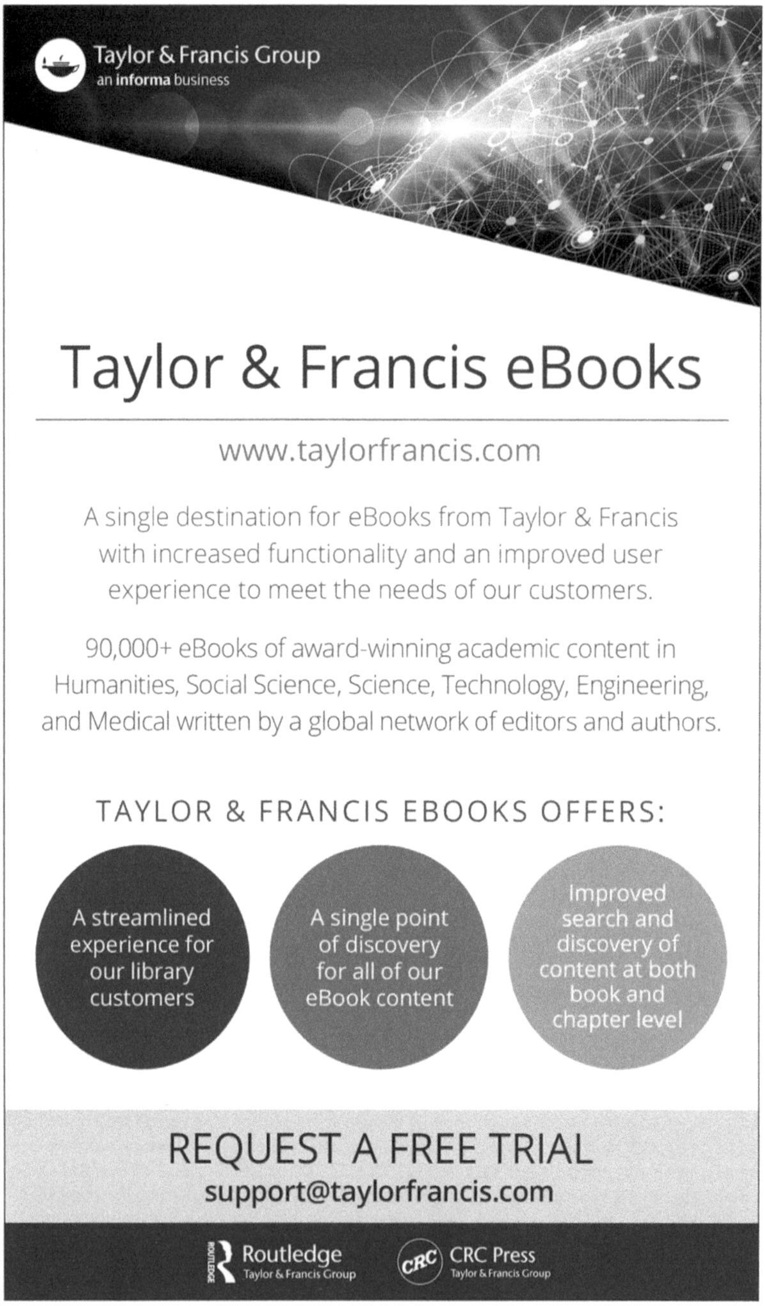
Taylor & Francis Group
an informa business
Taylor & Francis eBooks
www.taylorfrancis.com
A single destination for eBooks from Taylor & Francis with increased functionality and an improved user experience to meet the needs of our customers.
90,000+ eBooks of award-winning academic content in Humanities, Social Science, Science, Technology, Engineering, and Medical written by a global network of editors and authors.
TAYLOR & FRANCIS EBOOKS OFFERS:
A streamlined experience for our library customers
A single point of discovery for all of our eBook content
Improved search and discovery of content at both book and chapter level
REQUEST A FREE TRIAL
support@taylorfrancis.com
Routledge
Taylor & Francis Group
CRC Press
Taylor & Francis Group